Portuguese Landscape Architecture Education, Heritage and Research

In 2022, the Landscape Architecture course in [illegible] 80 years of existence. This edited collection, *Portuguese Landscape Architecture Education, Heritage and Research*, commemorates this important milestone by bringing together some of the most respected names in Portuguese Landscape Architecture.

Although the book's content is targeted at the assessment of the Portuguese history and influence, the themes under analysis are all-encompassing within the major fields, namely pedagogy; heritage; theory and methods; and design and landscape planning and management. The book seeks to address several research questions, including

- How has landscape architecture evolved in Portugal and how has it been revealed in the different disciplinary areas and educational institutions, particularly considering the great challenges of today?
- What legacy did Cabral, Sousa da Câmara and the first generation of landscape architects leave us that can be identified in the theory and practice of research projects, recent or ongoing, carried out by Portuguese landscape architects?
- How has the education, research and practice of Landscape Architecture in Portugal been influenced or reflected by the exchange of knowledge with other countries?

This book will be of interest to researchers and students, as it encompasses an extensive contribution to the field of Landscape Architecture studies, aiming to impact both on the theory and practice of the discipline.

Maria Matos Silva, Assistant Professor of Landscape Architecture (LA) at Instituto Superior de Agronomia (ISA-UL), is currently coordinating the master's programme in LA (2022–). She has a degree in LA from ISA-UL (2007), a master's degree (2010) and PhD (2016) from the University of Barcelona, where she focused on urban and public space design. She has been a research associate of Centro de Investigação em Arquitectura, Urbanismo e Design (CIAUD) at the Faculty of Architecture, UL since 2016. She was awarded an Honourable Mention in the Scientific Prize of the UL/CGD (2019) and is co-PI of the "MetroPublicNet" FCT research project.

Cristina Castel-Branco is a landscape architect and professor at the University of Lisbon. As a Fulbright student, she received an MLA from UMass and obtained her PhD in the history of garden art. She has published on subjects in Landscape Architecture history, restoration, and ecological design. She is founder and the first president of the Historic Gardens Association; director of the Botanical Garden of Ajuda (1997–2002) and is a member of the UNESCO-ICOMOS cultural landscapes. She maintains a professional practice in Lisbon, the ACB studio, which has been prize-winning since 1990. She was granted the award of Officier des Arts et des Lettres in 2015 and the Japanese Government Praise of Merit in 2020.

João Ferreira Nunes is a landscape architect with a great passion for drawing and understanding the processes of the world. The founder of PROAP, his professional and teaching activities intertwine and contribute to the development of new lines of research and experimentation. He is full professor at the Accademia di Architettura in Mendrisio, Switzerland and a visiting professor at international universities.

Luís Paulo Ribeiro, Assistant Professor of Landscape Architecture (LA) at Instituto Superior de Agronomia (ISA-UL), has been lecturing since 1999. He obtained his degree in LA at ISA-UL, and his PhD from the University of Massachusetts, USA (1997). He is a founding partner of TOPIARIS (1988), an expert in the restoration of historical gardens and a coordinator of related projects, such as the Tropical Botanical Garden of Lisbon. He is head of the master's and PhD programs in the LA department and a regular invited lecturer at the University of Massachusetts, where he was awarded the 6th Fábos Medal for the excellence of his work on Greenway planning and design.

Teresa Andresen is a landscape architect and agricultural engineer (CLAP, 1982; EA, 1982). She has been a professor in higher education for 30 years and was a member of the Scientific Council of the European Environment Agency from 2002 to 2008; president of the European Foundation for Landscape Architecture from 2004 to 2007; vice president of the International Federation of Landscape Architects (2007); and Portuguese Expert at the Permanent Commission for World Heritage (UNESCO) from 2014 to 2017. She has been a member of the National Council for Environment and Sustainable Development since 2013 and president of the Board of AJH–Portuguese Association of Historic Gardens since 2017.

Project Thinking on Design
Series Editors: Alessia Allegri, Ljiljana Cavic and João Pedro Costa

Portuguese Landscape Architecture Education, Heritage and Research
80 Years of History
Edited by Maria Matos Silva, Cristina Castel-Branco, João Ferreira Nunes, Luís Paulo Ribeiro and Teresa Andresen

Portuguese Landscape Architecture Education, Heritage and Research

80 Years of History

Edited by Maria Matos Silva, Cristina Castel-Branco, João Ferreira Nunes, Luís Paulo Ribeiro and Teresa Andresen

Routledge
Taylor & Francis Group
LONDON AND NEW YORK
ULISBOA | UNIVERSIDADE DE LISBOA
FACULDADE DE ARQUITETURA
CIAUD Centro de Investigação em Arquitetura Urbanismo e Design
fct Fundação para a Ciência e a Tecnologia

Designed cover image: Alessia Allegri

First published 2025
by Routledge
4 Park Square, Milton Park, Abingdon, Oxon OX14 4RN

and by Routledge
605 Third Avenue, New York, NY 10158

Routledge is an imprint of the Taylor & Francis Group, an informa business

This work is financed by national funds through FCT—Fundação para a Ciência e a Tecnologia, I.P., under the Strategic Project with the references UIDB/04008/2020 and UIDP/04008/2020.

British Library Cataloguing-in-Publication Data
A catalogue record for this book is available from the British Library

Library of Congress Cataloging-in-Publication Data
Names: Silva, Maria Matos, editor. | Andresen, Teresa, editor. | Castel-Branco, Cristina, editor. | Ribeiro, Luis Paulo Faria, editor. | Nunes, João Ferreira, 1960– editor.
Title: Portuguese landscape architecture education, heritage and research : 80 years of history / edited by Maria Matos Silva, Teresa Andresen, Cristina Castel-Branco, Luís Paulo Ribeiro and João Ferreira Nunes.
Description: Abingdon, Oxon ; New York, NY : Routledge, 2025. | Series: Project thinking on design | Includes bibliographical references and index.
Identifiers: LCCN 2024007008 (print) | LCCN 2024007009 (ebook) | ISBN 9781032592886 (hardback) | ISBN 9781032592862 (paperback) | ISBN 9781003454007 (ebook)
Subjects: LCSH: Landscape architecture—Study and teaching—Portugal. | Landscape architecture—Study and teaching—Portugal—History. | Landscape architecture—Portugal. | Landscape architecture—Portugal—History.
Classification: LCC SB469.43.P8 P67 2025 (print) | LCC SB469.43.P8 (ebook) | DDC 712.09469—dc23/eng/20240423
LC record available at https://lccn.loc.gov/2024007008
LC ebook record available at https://lccn.loc.gov/2024007009

ISBN: 978-1-032-59288-6 (hbk)
ISBN: 978-1-032-59286-2 (pbk)
ISBN: 978-1-003-45400-7 (ebk)

DOI: 10.4324/9781003454007

Contents

Illustrations

Figures

Tables

Contributors

Inês Adagói is a PhD student in Landscape Architecture, since 2018, at the School of Agriculture, University of Lisbon (ISA, UL). She is currently studying landscape planning of low-density rural landscapes in Portugal and ways to implement sustainable development measures in these territories. She is also collaborating in the "SCAPEFIRE" project, funded by the Foundation for Science and Technology (ref. PCIF/MOS/0046/2017).

Teresa Andresen is a landscape architect and agricultural engineer (CLAP, 1982; EA, 1982). She has been a professor in higher education for 30 years and was a member of the Scientific Council of the European Environment Agency from 2002 to 2008; president of the European Foundation for Landscape Architecture from 2004 to 2007; vice president of the International Federation of Landscape Architects (2007); and Portuguese Expert at the Permanent Commission for World Heritage (UNESCO) from 2014 to 2017. She has been a member of the National Council for Environment and Sustainable Development since 2013 and president of the Board of AJH–Portuguese Association of Historic Gardens since 2017.

Leonor T. Barata is an environmental engineer from Nova School of Science & Technology (FCT, UNL) and has a master's in Landscape Architecture from the School of Agriculture, University of Lisbon (ISA, UL). Between 2003 and 2009 she coordinated and advised several environmental impact studies, at which point she also began collaborating on Landscape Architecture projects until recently. She is a researcher at Green and Blue Infrastructure Thematic Line (LEAF R&D Unit) and, since 2014, she is a PhD student in agro-ecological farm landscape design.

Ana Beja da Costa, Landscape Architect, Postdoctoral Research Fellow in the MetroPublicNet research project, at the School of Architecture, Universidade de Lisboa. PhD in Landscape Architecture and urban ecology from the School of Agriculture, Universidade de Lisboa (2020). She participated in several research projects on landscape design and ecology applied to human settlements, namely in Timor-Leste, India, Ghana and Mozambique. In parallel, has continuously practiced as a landscape architect in Belgium, the Netherlands and Portugal, demonstrating a continued interest in interdisciplinary practice possibilities. She holds a postgraduate master's degree in human settlements (KU Leuven, 2008) and master's degree in Landscape Architecture (School of Agriculture, ULisboa, 2005).

Fátima Bernardo is an assistant professor at the Department of Psychology at Évora University where, since 2000, she teaches courses across a range of fields within the field of environmental psychology, social psychology, psychology of groups and group dynamic. She is a researcher in the Center for innovation in Territory, Urbanism and Architecture (CITUA-IST-UL), Technical University of Lisbon, Portugal and has a PhD in psychology from Évora University, Portugal. Her main research interests and publications are environmental psychology (risk perception, place identity, landscape perception) and social psychology (social identity, perceiving groups). She is member of IAPS (International Association People-Environment Studies) and Psicamb (Environmental Psychology Association).

Miguel Brilhante has a master's in biology of plant resources (FCUL) and received the award of the "Best Student" by ARALAB. Since 2021, he is doing his PhD at ISA/ULisboa. Since 2017, he worked in some research topics, namely regarding the Macaronesian and African flora and botanic gardens.

Raquel Carvalho graduated Landscape Architecture at Technical University of Lisbon (Higher Institute of Agronomy–ISA) in 2005. Since 2005 she collaborates at ACB Landscape Architecture studio and has continuously practiced as a landscape architect, dealing with public and private gardens and restoration of historic gardens. She is a PhD in Landscape Architecture and urban ecology (University of Porto) with a full doctoral scholarship attributed by the Portuguese Foundation for Science and Technology (FCT). Her research interests focus heritage landscapes as well as contemporaneous projects, as a result of her long professional experience in studio. Since 2020 she has been an invited assistant professor at the Higher Institute of Agronomy.

Cristina Castel-Branco is a landscape architect and professor at the University of Lisbon. As a Fulbright student, she received an MLA from UMass and obtained her PhD in the history of garden art. She has published on subjects in Landscape Architecture history, restoration and ecological design. She is founder and the first president of the Historic Gardens Association; director of the Botanical Garden of Ajuda (1997–2002) and is a member of the UNESCO-ICOMOS cultural landscapes. She maintains a professional practice in Lisbon, the ACB studio, which has been prize-winning since 1990. She was granted the award of Officier des Arts et des Lettres in 2015 and the Japanese Government Praise of Merit in 2020.

Francisco Castro Rego, is a former professor at the University of Lisbon (ISA/ULisboa), a senior researcher in fire ecology, with a PhD in forestry and wildlife and range management from Idaho University (EUA). For more than 20 years he was Scientific Coordinator of the CEABN-InBIO at ISA/ULisboa. From 2005 to 2007 he was director of the Forest National Services, and in 2001 he was an advisor at the Ministry of Agriculture. He has held various directorships in national and international organizations such European Forest Institute. He has vast experience on coordinating projects on the topics of landscape and forest management and fire ecology.

Ana Raquel Cunha is a student of the PhD program in Landscape Architecture and urban ecology (LINK) at ISA/ULisboa. Her study main topic is "Botanic Gardens as Living Museums: revealing the historical and natural heritage of the Ajuda Botanical Garden", with a FCT scholarship (2020/04824/BD). She was a research fellow in two important projects: "LX GARDENS" and "PORBIOTA", mobilizing ISA biological collections.

Natália Cunha has a PhD in Landscape Architecture from the School of Agriculture, University of Lisbon. She collaborated as teaching assistant in Instituto Superior Técnico, from 2006 to 2012. Since 2002, she has been an integrated researcher in ISA, in the Green and Blue Infrastructure Thematic Line (LEAF R&D unit). Her experience focuses on the ecologically-based landscape planning, established on sustainable land management principles, and an integrative, and transdisciplinary methodologies that bring together the fundamental areas of ecological and social sciences. Her research explores green infrastructures, ecological networks and land morphology models.

Susana Dias, is a biologist, with a PhD in forestry, where one topic was the link between landscape management options and bird species conservation. She is a professor at the University of Lisbon (ISA/ULisboa), and she is on the directive board of the CEABN-InBIO. She has been involved in several research projects, at national and international levels, for the past 20 years. Since 2018, Susana has been the executive coordinator of three protocols with municipalities, aiming at assessing the ecosystem services of street trees to support decision planning.

Inês Duarte is a landscape architect and has a master's in management and nature conservation and a PhD in forestry and natural resources. She is a researcher in the CEABN-InBIO at ISA/ULisboa, in the topics of landscape ecology, nature conservation and ecosystem services. She has been involved in several research projects at national and international levels. She also lectures on design in cultural and metropolitan landscapes, environmental planning and Landscape Architecture at the University of Algarve.

Gonçalo Duarte Gomes is a chartered landscape architect with a professional career of 20 years in the private and public sectors, in different areas of activity. He is a researcher at the Center for Art History and Artistic Research at the University of Évora and a guest speaker and moderator at several national and international events, with research articles published in several countries. He is the author of the book, "A doce melodia de um coro de mudos", and collaborates regularly with the media and various publications.

Paulo Farinha-Marques is an associate professor of Landscape Architecture in the faculty of science of the University of Porto. He graduated in Landscape Architecture in Lisbon and obtained a PhD in the faculty of architectural studies of the University of Sheffield. Since then, he teaches and produces Landscape Architecture projects at various scales. His main areas of interest are the planning and design of public outdoor spaces with special emphasis on urban green networks, green corridors, parks, gardens, green squares, green streets and designed woodlands. He currently coordinates a special landscape, rich in designed outdoor spaces and cultivated biodiversity: the Botanical Garden of the University of Porto.

Diana T. Fernandes has a PhD in Landscape Architecture and urban ecology (2021) and a master's degree in Landscape Architecture, by the faculty of sciences, Univ. Porto (2013). As a member of Water Resources Systems' Research Team, at Centro Interdisciplinar de Investigação Marinha e Ambiental, a collaborator in a river restoration company, E.Rio, and a member of the managing board of Centro Ibérico de Restauro Fluvial, she develops her work mainly in river rehabilitation design and planning at a local and regional scale. She also has been involved in several research projects, regarding green infrastructure and spatial planning.

João Ferreira Nunes is a landscape architect with a great passion for drawing and understanding the processes of the world. The founder of PROAP, his professional and teaching activities intertwine and contribute to the development of new lines of research and experimentation. He is full professor at the Accademia di Architettura in Mendrisio, Switzerland, and a visiting professor at international universities.

Ana Júlia Francisco is an agriculture engineer and a specialist in urban arboriculture at Lisbon Municipal Council (CML). She coordinated the Garden Division at CML between 2006 and 2011. Since then, she coordinates the Division of Maintenance and Requalification of the Green Structure (DMREV, division of the Municipal Directorate for Environment, Green Spaces, Climate and Energy/ Department of Green Structure–DMAEVCE/DEV), dedicated to the management of the city's trees, either under municipal management or in support of parish councils.

Luísa Franco has a PhD in Landscape Architecture from the Instituto Superior de Agronomia, University of Lisbon (ISA, UL). She has developed research in green infrastructures and landscape and land use planning since 2011 and has participated in consulting projects for public entities. Since 2021 she has been a junior researcher in the LEAF R&D unit (Green and Blue Infrastructures thematic line). Her research and professional interests are geographical information systems and its application to landscape and land use planning, evaluation of land use planning policies, planning of green infrastructures in urban and rural areas and planning for the prevention of rural fires.

Davide Gaião, is an agronomic engineer, with a degree in ecology and landscape and a master's in agronomic engineering from the University of Trás-os-Montes and Alto Douro. Since 2015 he has been an assistant professor at the Agrarian School of Polytechnic Institute of Viseu, and he collaborates with the CEABN-InBio at ISA/ULisboa in projects related with urban ecology. He has been working on urban tree inventory in several Portuguese cities, as well as carrying out biomechanical and phytosanitary assessments on public trees, studying pests and diseases of trees and agricultural crops.

Anubhav Goyal graduated in 2009 in architecture from Rajasthan University–Aayojan School of Architecture at Jaipur and completed the master's in urban planning in 2011 from Malaviya National Institute of Technology with post-graduation in project management from National Institute of Construction Management and Research–Pune. Goyal was licenced in applications of remote sensing & GIS for natural resources in 2015 by Indian Institute of Remote Sensing and is a research candidate in Universidade de Lisboa, Centro de Investigação em Arquitetura Urbanismo e Design-CIAUD, *formaurbis* LAB, Portugal. Goyal works in the area(s) of humanities with emphasis on urbanism and architectural design.

Maria da Graça Saraiva, Associate Professor at the School of Architecture of the University of Lisbon (UL), has been retired since 2013. She trained as a landscape architect and agronomist engineer at ISA, with a master's degree in urban and regional planning and a PhD in Landscape Architecture at UL. Researcher at CIAUD/FA/UL (Research Centre of Architecture, Urbanism and Design), she has been participating in European and national research projects on landscape and spatial planning, landscape perception and sustainability river restoration and natural resources planning, urban forestry and vegetation.

Maria João Fonseca is a landscape architect and researcher. She has been working on the Gonçalo Ribeiro Telles' archive since 2019 (until Feb 2022 as a post-doctoral fellow in the ROSSIO Infrastructure project), coordinates the citizen science project "Germinar um banco de sementes" [Sprouting a seedbank] and is a member of the DGArtes/Ministry of Culture Evaluation Committee. She has a PhD in architecture (ISCTE-IUL), graduated (Instituto Superior de Agronomia—UTL), and she has master's degree (University of Greenwich) in Landscape Architecture. She holds postgraduate studies in "Historical Archival Science" and "Emerging Cultures and Discourses: from criticism to artistic manifestations" (FCSH/Un. Nova de Lisboa).

Maria José Curado is an assistant professor at the faculty of sciences, Univ. Porto, in Landscape Architecture courses. She's a member of the Research Centre in Biodiversity and Genetic Resources developing its research in the field of landscape ecology, land use planning and landscape management. She has a degree in Landscape Architecture from ISA, Univ. Lisbon Technician (1994), master's in planning and urban environment project, Univ. Porto (1998) and a PhD in environmental applied sciences, Univ. Aveiro (2003). Before joining the academic career, she developed the professional activity of Landscape Architecture in a liberal professional system (1994–2004).

Isabel Loupa Ramos, is a landscape architect who holds a master's in human geography and a PhD in environmental engineering. She teaches courses across a range of fields within the landscape ecology and urban planning domain at Instituto Superior Técnico, University of Lisbon. She carries out her research activities at CITUA (http://citua.tecnico.ulisboa.pt) at the interface between cultural and natural processes and urban and rural landscapes. She focuses on the integration of landscape and biodiversity into spatial planning, using green infrastructure planning as pathway for sustainability transitions. She is an elected representative of the European Chapter of Landscape Ecology (IALE-Europe).

Manuela R. Magalhães has a degree and PhD in Landscape Architecture by the School of Agriculture, University of Lisbon (ISA, UL). She was a Superior Technique of Public Administration (1971–1988), in the area of land use planning; liberal professional (1971–1988) in the area of Landscape Architecture plans and projects; assistant and assistant professor at ISA/University of Lisbon (1988–2013); guest professor at Instituto Superior Técnico/Technical university of Lisbon (1997–2011); and founder and coordinator of the former Centre for Landscape Architecture Studies "Professor Caldeira Cabral" currently incorporated in LEAF/ISA/ULisboa. She is also the coordinator of several scientific projects.

Rodrigo Maia has a PhD in civil engineering by the faculty of engineering, Univ. Porto. He is an associate professor at FEUP, vice president of European Water Resources Association and team leader of the Research Team of Water Resources Systems at Centro Interdisciplinar de Investigação Marinha e Ambiental, coordinating projects on hydrology, extreme events, water resources management, fluid mechanics and fluvial rehabilitation. He's involved in research on international rivers basins, namely of the Iberian, among others national and international projects, and he was involved in the creation of a Drought Early Warning Management System for the Portuguese National Water Authority.

Bruno Marques is the associate dean (Academic Development) and senior lecturer in Landscape Architecture at Victoria University of Wellington, Faculty of Architecture and Design Innovation. His main research interests relate to integrating Indigenous methods in participatory design and place-making in landscape rehabilitation and ecosystem services.

Isabel Martinho da Silva graduated in Landscape Architecture (1988) at the High Institute of Agronomy of the Lisbon University and completed a master's in Landscape Architecture (1996) and a PhD in renewable natural resources studies (2001) at the University of Arizona, USA. Currently, she is an assistant professor of Landscape Architecture at the University of Porto and a senior researcher in the group LPDM (Landscape Planning, Design, and Management) in the Biopolis Association–CIBIO. She has been developing research in landscape change, sustainable design, urban agriculture, public perception, and environmental justice.

Maria Matos Silva, Assistant Professor of Landscape Architecture (LA) at Instituto Superior de Agronomia (ISA-UL), is currently coordinating the master's programme in LA (2022–). She has a degree in LA from ISA-UL (2007), a master's degree (2010) and PhD (2016) from the University of Barcelona, where she focused on urban and public space design. She has been a research associate of Centro de Investigação em Arquitectura, Urbanismo e Design (CIAUD) at the Faculty of Architecture, UL, since 2016. She was awarded an Honourable Mention in the Scientific Prize of the UL/CGD (2019) and is co-PI of the "Metro-PublicNet" FCT research project.

Jacqueline McIntosh is a senior lecturer in architecture at the Wellington School of Architecture of Victoria University of Wellington. Current research revolves around investigating therapeutic and rehabilitative built environments. This multidisciplinary and multi-institutional bicultural research focusses on the design of the built environment for wellness including social and cultural inclusion.

Ana Müller is a PhD student in Landscape Architecture at LEAF R&D research centre (Green and Blue Infrastructures thematic line), in the School of Agriculture, University of Lisbon (ISA, UL). She has a master's in Landscape Architecture (2011) from ISA, University of Lisbon, with the dissertation "Transformations of the Rural Landscape of Northern Portugal". Since 2003 she has developed research in soft mobility, green infrastructures, and rural and urban land use planning.

Ludovica Nardella holds a master's degree in architecture and civil engineering from the Faculty of Architecture of the University of Naples Federico II, in Italy. She has been enrolled in the National Council of Architects, Landscape Planners, and Conservationists since 1989. As a freelance architect, she worked on restoration projects of historical buildings and architecture projects. Currently, she teaches drawing and history of art at the State Scientific High School Coluccio Salutati in Montecatini Terme, Italy. She is now attending the LINK doctoral program in Landscape Architecture and urban ecology, at the Instituto Superior de Agronomia in Lisbon.

Leónia Nunes, is a forest engineer, with a PhD in forest sciences. She is a researcher at the CEABN-InBIO at ISA/ULisboa, where she develops research in the silviculture of mixed forests, spatial structure of forest, modelling the growth of ecosystems and national forest inventories. She has been involved in several research projects and recently she has been collaborating with municipalities in the estimation of urban environmental benefits. She is a member of the directive board of the Forest Stewardship Council (FSC) Portugal and vice president of the Portuguese Society of Forest Sciences.

Luís Paulo Ribeiro, Assistant Professor of Landscape Architecture (LA) at Instituto Superior de Agronomia (ISA-UL), has been lecturing since 1999. He obtained his degree in LA at ISA-UL, and his PhD from the University of Massachusetts, USA (1997). He is a founding partner of TOPIARIS (1988), an expert in the restoration of historical gardens and a coordinator of related projects, such as the Tropical Botanical Garden of Lisbon. He is head of the master's and PhD programs in the LA department and a regular invited lecturer at the University of Massachusetts, where he was awarded the 6th Fábos Medal for the excellence of his work on Greenway planning and design.

Selma B. Pena develops research in the School of Agriculture, University of Lisbon (ISA, UL), in the actual LEAF R&D unit, since 2004. She is head of the Green and Blue Infrastructure Thematic Line (since 2020), from LEAF R&D Unit. Her research area focuses on landscape and land use planning, soil and water conservation, landscape stability assessment, GIS methodologies integrating policies and landscape planning tools. She also teaches the landscape planning courses from the master's and graduates Landscape Architecture courses from ISA/ULisboa.

Eduardo Costa Pinto is a partner at GETOUT Landscape Architecture and holds a degree in Landscape Architecture, by *Instituto Superior de Agronomia*, University of Lisbon. Since 2007, he has worked in several studios in Portugal and Italy, namely: D-RECTA Urban I, *Promontório* Architects, PROAP *Italia*, or Campos Costa Architects. Author/co-author of the following research projects: "Seven Circles", funded by the General Directorate of Arts (DGARTES) and Millennium BCP Foundation (MBCPF); "*O Mais Comprido Museu do Mundo*", financed by the MBCPF and with the High Sponsorship of the Presidency of the Portuguese Republic; "*Árvores sem Raízes*" radio podcast, at Rádio Antecâmara, with the support of DGARTES.

Teresa Portela Marques graduated in Landscape Architecture and PhD at the Higher Institute of Agronomy (University of Lisbon), is an assistant professor in the Landscape Architecture courses at the Faculty of Sciences, University of Porto, and a member of the Research Centre for Biodiversity and Genetic Resources (CIBIO/INBIO). Her research is mostly developed in the context of the history and critique of landscapes of amenity, particularly those designed in the 19th century and first decades of the 20th century, as well as on the conservation of landscapes of historic and cultural value; she is the author of several publications and projects in this field.

Maria Manuel Romeiras is a professor at ISA/ULisboa. She has received some honors and/or awards. Since 2018, she is corresponding member of the Academy of Sciences of Lisbon and in 2019, she was distinguished by Ciência Viva in the book *Women in Science*. Presently, she is the curator of the "Tapada da Ajuda Botanical Park" (ISA/ULisboa).

João Rafael Santos, Architect, has a PhD in urbanism and is an assistant professor of architecture and urbanism on master's and PhD levels at Lisbon School of Architecture, Universidade de Lisboa with a postdoctoral scholarship at The University of Tokyo. He is the coordinator of URBinLAB research group, and his interests include urban and territorial design and spatial planning, especially in the field of metropolitan studies with a focus on the relationship of infrastructure with public space and socio-spatial development. He is the coordinator of the FCT-funded research project "MetroPublicNet: A Metropolitan Public Space Network" to support the robust, low-carbon and cohesive city. He won the Metrópoles Ciência Prize in 2016 for his research on Lisbon Metropolitan Area.

João Silva is a PhD fellow at the School of Agriculture, University of Lisbon (ISA, UL). His current research focus on the relationships between different farming systems, biodiversity, and ecosystem services. He is a researcher member of the group "Green & Blue Infrastructures" of the R&D Unit "LEAF–Linking Landscape, Environment, Agriculture and Food", since 2010.

Ana Luísa Soares is a landscape architect and professor at the University of Lisbon (ISA/ULisboa). She is a researcher at the Centre for Applied Ecology "Prof. Baeta Neves" (CEABN-InBIO), and recently she has been appointed Pro-Rector of ULisboa. From 2009 to 2014 she was vice president of the Management Board of ISA, with the responsibility over the ISA Campus ISA and the Botanical Garden of Ajuda patrimony. She has participated in several research projects and was the coordinator of the project "LX GARDENS: Lisbon's Historic Gardens and Parks". Since 2019 she is the director of the Botanical Garden of Ajuda, ISA/ULisboa.

Sónia Talhé Azambuja obtained a degree in Landscape Architecture from the School of Agriculture (ISA) and pursued studies at the School of Arts and Humanities of the University of Lisbon, concluding her master's in Art and PhD in history of art. She currently holds the position of an assistant professor of Landscape Architecture at the University of Lisbon and at the University of Algarve. She is the author of two books on the symbolism of nature, the President of the Association of Friends of the Botanical Garden of Ajuda-AAJBA, and an International Member of ICOMOS/IFLA International Scientific Committee on Cultural Landscapes (2009–2019).

Domingas Vasconcelos is a PhD student in Landscape Architecture and urban ecology (University of Lisbon/University of Porto/University of Coimbra, Portugal). She graduated in Architecture in the University of Porto–Faculty of Architecture (1990) and obtained a master's degree in art history from the Faculty of Arts and Humanities (2004). Since 1992 she has been working as an architect at Porto City Council, in the Urban Studies Division (until 1999) and in the Cultural Heritage Division (since 2000). She supports professional work focused on the architectural and urban heritage of Porto with research on heritage and sustainability.

Lígia Vaz de Figueiredo is trained as a landscape architect, with an M.Sc. in urban and environmental regeneration and a PhD in fluvial restoration and management. She collaborates with CIAUD, the Research Centre for Architecture, Urbanism, and Design, Lisbon School of Architecture, Universidade de Lisboa (https://ciaud.fa.ulisboa.pt/index.php/pt/). She has been participating at European and national research projects, and her research focuses on integrating ecological and social issues in the context of fluvial landscape, landscape assessment, and public participation regarding fluvial restoration processes.

Wim Wambecq is an architect and urbanist with a PhD in engineering science, architecture at KU Leuven, Belgium. He is a visiting professor at the Department of Architecture of the University of Leuven (KU Leuven), where he teaches architecture from urban design perspective (public infrastructure) and landscape urbanism studios in European and Southeast African development contexts. His doctoral dissertation "Forest Urbanism in the Dispersed Flemish Territory" won the first prize at the third Manuel De Solà-Morales European Award in 2021 (ETSAB Barcelona School of Architecture & UPC Polytechnic University of Catalonia). He is also the co-founder of the interdisciplinary architecture, landscape and urban design Atelier Midi.

Acknowledgements

The editors wish to acknowledge all authors who have contributed to this book that reflects the diversity and depth of contemporary research regarding Portuguese Landscape Architecture.

A special thank you to the Scientific Committee, whose insightful critiques and constructive feedback have raised the quality of this work, and to our meta-reviewer Anubhav Goyal, who facilitated a coherent and smooth process all the way from the call for abstracts until the final submission.

We further like to recognize the supporting institutions of IFLA (International Federation of Landscape Architects) and APAP (Associação Portuguesa de Arquitectos Paisagistas), as well as the funding support from CIAUD (Research Centre for Architecture, Urbanism and Design, Lisbon School of Architecture, University of Lisbon) and FCT (Portuguese Foundation for Science and Technology).

Finally, a heartfelt acknowledgement to all those who have played a role in the affirmation of Landscape Architecture in Portugal over the past 80 years.

Foreword I

Bruno Marques[1]

Throughout its history, Landscape Architecture has grown from focusing on gardens and aesthetics to a multidisciplinary field that addresses ecological, social, and cultural considerations. Over the last century, the profession expanded to include urban design, ecological restoration, and sustainable development. Today, Landscape Architecture is a well-established and recognised profession practised in various parts of the world, combining elements of design, horticulture, ecology, and planning to create outdoor spaces that are functional and sustainable. As a profession, we play a vital role in shaping the built environment and promoting resilient landscapes. While the specific approaches and practices may vary depending on cultural, environmental, and social contexts, the fundamental principles and goals of Landscape Architecture remain consistent. In the rich tapestry of Landscape Architecture, Portugal emerges as a captivating thread, weaving heritage, education, and evolving perspectives of this dynamic discipline over the last 80 years.

The connection to the world stage of the profession in Portugal surges through Francisco Caldeira Cabral, who became the fifth president of the International Federation of Landscape Architects (IFLA) between 1962 and 1966. IFLA was officially founded in 1948 in Cambridge, England, marking the establishment of the first worldwide organisation dedicated to Landscape Architecture. The initial membership included representatives from 15 countries, and Sir Geoffrey Jellicoe became the first president of IFLA. In its early years, IFLA focused on facilitating communication and cooperation between national Landscape Architecture associations, which the Associação Portuguesa dos Arquitectos Paisagistas (or Portuguese Association of Landscape Architects) is a member of. Headquartered in the beautiful Potager du Roi in Versailles, France, IFLA keeps working to promote the profession and exchange knowledge and experiences among professionals from different countries.

As the first Portuguese landscape architect, Cabral was known as a multilingual, multi-talented, constant traveller and an eternal student. He would adventure himself throughout Europe and Africa with students en route to IFLA Councils and Congresses and, with that, explore magnificent landscapes and use them as a living lab of learning. Cabral taught students to look at landscape; to see the layers of history, of geology, of human use; and to understand the way it lives and sustains the lives of those who have sprung from and passed through it, but most importantly, he taught how to maintain it alive for future generations. He was not only clever but also wise.

While the world has changed a lot since Cabral's presidency in IFLA, the role of landscape architects has evolved beyond designing parks and gardens. Landscape Architecture now encompasses the entire environment, from urban to regional areas, including tourism-related green spaces, the preservation of national parks, and the recovery of various forms of heritage. Landscape architects are specially trained to contribute to strengthening ecological and social resilience. Currently, our profession faces high expectations regarding involvement in large-scale projects, creating new rural and urban areas while combating biodiversity loss and adapting to climate change. Resilience is the response we must aim for.

Landscape Architecture has a rich history that spans centuries, evolving in response to cultural, social, and environmental changes. The ongoing environmental changes and their widespread effects are daily reminders that nature-based solutions and Landscape Architecture are the way forward. We are now facing critical challenges, including the depletion of natural resources and loss of biodiversity, the impact of climate change and rising sea levels, increasing population and urbanisation, outdated infrastructure, the expansion of digitalisation and emerging technologies, and social segregation and its effects on mental health and well-being. These are some of the difficulties that professionals and academia must confront.

To address our global challenges, sharing knowledge and best practices allows us to reflect positively on the past to understand better our present actions and grow, learn, and gain energy to embrace the future. This book—titled "Portuguese Landscape Architecture Education, Heritage and Research" and edited by Maria Matos Silva, Cristina Castel-Branco, Luís Paulo Ribeiro, João Ferreira Nunes, and Teresa Andresen—provides the rich context of Portugal's Landscape Architecture journey, which unfolds across three distinct parts: Heritage and history of Portuguese Landscape Architecture; A Portuguese context on education and pedagogy in Landscape Architecture; and, Portuguese perspectives on theory and methods in Landscape Architecture. In each section, the editors meticulously unveil the shared wisdom embedded in the foundation of the discipline and its promising trajectory into the future.

The narrative begins with an overview, highlighting the expectations set for the enduring Landscape Architecture legacy in Portugal. It becomes evident that this is not merely a retrospective glance but a forward-looking examination, urging constant revisitation and a commitment to growth, self-criticism, and adaptability. As the book unfolds, it becomes apparent that Portuguese Landscape Architecture has become a reference beyond its borders, notably in the European context. Through the legacy of the past 80 years, the book unravels the profound implications of the theoretical contributions and their instrumental role in promoting the discipline. The visionary approaches and exquisite personalities of many leading and some less-known Portuguese landscape architects not only raised ecological awareness but also shaped educational programmes to address the threats posed by industrial development and urban expansion.

The narrative seamlessly transitions to present chapters that contribute to understanding the local landscape and extend the conversation to a broader global discourse on Landscape Architecture education and professional practice, from the restoration of historic gardens to the significance of urban parks and gardens, green and blue infrastructure, integration of Landscape Architecture into urban planning and environmental management and the exploration of socio-cultural dimensions. It equally delves beyond national borders and explores issues related to coastal management and the importance of traditional knowledge. This book is not just a critical reflection but a call to action.

In this book, the reader is left with a profound appreciation for the multifaceted layers that shape Portuguese Landscape Architecture and its implications worldwide. The collective wisdom, historical narratives, and forward-looking perspectives presented in this book serve as a testament to the past and a guidepost for a more informed, consistent, varied, and ambitious future for Portuguese Landscape Architecture. Throughout, we are reminded that Landscape Architecture, like the landscapes it shapes, is a dynamic and resilient force that continuously adapts to the challenges and opportunities of the world around us.

In conclusion, this book serves as a beacon, illuminating the path for future critical reflections and discussions within and beyond borders. As we navigate the complexities of the present and prepare for an ambitious future, Portuguese Landscape Architecture stands as a testament to the enduring spirit of those who will follow. Landscape architects are pivotal in shaping a prosperous future characterised by resilience, transformation, and long-term sustainability. By collaborating with diverse professionals and embracing innovative approaches, we can tackle climate action, protect biodiversity, enhance ecosystems, promote health and well-being, support community participation, embrace technology and evidence-based design, improve food security, design resilient landscapes, and incorporate traditional practices. Landscape architects and organisations like IFLA must take a stand and work collectively to address these challenges, ensuring a better and more sustainable future for all, and this book is a true reflection of that.

Note

1 Dr Marques is the 23rd president of the International Federation of Landscape Architects (IFLA), and the associate dean, Faculty of Architecture and Design Innovation, Victoria University of Wellington, New Zealand.

Foreword II

João Ceregeiro[1]

> At the same time, around the first of October or later, the elm trees are at the height of their autumnal beauty, large yellow-brown masses, fresh from the September oven, hanging over the road. Their leaves are at full maturity. I wonder if there is an equivalent maturity in the lives of the men who live in its shadow.
>
> Henry David Thoreau, *Wild Apples/Autumnal Tints*

There are dates that present themselves as opportunities for reflection, to look back and around, to help us identify the present and better position ourselves for the future. We do it with ourselves, and in a mature gesture, landscape architects have done it once again, on the occasion of the celebration of the 80th anniversary of the founding of the Course of Landscape Architecture in Portugal, crossing fires in 2022 with the Portuguese Association of Landscape Architects–APAP, in the celebration of 80–100, opening a broad cycle of workshops over 2 years.

In general, the production of literature and information disseminated around Landscape Architecture issues is scarce and therefore essential. Whether to the general public or in more restricted contexts, the dissemination of worked information, even from academia, as in this case, helps us to understand and prepare the general state of the profession, involving curricular dynamics, with a direct influence on professional operations and the institutional context of Landscape Architecture.

We need to reverse the Cassandra myth and publicise Landscape Architecture wherever possible, in written and oral form, on social networks, in the various speciality forums or in the general press. It is an imperative of conscience, if not a professional obligation for everyone, to expose the profession and its challenges to the community, to make it present, necessary and respected.

It was by explaining and provoking change in the political and social status quo that our masters Francisco Caldeira Cabral and later Gonçalo Ribeiro Telles and their colleagues from the first generation paved the way in Portugal for Landscape Architecture to exist today as a recognised profession, operating according to the ecological principles that regulate biophysical factors in the territory and involving nature in the narrative of their proposals. This is what F. L. Olmsted did in the embryonic phase of the profession, in pedagogical intervention and constant struggle, as a conservationist, hygienist and militant of the values of modern man. From this point of view, Landscape Architecture is part of a cultural, social and political process in constant mutation, considering the permanent transformation of systems.

However, in a world in deep imbalance, guided by a hermetic plutocracy, the word mutation can be understood as a profound social, economic and cultural revolution, preventing nature from taking over the process or the depletion of natural resources as a result of current levels of consumption.

Between reissues and new themes, APAP has recently supported the launch of works by colleagues, enriching the theoretical debate, the substance of the profession and its legacy, reflecting through their knowledge, the capacity of Landscape Architecture to involve diverse themes, but always complementary to each other. Without prejudice to many others, I would mention "Arquitectura Paisagista ou A Organização do Espaço nas Paisagens" by Ilídio Alves de Araújo, "Das Pedras Pão" by Henrique Pereira dos Santos, or the 3rd edition of "Árvore em Portugal" by Francisco Caldeira Cabral and Gonçalo Ribeiro Telles.

The backdrop to "Portuguese Landscape Architecture Education, Heritage and Research: 80 Years of History" is ecological thinking, and its close relationship with the practice of the profession in its multiple dimensions and scales reminds us that there is a simple order in the things around us that is difficult to detect, reproduce and pass on to a cognitive dimension. This reality, once touched, definitely involves us, inducing a commitment to attention and vigilance in preserving this open and global relationship, involving the non-conformism and creative capacity of Landscape Architecture.

Nations need a humanising benchmark that dismantles the sectoral interests that sustain the neoliberal varnish that has tinged public policies for decades, impervious to the implementation of the ecological precepts of the environment, the landscape and its management. These are antagonistic times, resistant to anything that might involve sharing the wealth of the territories of knowledge in the global space. We are moving in the opposite direction to what we want.

Landscape Architecture's strength of being able to touch on so many technical fields is also its weakness, in a culture of specialities and specialisations that is not very keen on integrated visions. Landscape Architecture's proposals are very likely to affect vested corporate and sectoral interests. Promoting complementary training in political sciences for recent Landscape Architecture graduates could be a way of making up for the landscape's lack of presence in the decision-making arena.

APAP's involvement in the National Architecture and Landscape Policy has made it possible to exploit the wording of the European Landscape Convention, converging new ideas to raise political awareness and the ability to bring together what is essential in safeguarding landscape, architecture and cultural heritage. Special attention is currently being paid to the need to combat environmental and landscape illiteracy. In this context, a curriculum support programme has been set up at secondary school level. Interestingly, this concern has now also been raised by the IFLA (International Federation of Landscape Architects), which has been spreading the word among national organisations about the need to implement programmes of this nature.

Landscape Architecture was born in Portugal as a result of Francisco Caldeira Cabral's vision, his perception of the moment, his personal involvement in an official state project, the creation of the disciplinary body of the first course and consequently the training of the first generation of landscape architects strategically placed in the state apparatus. With the profession underway, structured environmental, land use planning and nature protection policies emerge. International reference projects were realised and publications were made that were essential to teaching.

The production of information and the work space of Landscape Architecture in Portugal has gone through some brilliant decades, and at a glance I can identify some achievements that place our profession in the front ranks internationally: Lisbon's Green Structure (1999), a pioneer of its kind; the Algarve's Landscape Planning Plan (1969), a forerunner of Ian McHarg's method; the National Ecological Reserve–REN (1983), unprecedented as a planning instrument; Gonçalo Ribeiro Telles, awarded the Geoffrey Jellicoe Prize (2013); the creation of the National Network of Protected Areas (1993); or the opening of five Landscape Architecture schools for a population of 10 million inhabitants.

In 1953, Cabral created the Centre for the Study of Landscape Architecture–CEAP, as a space for research, work and representation of the profession. It was through CEAP, a forerunner of the professional association, that the profession was represented at the IFLA, founded in 1948, through meetings and congresses, with the work carried out by the first landscape architects.

In March 1976, the statutes were approved, and the APAP was founded by colleagues who were aware of the importance of associative work to defend and publicise the profession.

With public utility status since 1995, APAP currently has 1,500 members, has a professional code of ethics, represents IFLA–Europe, which is responsible for evaluating schools' curriculum structures and maintains its goal of legitimate recognition as a professional association under public law.

The last few boards have made an effort to make the professional association more operational and more visible, promoting vigilant action on the rights of the profession and its members and focusing on the definitive recognition of the profession.

Today, the fate of outdoor spaces and the landscape in general involves other players and professions. The room for manoeuvre is more restricted, but the landscape is still available to receive us, possibly attentive to its virtues, whims and complexities.

Note

1 148-APAP

Introduction

Three principles of ecological thinking for an interpretation of Portuguese Landscape Architecture education, heritage, and research

Maria Matos Silva

Jardim da Europa à beira-mar plantado
de loiros e de acácias olorosas;
de fontes e de arroios serpeado,
rasgado por torrentes alterosas,
onde num cerro erguido e requeimado
se casam em festões jasmins e rosas;
balsa virente de eternal magia
onde as aves gorgeiam noite e dia

Tomás Ribeiro, extract from
the poem "A Portugal" (1862)

Portugal, so-called "garden by the sea" ("Jardim . . . à beira-mar plantado", Tomás Ribeiro 1862), has one of the oldest European teaching traditions in Landscape Architecture. While its antiquity enables a particular degree of maturity, what is of most interest is what these 80 years of history have given us, namely accumulated knowledge and experience. There is therefore reason enough to save some time and reflect on this extension in time, questioning the roots of Portuguese Landscape Architecture education and practice and how it has evolved, if there is a clear and unequivocal acknowledgement of the discipline today and most importantly what can we envision for the next 80 years?

In this search around the hypothesis of a particular Portuguese Landscape Architecture identity, the legacy of landscape architects who dared to dream beyond conventional boundaries is a constant presence. Indeed, over the course of eight decades, Portuguese Landscape Architecture has absorbed the traits of knowledge, creativity and courage of the pioneers of the profession, as well as the strong personalities that followed them, leaving indelible marks on the national and international arenas. Fortunately, these marks have been cultivated, have flourished and have given fruits, particularly through Landscape Architecture education which, from north to south of the country, has been maturing with continuous cycles of unwavering dedication.

In this book you may find specific highlights on heritage and history of Portuguese Landscape Architecture, on the Portuguese context of Landscape Architecture education and pedagogy and on Portuguese perspectives on theory and methods in Landscape Architecture, each corresponding to the three parts that

DOI: 10.4324/9781003454007-1

structure the publication. Three different themes where the resonance of shared wisdom is revealed not only in the clear and robust foundation from which the discipline emerged, but also in the recognition of a Portuguese Landscape Architecture journey of consistent and rich growth with a promising future.

Although this book brings together a particularly selected set of scientific articles (See Methodology and review process), the scope of the presented results is not limited merely to the academia. The consolidation of research lines specifically related to landscape design, planning and management is unravelled over the course of the chapters, showing a clear and enmeshed connection between educational and professional activities, furthermore elucidating how Portuguese Landscape Architecture is at the forefront of the theory and methods specific to the discipline, namely in the fields of sustainability, resilience and interdisciplinarity applied to the architecture of complex landscape systems. It is a book that uses the past to explore the present and prospect the future. It is the celebration of an enriching dynamic between intellectual heritage and contemporary thought whose outreach in results, reflections, criticisms, challenges or prospects goes beyond the exclusive scope of the Landscape Architecture discipline.

Flipping through the chapters of "Portuguese Landscape Architecture Education, Heritage and Research: 80 Years of History" is to be transported through the fundamentals of ecological thinking, which is here considered as the basis of Landscape Architecture education and a powerful tool of Landscape Architecture practice. Three main principles of ecological thinking are therefore highlighted as a proposed framework to interpret these past 80 years and prospect the 80 years to come, more specifically: 1) the multiplicity, diversity and interconnectivity of relations; 2) the open-ended processes of cycles and systems and 3) uncertain and variable dynamics as a driving force for positive change.

Very much like the human being, landscapes are healthier and more resilient the stronger, safer, wider and more complex and multiple their anatomies and their interconnections are, both internally and externally from their systems. An ecological thinking principle that constantly defies existing trends in favour of any type of mono-culture (culture here applied in all its meanings, from the breading plants or animals to education and civilisational values). As manifested in the advanced research presented in this book, Landscape Architecture in Portugal has been developing openly and far from a self-absorbed path. It has namely grown and refined as a result of different types of internationalisations, either through direct, individual, exchange of knowledge and experience (such as the case of Cabral and Porcinai in Chapter 4, "Landscape Architecture education in Portugal and Italy"), or through long-ranged international collaborations between universities (as confirmed in Chapter 11, "The legacy of three generations of architects-professors" or Chapter 19, "Multiscalarity, interdisciplinarity, and research-by-design towards a metropolitan landscape") or by expanding national offices abroad (such as mentioned in Chapter 11, "The legacy of three generations of architects-professors"). Indeed, the multiplicity, diversity and interconnectivity of relations as a primary ecological category that must be prioritized, encompasses not only the relations with or within ecosystems, but also the convivial and cultural relations from within.

Relations that further encompass different ways of seeing and experiencing time, space and places (Barata, 2022).

Understanding and respecting natural cycles, such as the water cycle, nutrient cycles or energy flow, together with the acknowledgement of their "open-ended" dynamics, is forcefully attached to a strong and particularly intimate relation with the time factor. Time, being the underlying engine of any process, is thus instinctively seen as value of itself within this "Ars cooperative Naturae" (the art to convince nature to collaborate with us), as Cabral so straightforwardly described the discipline back in 1961 (Andresen, 2001). Later in 1984 also J. B. Jackson referred that "the act of designing Landscape is a process of manipulating time" (1984). In the same line of reasoning, João Ferreira Nunes further develops upon the idea of "domestication" when describing his way of working and designing with natural processes for the benefit of Man (Nunes et al., 2011), more so recognizing the existence and the need of compliance with the multiple "times" existing as both constituents and actors in the landscape, together with the unrepeatability of moments (Costa, 2021). Overall, as a Human artefact that only exists in tandem with time, Landscape is therefore forever unfinished, making landscape design an eternal open-ended work. As a result, long-term thinking as a 'modus operandi', alongside a permanent elasticity of thought between scales and the times of different processes, is fundamentally ingrained to a Landscape Architecture endeavour. The long-term thinking that is crucial for planned adaptation (Costa, 2013), as well as the long-term thinking that takes advantage of accumulated knowledge, our common memory, to prospect and plan about the forever uncertain future (Matos Silva, 2020). The presented overview of Portuguese Landscape Architecture Education, Heritage and Research" has brought great expectations to a long future of Landscape Architecture in Portugal. Yet as conveyed in the overall reading, the themes here gathered under scrutiny envision distant timeframes. An approach that implies constant and never-ending revisitation and challenging as an established methodology. A process of growth in maturity, which necessarily implies self-criticism, the acceptance of delusions and the willingness to change.

Lastly, also latent throughout the research conveyed in each chapter, is Landscape Architecture's foundation on uncertain and variable dynamics as a driving force for positive change. In various statements Cabral mentions that the fundamental condition of life is variety and balance (Cabral & Ribeiro Telles, 1999 [1960]) arguing that the greater the variety (also of relations, as mentioned earlier), the more stable is the balance. António Christofoletti named this notion of "equilibrium" interpreted as a dynamic balance occurring in an open system in which the import and export of energy and matter are equated through adjustments in the forms of the system itself (1936). The concept of "homeostasis", defined by Walter Cannon (1929) and applied to Landscape by Cabral, also introduced a dynamic perspective in the way of looking at the world, challenging the notions of frontiers or limits within which Man can act in order to maintain the resilience of systems (AA.VV., 2010). Indeed, the quest proposed by the discipline's pioneers for the promotion of an ongoing dynamic equilibrium, as in the ongoing search for balance in the settling of in/out exchanges within its constituent open-ended cycles and systems, is here considered as the engine that feeds Landscape Architecture greatest creative achievements. This restlessness of continuously wanting to improve the equation on the management of natural resources and overall energy flows towards a result that optimally should tend to zero is clearly evidenced in recent Landscape Architecture endeavours that artistically used cultural and natural systems to solve complex problems such as flooding, water scarcity, mobility, biodiversity loss or resource management. Consider, for instance, the sustainable drainage systems in Asprela Park at Oporto (2022); the combination of active mobility systems with a pre-established car traffic network at the "water plaza" Parque Gonçalo Ribeiro Telles (2021); the management plan of Parque das Serras do Porto (2018); the planting plan composed of dune flora at Praia dos Pescadores, Tagus Linear Park at Vila Franca de Xira (2012) and the retention basin as part of the open-air amphitheatre of Quinta das Lagrimas, Coimbra (2008), just to name a handful of examples. As demonstrated throughout the book, Portuguese Landscape Architecture education and practice is a reference outside its frontiers, and namely in the European context, not so much by the age of the course but mostly due to its inherent characteristics alongside the quality of the outcomes produced over time. The conclusions that arise out of the presented range of scientific articles may therefore further contribute towards a comprehensive study of Landscape Architecture education and practice in Europe. Indeed, considering that history informs the present and that the present prepares for the future, this book is an important milestone not only for self-awareness and assessment, providing new perspectives from within, but also as a means which enables scrutiny by colleagues across borders. In other words, it is a book that may lead the way for further critical reasonings that, by challenging what appears to be stable and predictable, may significantly contribute towards a more informed, consistent, varied and ambitious assessment regarding Portuguese Landscape Architecture, its present day and future to come.

Coupled to this disciplinary way of thinking and acting founded in ecology is a strong sense of responsibility. A weight that mobilises in the face of the long announced, and persisting, ecological crisis. An ecological crisis that is not only environmental, but also a social and mental crisis, as argued by Félix Guattari in 1989 (Guattari) and reinforced by the widespread demands of contemporary youth. This urge to contribute towards a better future comes with the recognition of the value of an active role of Landscape Architecture in the decision-making of a new global society, adding to the debate on hypotheses, proposals and strategies associated with alternative social and economic models. Yet in doing so, Landscape Architecture education and practice must be revisited in light of today's reality and its present-day major concerns, not only regarding the environment, but also in matters such as social justice, inclusion and mental health. More specifically it must focus on issues related to climate change, environmental equity, deceleration and degrow, salubrity, soil loss, carbon capture, ecological footprint, active mobility, spatial cohesion, sustainable tourism and cultural and natural heritage, among others.

For all the aforementioned, "Portuguese Landscape Architecture Education, Heritage and Research: 80 Years of History" constitutes both an exclamation and a critical introspection. In an era with a parallel virtual world, in which daily lives are dominated by virtual interactions and relations, it is important to re-empower Landscape Architecture cultural, aesthetic and ecological competences. Not only by protecting and revealing the existing, but also by creating and producing new landscapes, with new ecologies, cultures and aesthetic values that enable new or improved connections (with our own nature and between natures). This book is therefore one more effort towards the need for Landscape Architecture to embrace responsibilities and be at its best, namely by evidencing and promoting the core values of its academic reasonings, professional technical competences and political courage. Facing the value of the testimony of the indelible spirit of this large family that built the 80 years of Portuguese Landscape Architecture education and practice, especially those who dared to dream, create, and redefine the many possibilities, it is important to emphasise that time didn't stop and that new dreams are needed.

Reference list

AA.VV. (2010). *Caracterização da Arquitectura Paisagista em Portugal*. Instituto Superior de Agronomia, Universidade Técnica de Lisboa. Available at: https://apap.pt/wp-content/uploads/2017/08/Caracterizacao-da-Arquitectura-Paisagista-em-Portugal.pdf [Accessed 13th November 2023].

Andresen, T. (2001). *Francisco Cadeira Cabral*. United Kingdom: LDT Monographs No. 3.

Barata, A. (2022). *Para viver em qualquer mundo. Nós, os lugares e as coisas*. Lisboa: Documenta.

Cabral, F.C., & Ribeiro Telles, G. (1999 [1960]). *A Árvore em Portugal*. Lisboa: Assírio & Alvim.

Cannon, W.B. (1929). Organization for physiological homeostasis. *Physiological Reviews*, IX, 399–431.

Christofoletti, A. (1936). *Geomorfologia*. São Paulo, Brazil: Universidade de São Paulo, Editora Edgard Blücher Ltda.

Costa, J.P. (2013). *Urbanismo e Adaptação às Alterações Climáticas, as frentes de água*. Lisboa: Portugal, Livros Horizonte.

Costa, R.S. (2021). *Rising Series*. Available at: www.ruisoarescosta.com/.

Guattari, F. (2000 [1989]). *The Three Ecologies*. London and New Brunswick, NJ: Original from the University of Michigan, Athlone Press.

Jackson, J.B. (1984). *Discovering the Vernacular Landscape*. New Haven and London: Yale University Press.

Matos Silva, M. (2020). *Public Spaces for Water. A Design Notebook*. London, UK: CRC Press Routledge Taylor & Francis Group.

Nunes, J., Ribas, C., Zollo Espejo, D., Zoilo, I., & Campos, T.T. (2011). *Concursos perdidos—Arquitectura Paisagista = Lost competitions—Landscape Architecture*. Trento, Italy: List—Laboratorio Internazionale Editoriale.

Ribeiro, T. (1862). *A Portugal. D. Jayme ou A Dominação de Castella*. Rio de Janeiro: Typ. Portugal e Brasil de J.P. da Silva Rocha.

Methodology and review process

Maria Matos Silva and Anubhav Goyal

Introduction

In the year 2022, Portuguese education in Landscape Architecture fulfilled 80 years of existence. To celebrate this important milestone, several events were carried out under the overall guidance of the executive committee composed of Cristina Castel-Branco, Teresa Andresen, Luís Paulo Ribeiro, and João Ferreira Nunes. These celebrations envisioned not only the promotion of the discipline of Landscape Architecture as one of the first to apply the principles of ecology to its professional practice, but also the honouring of two of the most prominent Landscape Architecture personalities that enabled the creation of the Landscape Architecture course in Portugal, namely Francisco Caldeira Cabral (1908–1992) and Manuel de Sousa da Câmara (1929–1992). These activities included: 1) an international conference "80 years of Education in Landscape Architecture: Vanguard teaching for a pioneering practice in sustainability" (coordinated by Ana Luísa Soares), which had the participation of referenced personalities related to Landscape Architecture such as Carl Steinitz (Harvard Graduate School of Design and University College London, and Honoris Causa from Instituto Superior de Agronomia, University of Lisbon, 2023), Richard Stiles (Vienna University of Technology), James Wescoat (Massachusetts Institute of Technology), Lilli Lička (University of Natural Resources and Life Sciences, BOKU, Vienna), and Norbert Kühn (Technical University of Berlin); 2) an itinerant exhibition "80 Years of Landscape Architecture Teaching in Portugal (1942–2022): Art and Ecology" (curated by Sónia Talhé Azambuja) that travelled among the existing Portuguese universities with Landscape Architecture degrees; and 3) the inauguration of a tree-lined avenue at Tapada da Ajuda in honour of the founder of Landscape Architecture in Portugal, Francisco Caldeira Cabral (coordinated by Cristina Castel-Branco). This book is also formulated as part of these celebrations.

Aim and objectives

The primary objective of this publication is to provide insight upon the questions of "How has Landscape Architecture education evolved in Portugal throughout these 80 years?"; "What enduring influences from Cabral and Sousa da Câmara education programme can be identified in the current curriculum, or in recent or ongoing

DOI: 10.4324/9781003454007-2

research projects and initiatives?"; "What important milestones, throughout these 80 years, can be considered relevant for the identity and future outcomes of Landscape Architecture in Portugal?"; "What has been the relevance of the teachings of the first generation of landscape architects today, namely facing the great contemporary challenges (ecological, climatic, social, economic)?"; and "How has education in Landscape Architecture in Portugal been reflected in an exchange of knowledge with other countries?". Based on these general and encompassing questions, alongside associated reflections, the scope of the "call for papers" is predefined under three main thematic blocks—Heritage and History, Education and Pedagogy, and Theory and Methods—entailing academic theories and concepts that outline the subject matter and offer new insights regarding the past, the present, and the future of Landscape Architecture in Portugal.

The editors of this scientific publication, responsible for the quality and coherence of the outcome, include the members of the previously mentioned executive committee of the joint celebrations[1] and Maria Matos Silva who was put in charge of the publication overall coordination.

Structure

The encompassing scope of research chapters is organised in three parts: Part I—Heritage and history of Portuguese Landscape Architecture; Part II—A Portuguese context on education and pedagogy in Landscape Architecture; and Part III—Portuguese perspectives on theory and methods in Landscape Architecture.

In the first part evidence is presented on the pioneering endeavours of the first generation of Portuguese landscape architects and their role in the establishment and identity of a new practice-oriented profession with strong cultural, ecological, and aesthetical principles of action. More specifically, personalities such as Francisco Caldeira Cabral, the founder of Landscape Architecture education in Portugal, and Gonçalo Ribeiro Telles (Sir Geoffrey Jellicoe award 2013), well-known for its determinant political achievements, are mentioned as the pillars from which Portuguese Landscape Architecture education, research, and practice developed from.

The second part, dedicated to education and pedagogy in the specific Portuguese context, highlights not only the relevance of outstanding professors and their talent in the art of teaching, but also the educational opportunities that come from overcoming frontiers and connecting disciplines. It is furthermore reinforced the fundamental importance to base education on a solid and clear identity framework, as well as to continuously assess learning processes so that to endure the rightful high expectations that the future has for the profession of Landscape Architecture.

The third and final part comprises a sample of Portuguese perspectives on theory and methods in Landscape Architecture, including research on the historical evolution, future management, and governance prospects of Portuguese landscapes, from north to south, from specific climatic regions and river's watersheds to differentiated ecologies in urban contexts.

Operational processes: Selection and review of submitted manuscripts

Preparatory actions

At an early planning stage, to assess the relevance and interest of a scientific publication as this one, encompassing the aforementioned characteristics, editors enquired the national research centres with significant and compelling research work in the area of Landscape Architecture, more specifically: CHAIA (Centro de História da Arte e Investigação Artística) of Évora University; CIBIO-InBIO (Research Network in Biodiversity and Evolutionary Biology) of the Faculty of Sciences, University of Porto; and LEAF (Linking Landscape, Environment, Agriculture and Food) and CEABN (Centro de Ecologia Aplicada Prof. Baeta Neves) of Instituto Superior de Agronomia, University of Lisbon.

In view of the positive general interest shown by the current national Landscape Architecture academy, a scientific committee was set up encompassing both national and international experts. In charge of ensuring the pertinence and quality of received entries, the scientific committee entailed a total of 24 experts from 10 different nationalities: Brazil, France, Germany, Hungary, Italy, New Zealand, Portugal, Spain, Türkiye, and United States of America (Table 0.1).

Table 0.1 Names and affiliations of contributing members of the scientific committee.

	Name	**Affiliation**
1	Albert Fekete	Hungarian University of Agriculture and Life Sciences (MATE), Hungary
2	Ana Beja da Costa	University of Lisbon, School of Architecture, Portugal
3	Anatole Tchikine	Curator of Rare Books, Dumbarton Oaks Research Library and Collection, Trustees for Harvard University
4	Andre Santos Nouri	Bilkent University, Turkey
5	Anette Freytag	Rutgers University, USA
6	Bruno Marques	Victoria University of Wellington, New Zealand
7	Camila Gomes Sant' Anna	Federal University of Goiás, Brazil
8	Carlos Terra	Federal University of Rio de Janeiro, Brazil
9	Cecilia Herzog	Pontifical Catholic University of Rio de Janeiro, Brazil
10	Christian Kuepfer	University of Nürtingen-Geislingen, Germany
11	Cláudia Fernandes	University of Porto, Faculty of Sciences, Portugal
12	Desidério Batista	University of Algarve, Faculty of Sciences and Technology, Portugal
13	Jorge Cancela	University of Lisbon, School of Architecture, Portugal
14	Leonor Matos Silva	University Institute of Lisbon (Dinâmia'CET), Portugal
15	Lucia Costa	Federal University of Rio de Janeiro, Brazil
16	Luigi Zangheri	Accademia delle Arti del Disegno, Italy
17	Marta Enokibara	São Paulo State University (UNESP), Brazil
18	Pedro Arsénio	University of Lisbon, School of Agriculture, Portugal
19	Pedro Calaza	Galician Agriculture Engineers College, Spain
20	Robert Ryan	University of Massachusetts Amherst, USA
21	Stéphanie de Courtois	Ecole Superieur de Paysage de Versailles, France
22	Teresa Amaro Alfaiate	University of Lisbon, School of Agriculture, Portugal
23	Teresa do Paço	University of Lisbon, School of Agriculture, Portugal
24	Tessa Matteini	University of Florence, Italy

The first call for abstracts was launched in the end of May, alongside the dissemination of an online webpage[2] that enabled updated information, namely regarding scope of the envisioned publication, editors, institutional supports, scientific committee members, and important deadlines.

To facilitate the coordination of the manuscript entries and the encompassing review process involving authors and the scientific committee, an established web-based interface was used. Furthermore, the role of the meta-reviewer was put in action as the overall external manager of the online platform and encompassing peer-review process, adding an extra layer of quality control. More specifically, the Conference Management Toolkit (CMT) platform[3] was used, and the role of meta-reviewer is assumed by Anubhav Goyal, co-author of this chapter and doctoral candidate at the Lisbon School of Architecture, University of Lisbon.

Peer review

The conducted peer-review process included two separate phases, one dedicated to the assessment of the submitted abstracts and a second phase focused on the assessment of the full manuscripts (Figure 0.1). In each phase, entries were assessed by two independent members of the scientific committee in a double-blind review process—i.e., a review process in which both the reviewer and author identities are concealed from each other. As the scientific committee members were publicly identified with names and affiliations, namely in the mentioned webpage of the encompassing celebrations, potential conflicts of interest were tackled by inquiring authors on such possibility as a mandatory question within the submission form as well as by a cross-verification by the meta-reviewer. This way, two members of scientific committee assured the peer review of each abstract and full manuscript,

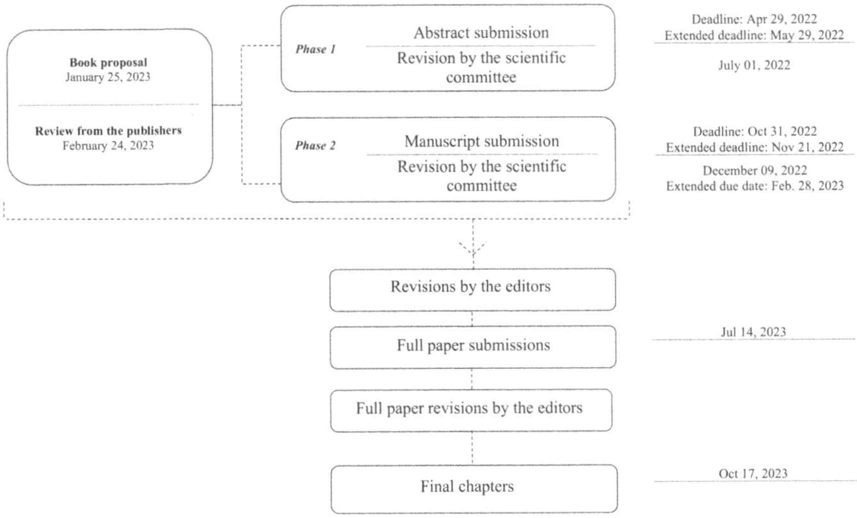

Figure 0.1 Methodology diagram.

certifying that its content met the requested high academic and scientific standards alongside adequate quality of writing. Whenever useful, the scientific committee provided additional feedback and suggestions regarding the given content in order to engage the authors with their best efforts and further contribute to the overall success of the envisioned publication. The selection process of abstracts and full manuscripts was therefore carried out in a homogeneous and horizontal manner, bearing in mind the main proposed objective highlighted in the "Call for Abstracts".

The call for abstracts resulted in 25 entries, from which 19 were selected to be develop into full manuscripts. After the systematic double-blind peer review of the abstracts, the results and accompanying assessment report was shared to contributing authors. The authors of the selected abstracts were further informed to develop and upload complete manuscripts in the online management platform by October 31, 2022, taking into consideration the remarks provided by the scientific committee and editors. Until the extended date, November 21, 2022, full manuscripts were uploaded in the online management platform, and by February 28, 2023, all manuscripts were subsequently reviewed by the scientific committee and book editors.

Review process: Assessment quality

Assessment questionnaire

The stages of review process and manuscript selection was made explicit from the start, being an exhaustive procedure based on principles of relevance and scientific quality, more specifically: originality, addition of value to the field, relevant bibliography, results, and implication for practice and research. This outline ensured the review process to be comprehensive, thorough, and objective. A realistic timeframe, with mandatory deadlines for the development of reviews was furthermore pre-established.

In addition, a clear and concise questionnaire was formulated, with mutually exclusive and exhaustive options as well as space for "comments for authors" and "comments for editors". The questionnaire covered various aspects such as the content quality, submission categorization, value added to the field, reviewer familiarity with the subject matter, submission length appropriateness, and a section for comments. Each section provides a spectrum of options (Table 0.2). This review questionnaire, within the online management platform, was provided for reviewers to fill in, not only as an additional guideline for assessment quality, but also as a form whose result would further assist authors in the improvement of the provided manuscripts. Through the online management platform, reviewers could also provide comments to authors and editors directly in the provided document in pdf or word formats. Overall, the questionnaire served as a comprehensive tool to help reviewers to systematically evaluate the submissions while providing clear feedback to both authors and the editors for improvement purposes.

Table 0.2 Questionnaire.

The questionnaire
1. Recommendation: Type: Options • Reject: Content inappropriate to the proposed book or has little merit • Probable Reject: Basic flaws in content or presentation or very poorly written • Marginal Tend to Reject: Not as badly flawed; major effort necessary to make acceptable but content well-covered in literature already. • Marginal Tend to Accept: Content has merit, but accuracy, clarity, completeness, and/or writing should and could be improved in time. • Clear Accept: Content, presentation, and writing meet professional norms; improvements may be advisable but acceptable as is. • Must Accept: Candidate for outstanding submission. Suggested improvements still appropriate
2. Submission Categorization: Type: Options • Highly theoretical • Tends towards theoretical • Balanced theory and practice • Tends toward practical • Highly practical
3. Overall Value Added to the Field: (Check as many as appropriate) Type: Options • New information • Valuable confirmation of present knowledge • Clarity to present understanding • New perspective, issue, or problem definition • Not much • Other
4. Reviewer Familiarity with Subject Matter: (Relates to the confidence you have in your review) Type: Options • Low • Moderate • High
5. Is the submission length appropriate: Type: Options • Yes • No • Unsure
6. Comments for the Authors: (Constructive comments to the author(s) would be appreciated) Type: Comment
7. Comments for the Editorial Board (authors will not see these comments): Note: Reasons must be included for all submissions, because they help us determine what to do when reviewers disagree with each other. Type: Comment

Review and assessment of publishing house

Considering the pertinence and scientific quality of abstracts accepted by the scientific committee, a book proposal was submitted to Taylor & Francis publishing house under the series "Project Thinking on Design" introduced by the Research

Centre for Architecture, Urbanism and Design (CIAUD). The application outlined the book's structure in three parts, corresponding to the three major themes initially proposed for authors in the "call for abstracts" (Heritage and History, Education and Pedagogy, and Theory and Methods), furthermore encompassing original manuscripts from each editor as well as introducing and concluding chapters.

The book application, which included the abstracts previously reviewed by the scientific committee, was subjected to an additional external peer review and assessment by Taylor & Francis publishing house. The evaluation resulted in a positive recommendation for publication with suggested revisions, namely regarding the scope and goals of overall intended outcomes. Editors acknowledged the conveyed suggestions, having considered the highlighted critiques in the conducted reviews of each submitted manuscript as well as in the development of their own original articles.

Results

How has the genesis of Landscape Architecture in Portugal influenced the practice and pedagogy of contemporary Portuguese Landscape Architecture? Is there a distinct identity of Portuguese Landscape Architecture? What professional and research futures may emerge from the discipline of Landscape Architecture and the accumulated knowledge of its maturity? The challenge put forward to the contributing authors was exactly to enrich this ongoing discussion. Does the past still hold lessons to be learned? How do we sustain and empower the profession? What future tendencies, in the ways we think and act, can contribute to an ever-enhancing process of what it is to be a landscape architect? What are the possible futures of Portuguese Landscape Architecture?

This book is planned to explore, examine, and develop upon these pre-defined queries, delving deeply into the proposed theme, presenting original research, analysis, and insights aimed at contributing to existing knowledge within the field. Indeed, the book contributes to strengthening the discipline's theoretical foundations by granting contemporary reasonings to its collective memory, serving the purpose to deepen the reflection upon new and improved educational models. Being an international publication, written in English, the book also enables peers from all over the world to attest and validate on the importance of what is being discussed, furthermore facilitating the dissemination of the history and culture specific to Portuguese Landscape Architecture.

The joint effort of 46 contributors, including the manuscript authors and the editors, provides a diverse and comprehensive perspective on the theme. A combination of various viewpoints and expertise that create a rich and well-rounded exploration of the subject which may serve as an essential resource for students, educators, and researchers; subtly counteracting the identified vulnerability of a potential "lack of a comprehensive history of landscape architecture in Portugal" (Chapter 5, "Landscape Architecture in Portugal: The quest for a needed memory"). Overall, the presented research evidences an encompassing self-assessment on the history and contemporary endeavours of Portuguese Landscape Architecture in the fields of education, heritage, and research. It furthermore encourages for its main findings to be critically scrutinized by peers from all over the world that, although maybe unfamiliar to the Portuguese context, carry equivalent ambitions for the future of education, heritage, and research in Landscape Architecture—altogether united in favour of ecologically rich, socially healthy, resilient, and beautiful landscapes.

Notes

1 Instituto Superior de Agronomia (ISA), Universidade de Lisboa. SOBRE—80 anos de Ensino em Arquitectura Paisagista // Comissão Executiva. [online] Available at: http://80anosap.isa.ulisboa.pt/sobre-2/ [Accessed 10 November 2023].
2 Instituto Superior de Agronomia (ISA), Universidade de Lisboa. Research in Landscape Architecture: Celebrating 80 Years of Landscape Architecture Education in Portugal—CALL for ABSTRACTS. [online] Available at: https://80anosap.isa.ulisboa.pt/research-in-landscape-architecture-celebrating-80-years-of-landscape-architecture-education-in-portugal-call-for-abstract/ [Accessed 11 October 2023].
3 Microsoft Conference Management Toolkit. [Online] Available at: https://cmt3.research.microsoft.com/User/Login [Accessed: October 11, 2023].

Part I

Heritage and history of Portuguese Landscape Architecture

A time frame for Landscape Architecture

Teresa Andresen

How often have we heard to say that heritage and history are different but inextricably linked? History is about the past while heritage is about a process and is both tangible and intangible. Heritage is both legacy and understanding on a narrative developed from a constructive discourse.

However, one cannot help to ask "Whose History?" and "Whose Heritage"? The editors tell us: heritage and history of Portuguese Landscape Architecture . . . and we ask "For Whom?". I'll let you know: for those who might share a desire to build a common understanding on Portuguese Landscape Architecture.

Dear Reader, the editors provided the timeframe: the last 80 years of Portuguese Landscape Architecture which carries the modern, the contemporary, and the future, the place, and the heritage of antient, multicultural and geo-biodiverse roots—eight decades that bring together the Western world from the brinks of an era to the ones of another, with many sub-eras in between, where war and disruption persist in the air.

Dear Reader, whether you are a landscape professional or not, if you are thinking that you are being invited to Utopia, please, go on and bear with me. I shall be thankful for I write for a continuous growth of my understanding as a creature of nature who has not given up to trust a viable practice for a physically, mentally and spiritually healthy landscape. The possibility of the preservation of the rights of the individual together with the likelihood of the achievement of collective work is for me an enduring challenge that encompasses the difficult, the best and the correct. I must warn you that your Writer—me—has her life of almost six decades been chained with Landscape Architecture to a point not to be clear about their differences. About 30 years ago, I began to understand and pay attention to the ideological side of Landscape Architecture and landscape architects as a group. Consciously or unconsciously, ideologies carry values and beliefs that tend to shape our views but also our misconceptions of the social reality (please, let me set aside the

DOI: 10.4324/9781003454007-3

political realm at this stage). However, they direct our decisions in the process of making sense of our actions and influencing the course of things.

Having shared these thoughts, dear Reader, I invite you to what I consider a first-time insight and an attempt by us to bring together a set of perspectives to a much-needed debate and well-timed inspiration for practice and research on Portuguese Landscape Architecture. From their reading, I propose a time frame for Landscape Architecture in Portugal divided into three moments: i) the first 30 years: from the 1940s to the 1970s, colonial and surviving authoritarianism; ii) the last 30 years of the 20th century: from the 1970s to the 2000s, European and learning to be democratic; and iii) the first two decades of the 21st century: from 2000s onwards, global and fighting dystopia.

Inspired by the 80 years of ongoing education in Landscape Architecture in Portugal, this book was envisioned "to present an overview of the Portuguese Landscape Architecture identity through a selected number of scientific articles gathered in three encompassing themes: Heritage and History, Education and Pedagogy, and Theory and Methods" (Introduction). For the first theme, the authors were invited

> to reflect on this extension in time, questioning the roots of Portuguese Landscape Architecture education and practice and how it has evolved, if there is a clear and unequivocal acknowledgement of the discipline today and most importantly what can we envision for the next 80 years?
>
> (Introduction)

I thank the editors and the authors.

The theme of "Heritage and history of Portuguese Landscape Architecture" counts eight contributions. Four of these are centred in the figures of the early players: "Lisbon city gardeners, from horticulture to Landscape Architecture (1850–1950)"; "Francisco Caldeira Cabral and the 'Panorama' magazine: Introducing a landscape conscience in tourist propaganda"; "Landscape Architecture education in Portugal and Italy. The Pioneering Approaches of Francisco Caldeira Cabral and Pietro Porcinai" and "Gonçalo Ribeiro Telles: Landscape architect, ecologist and politician in Portugal". The other four focus on practice and legacy: "Historic gardens as cultural heritage. From early debates to future initiatives"; "Landscape Architecture in Portugal: The quest for a needed memory"; "Thirty years of research in green infrastructure and landscape planning" and "From historical research to urban ecology: A new perspective on the evolution of landscape art".

The first 30 years: From the 1940s to the 1970s, colonial and surviving authoritarianism

Essentially, the first four decades of Landscape Architecture in Portugal can be identified with Francisco Caldeira Cabral (1908–1992). He is the founder and direct mentor of two generations of Portuguese landscape architects through the second half of the 20th century. He completed his degree in agronomy at the Higher Institute of Agronomy (*Instituto Superior de Agronomia, ISA*) in Lisbon in 1936 and went to Germany where he studied from 1936 to 1939 at the at the Institute for Garden Design (*Institut fur Gartengestaltung*) at the College of Agriculture (*Landwirtschaftlich-Gärtnerische Fakultät*) in Berlin. Cabral began lecturing in agronomy at ISA in 1940, and in 1942 he established the Autonomous Course in Landscape Architecture (*Curso Livre de Arquitectura Paisagista, CLAP*), an independent course in Landscape Architecture. In 1953 he inaugurated Landscape Architecture research at the Centro de Estudos de Arquitectura Paisagista (CEAP). From 1950 to the mid-1970s, Cabral achieved remarkable visibility through his involvement with the International Federation of Landscape Architects (IFLA), for which he served as both vice president (1960–1962) and president (1962–1966). At the same time, he found strategic placement for the young landscape architects in key positions in the central administration services and in the municipality of Lisbon. They became instrumental in contributing to an innovative and insightful approach to urban and rural problems in Portugal during the post-war period in the realms of design, planning and public policy and proved to have been a particularly effective and enduring enterprise.

In addition to teaching, he developed his own landscape design practice and performed other functions and activities as a member of the National Education Council and founder and president of the League for the Protection of Nature (*Liga para a Protecção da Natureza*). While a student in Berlin, Cabral had the collaboration of German experts on the project for the National Stadium in Portugal that marks the beginning of the practice of the Portuguese Landscape Architecture. His design work was mostly devoted to private estates and gardens, as well as a range of projects for public spaces, such as squares, gardens or parks. His activities at IFLA prevented him from taking on more ambitious projects until his retirement from ISA in 1975, after which he enjoyed almost two more decades dedicated to landscape design and planning in his office at Avenida da Liberdade, in Lisbon.

The Western roots of modern Landscape Architecture education are at the Royal Gardeners School (*Königliche Gärtnerlehranstalt*, KGL), in Potsdam (Sanssouci) and in Berlin (Botanical Garden), approved by Friedrich Wilhelm III, in 1823. KGL was founded and led by Peter Joseph Lenné (1789–1866) for three decades. It inaugurated the official teaching of horticulture and gardening in Prussia and became a centre for the dissemination of horticultural training, the practice of gardening and the development of artistic thought and theory. The innovative study programme at KGL emphasized the scientific, the practical and the artistic, a tripart training that became strongly connected with the design of parks and gardens. The school's excellent reputation for comprehensive training was renowned throughout Europe and even beyond. In 1903 the school was transferred to Berlin-Dahlem, the new location of the botanical garden. In 1929, a Chair of Garden Design was established at the College of Agriculture in Berlin by Erwin Barth (1880–1933). When Cabral began his study of Landscape Architecture in 1936, Barth's position of Chair of Garden Design had been assumed by Heinrich Friedrich Wiepking-Jürgensmann (1891–1973), at the College of Agriculture, in close collaboration with the architecture programme at the Technical University of Berlin-Charlottenburg. Cabral's 1936–1939 sojourn in Berlin attending the Institute for Garden Design coincided with the dominant years of National Socialism in Germany prior to the war. The school remained faithful to its ancient roots preparing a profession ready to move forward from landscape gardening and adapting to the challenges of society encompassing the realms of art, ecology, cultivation, building and sustainable development, an evolving school that represents the cradle of modern Landscape Architecture education in the Western world.

The impact of Cabral in the profession of Landscape Architecture in Portugal and worldwide in a little more than two decades has been summarized:

> i) created a curriculum and program of research in an entirely new disciplinary area; ii) introduced professional practice as a vital component of one's education; and iii) achieved remarkable international visibility during a period when authoritarian rule could have very well hindered that progression.
>
> (Andresen et al., 2023)

Barbara Birli has pointed out the uniqueness and pioneering experience of Cabral's influence on Portuguese Landscape Architecture by placing it in the context of other European educational courses: "Cabral may be seen as first example of knowledge transfer from one country to another and an early form of transnational cooperation in Landscape Architecture education with one university providing knowledge for the foundation of another university" (Birli, 2016, p. 156).

Cabral's role as an educator, a landscape expert and a conservationist is seminal and supports a practice based on an educational system that values common good, public interest in the productivity of resources, cultivation and conservation while also encompassing art, science, and technique, aesthetics and ecology as well as functionalism and a multiscale approach. The significance of the profession in Portugal matured while: i) reinforcing the autonomy and professionalization of the activity within several public institutions, both at the local and the national level, from urban public space to land use planning, to nature conservation; ii) pioneering the integration of ecological and cultural principles in landscape design, planning and management; and iii) working in multidisciplinary teams.

Dear Reader, our authors' contributions will be of your interest. They link the roots of Portuguese Landscape Architecture to the 19th century landscape gardening following the Western trends as in other countries and explore the personalities and achievements of Cabral and Ribeiro Telles.

The last 30 years of the 20th century: From the 1970s to the 2000s, European and building democracy

Cabral's active life mostly took place under the authoritarian regime of 'Estado Novo'. In April 1974, the democratic revolution broke out, introducing profound social, political and economic changes in Portugal. After his leave from ISA in 1975 he went on with teaching at the Évora University Institute, funded in 1973 and 6 years later renamed Évora University where his disciple Gonçalo Ribeiro Telles (1922–2020) was in charge of a new landscape programme (Andresen,

2001). Since the 1960s, Telles was recognized for his professional activity and was an emergent politician leader of the environmental cause. He and two others of Cabral's disciples—António Viana Barreto (1924–2012) and Ilídio de Araújo (1925–2015)—played a major role preparing for a new landscape planning framework in the period of 1976 to 1986 leading to Portugal's admission in the European Economic Community. In 1976, the National Service of Parks, Reserves and Landscape Heritage (Decree law No. 613/76 of July 27) was created under the leadership of landscape architect Fernando Pessoa. Telles served as Minister of Environmental Quality from 1981 to 1983 leaving office with the publication of the National Ecological Reserve (Decree law nº 321/83 of 5 July), the infrastructure for the integration of "areas essential to the ecological stability of the environment and the rational use of natural resources, with a view to the correct planning of the territory". In 1987 the Framework Law for the Environment (Law nº 11/87 of 7 April) was published with the indication of the following human environmental components: i) the landscape, ii) the natural and built heritage, and iii) the pollution (Article 17).

We can consider this period as the strong years of Portuguese Landscape Architecture in planning and public policies for spatial planning, housing and nature conservation, a process developed through the emergence of new disciplinary areas, such as environmental engineering or biophysical engineering that entered pioneering areas that so far had been of the realm of Landscape Architecture. Times were of crisis and recession and much pressure for urban growth and a fast abandonment of rural areas. Landscape design activity experienced a very low pace however a number of small Landscape Architecture offices emerged that today play a significant position in Portugal and in the international arena.

Two years after the democratic revolution of April 1974, Portuguese landscape architects funded the Portuguese Association of Landscape Architects (*Associação Portuguesa dos Arquitetos Paisagistas, APAP*) that in 1989 became a founding member of the European Landscape Architecture Foundation (EFLA), today, IFLA Europe. Another landmark in terms of Landscape Architecture education was the year of 1981. In ISA, CLAP became a full recognized degree in October under the visionary leadership of another of Cabral's disciples, Manuel de Sousa da Câmara (1929–1992), and at Évora University, in March, a new course was created with Telles and Cabral as founders and mentors. About 20 years later three more new programmes were created at the universities of Algarve (1999), Trás-os-Montes e Alto Douro (1999) and Porto (2001).

The 1990s witnessed a key event with a strong impact on Landscape Architecture in Portugal. We refer to the Expo '98, the official specialized World's Fair held in Lisbon under the theme "The Oceans, a Heritage for the Future". It received over 10 million visitors in 132 days, and 143 countries were represented, and it was the beginning of a new expansion of Lisbon towards the East. National and foreign Landscape Architecture offices had then a significant opportunity. In a way, Landscape Architecture recovered to a full potential of use of its competences adapted to the changes of time, space and society. In this decade the concept 'landscape' became appropriated by other disciplinary areas—from neuroscience to political science—including the media. This coincided with the growth of ecological awareness and the disciplinary realignment of ecology extending to urban ecology, ecological urbanism, ecological psychology . . . or landscape urbanism.

Dear Reader, the contributions from the authors focus diverse pertinent topics which are fundamental to the understanding and the debate of this time period and professional competencies, including practitioners, academicians and researchers. I refer to the emergence, since the 1980s, of the development of methodologies to approach cultural and natural heritage through historic gardens conservation and the promotion of green infrastructures—another topic highly representative of continuous Portuguese Landscape Architecture performance—and the more recent awareness of the importance of archives to support the history of Landscape Architecture in Portugal.

The first two decades of the 21st century: From 2000s onwards, global and fighting dystopia

Dear Reader, I leave you with our authors wishing you a fruitful reading and thinking, aware of our proactive and defensive attitudes. Professional boundaries are blurring in spite of technology and specialization. Information and communication technologies; artificial intelligence; platforms and applications; the increasing primacy of participatory, shared and adaptive decision-making processes at various levels of landscape governance; the multidisciplinary approach to landscape in design, management, conservation and planning and the desirable awareness of landscape-based policies are installed and moving fast. So is climate change, natural resources depletion, pollution, human population health, hate and anger. Sometimes, I wonder about the quests of the pioneers of Portuguese Landscape Architecture and feel inspired by their continuous commitment to a profession for the common good.

I wish to evoke Ken Fieldhouse (1948–2002), a landscape architect and town planner, and his commitment to promote a range of environmental publishing initiatives and focused on counteracting the scarce tendency for landscape architects to write. With Sheila Harvey, he was the editor of "Landscape Design Journal" and of the "Landscape Design Trust" monograph series from the Landscape Institute (UK). His untimely death led this project just to the publication of four monographs: Geoffrey Jellicoe (Harvey, 1998), Sylvia Crowe (Collens & Poell, 1999), Francisco Caldeira Cabral (Andresen, 2001) and Peter Shepheard (Dows, 2004). Thomas Church and Christopher Tunnard were to follow. The invitation of Ken Fieldhouse to write Cabral's monograph was a challenge, and only later I realized how well-timed it had been. I had the possibility to access Cabral's private archives and build a wide perspective on his work and ideas. The Gulbenkian Foundation (UK) and EDP supported the Trust for this purpose.

The arts' director, Manuel Costa Cabral, at the Gulbenkian Foundation invited me to prepare an exhibition at the Lisbon headquarters including the works and life of Cabral and the first generation of Portuguese landscape architects. The

Gulbenkian garden authored by Viana Barreto and Ribeiro Telles (1969), can be considered the work of maturity of Portuguese Landscape Architecture. The catalogue of the 2003 exhibition "From the National Stadium to the Gulbenkian Garden: Francisco Caldeira Cabral and the first generation of Portuguese landscape architects, 1940–1970" was the first comprehensive perspective on this time period. In October 2003, the Annual Conference of the European Council of Landscape Architecture Schools (ECLAS), "Landscape architecture and modernism: exploring the heritage and learning the lessons", met in Lisbon at the Gulbenkian Foundation, and it all resulted in a fruitful combination of efforts. I consider that two significant outcomes result from this project: i) the initiation of an archive of Portuguese Landscape Architecture from private firms at the so-called "Arquivo do Forte de Sacavém" under the Ministry of Culture and a photographic archive for the Arts Library at the foundation of all the drawings collected as part of the research process for the exhibition and catalogue from private and public institutions, such as Lisbon municipality, and a photographic archive of Landscape Architecture works that supported the exhibition and catalog (2002–2003); and ii) academic research in Portuguese Landscape Architecture.

On the last day of the exhibition, lasting negotiations were initiated between the general director of the archives and the owners or their heirs about the destiny of the materials assembled, surveyed and classified. Regarding research, it was significant in the creation in 2011 of LINK, a PhD programme in Landscape Architecture and urban ecology, a joint effort of ISA, Porto University and Coimbra University. I would like to emphasize two doctoral thesis that have been completed, "Contributions of Landscape Architecture to the public space of Lisbon (1940–1970)" by Teresa Camara (2015, later published in 2021) and "The German influence on the genesis of Landscape Architecture in Portugal" by Ana Catarina Antunes (2019). This research was preceded by a doctoral thesis, "From the landscape gardeners and horticulturists of Porto in the 19th century to modernism in landscape architecture in Portugal", by Teresa Portela Marques (2009), and a master thesis "The written legacy of Francisco Caldeira Cabral. Construction of theoretical thinking in landscape architecture" by Mariana Abranches (2014). The two last decades have witnessed an increase in publications whether monographic or scientific articles as well as books of contemporary Portuguese Landscape Architecture works.

The legacy of Landscape Architecture education and practice in Portugal is a long one in the European context. For the first three decades, Cabral and the profession were almost one. Having taught and practiced allowed Cabral to promote a generation of pioneers and future references of thought and practice for the following generations. The next decades brought in new areas of knowledge and practice, new Landscape Architecture programmes and new societal and environmental challenges. The profession grew to maturity and became part of globalization. The continuation of the profession will continue to adapt to the changes and to recreate working premises and techniques for an enduring healthy planet and life quality, daring "to dream beyond conventional boundaries" (Introduction).

Reference list

AAVV. (2003). *From the National Stadium to the Gulbenkian Garden: Francisco Caldeira Cabral and the first generation of Portuguese landscape architects, 1940–1970*. Coord. Teresa Andresen. Lisboa: Fundação Calouste Gulbenkian.

Andresen, T. (2001). *Francisco Caldeira Cabral*. United Kingdom: LDT Monographs No. 3 Editorial Board.

Andresen, T., Catarina Antunes, A., & Marques, T.P. (2023). Transmissions through landscape architecture education from Germany to Portugal. In Castel-Branco, C. (ed.), *Garden Transmissions: Word, Image, Experience, Future*. Turnhout, Belgium: Brepols Publishers.

Antunes, A.C. (2019). *A influência alemã na génese da Arquitetura Paisagista em Portugal*. Doctoral thesis, Faculdade de Ciências, Universidade do Porto. Repositório aberto Universidade do Porto. https://hdl.handle.net/10216/121946

Birli, B. (2016). *From Professional Training to Academic Discipline—The Role of International Cooperation in the Development of Landscape Architecture at Higher Education Institutions in Europe*. Dissertation, Technische Universität Wien, reposiTUm. https://doi.org/10.34726/hss.2016.38887.

Camara, T. (2015). *Contributos da Arquitetura Paisagista para o espaço público de Lisboa (1940–1970)*. Docoral thesis, Faculdade de Ciências, Universidade do Porto. Repositório aberto Universidade do Porto. https://hdl.handle.net/10216/85668

Camara, T. (2021). *Espaço público de Lisboa. Plano, projeto e obra da primeira geração de arquitetos paisagistas (1950-1970)*. Lisboa: Câmara Municipal de Lisboa.

Collens, G. & Poell, W. (1999). *Sylvia Crowe*. United Kingdom: LDT Monographs No. 2, Editorial Board.

Dows, A. (2004). *Peter Shepheard*. United Kingdom: LDT Monographs No. 4, Editorial Board.

Harvey, S. (1998). *Geoffrey Jellicoe*. United Kingdom: LDT Monographs No. 1, Editorial Board.

Marques, T.P. (2009). *Dos jardineiros paisagistas e horticultores do Porto de Oitocentos ao modernismo na arquitectura paisagista em Portugal*. Doctoral thesis, Universidade Técnica de Lisboa, Instituto Superior de Agronomia, Repositório da Universidade de Lisboa. http://hdl.handle.net/10400.5/2509

Pinto. Mariana Abranches. (2014). *O legado escrito de Francisco Caldeira Cabral: Construção de pensamento teórico em arquitetura paisagista*. Master's dissertation, Faculdade de Ciências, Universidade do Porto. Repositório aberto Universidade do Porto: https://hdl.handle.net/10216/78204

1 Historic gardens as cultural heritage. From early debates to future initiatives

Teresa Portela Marques and Maria José Curado

In 1988 a programme for the conservation of historic gardens was launched in Portugal, a joint initiative of the Portuguese Institute of Cultural Heritage (IPPC)[1] and the Portuguese Association of Landscape Architects, then chaired by Cabral (1908–1992). Thirty years after, in 2018, the European Year of Cultural Heritage, a meeting entitled "30 Years of Historic Gardens Restoration in Portugal" was held in Porto, a joint organisation of the Landscape Observatory of the Faculty of Sciences of the University of Porto (OP-FCUP) and the Portuguese Association of Historic Gardens (AJH) with the support of the Directorate-General of Cultural Heritage (DGPC) and the Portuguese Association of Landscape Architects (APAP). The meeting[2] brought together the promoters and protagonists of the programme, specialists in heritage and conservation of historic gardens in Portugal, and representatives of the DGPC and APAP, who analysed the initiatives implemented since the launch of the programme.

At the closing of the meeting, the document "Historic Gardens in Portugal and the Berlin Call to Action" was approved.[3] It appeals to competent organizations and all citizens to: 1) consider historic gardens as a central element of Portuguese and European cultural heritage; 2) consider including the UNESCO category of "cultural landscapes" in Portuguese legislation; 3) adopt greater and more effective funding to safeguard, restore, and promote Portuguese historic gardens; 4) consider historic gardens as a strategic resource for society, economy, culture, and the environment; 5) consider, as a collective task, the need to preserve historic gardens in order to transmit them for the enjoyment of future generations; 6) pay heed to the importance of stimulating stronger public involvement in safeguarding and transmission of Portuguese cultural heritage, manifested through gardens; and 7) encourage and support initiatives to facilitate the continuity of scientific research and the sharing and circulation of information. The document ends by stating the commitment that the Landscape Observatory (FCUP), in collaboration with AJH and APAP, constitutes itself as "a permanent platform for the collection of knowledge, training and coordination of 'cultural landscapes', which includes the category of 'gardens' as 'clearly defined landscapes, intentionally conceived and created by Man".[4]

All the concerns mentioned have been the focus of action by academia; associations, with emphasis on AJH and APAP; and individuals. The Landscape

DOI: 10.4324/9781003454007-4

Observatory (OP–FCUP) is currently surveying achievements in this field from 1988 to the present time.

The 2-year programme, initiated in 1988, aimed at an integrated and systematic conservation of historic gardens. APAP, particularly Cabral (then president), Teresa Andresen (general secretary), Ribeiro Telles, Edgar Fontes, Ilídio de Araújo, Aurora Carapinha, among others, supervised the work of young landscape architects. Although subjects related to the conservation and restoration of historic gardens were not, to date, an autonomous topic in the curricula of Landscape Architecture courses, the understanding of the artistic value and cultural significance of those gardens was, from an early stage, evident in the professional practice and academic teachings of the tutors.[5]

Actually, in 1943, Cabral regretted the lack of studies on Portuguese gardens and referred to the urgency of initiating them[6]

> coordinating all the scattered elements and collecting the largest possible number of data from all over the country. . .. From all of them it will be necessary, in the first place, to obtain plans that will allow us to thoroughly compare their layouts, photograph the most interesting aspects and record the plants species that are most commonly found there. At the same time, it will be necessary to collect as many historical references as possible which are scattered through publications and archives.[7]

Although his goal was primarily to investigate the existence of a 'Portuguese garden', to know the 'traditional characteristics' of the Portuguese 'old gardens' and apply them to modern gardens, it is certain that the importance given to the ancient fabric and to the historical sources of information was great, resembling, in this aspect, the principles that currently govern interventions in the historical fabric.

In the 1940s, in an intervention at Quinta da Agrela in Santo Tirso, Cabral showed a remarkable knowledge of the traditional spatial organisation of farms and estates, from their general layout to the use of vernacular materials and construction methods. The interventions proposed in 1950 for the historic Quinta of Sub-Serra, in Alhandra, and in 1954 for Quinta da Lameira, in Portalegre, reveal not only his erudition regarding the stylistic characteristics of the various periods of garden art and their contexts, but also his vision on the interventions in the historical fabric. According to Cabral, given the scarcity of elements for the "reconstitution of the gardens", knowledge of the fundamental characteristics of the "Portuguese garden" would allow interventions that would "respect the spirit" instead of having "the pretension of imitating what was there and has vanished forever". In the projects he developed for the two gardens, he chose to conserve and restore old elements characteristic of the "Portuguese garden" and to reinforce the vernacular character of the gardens by adding missing traditional elements and, eventually, modern elements, such as a lawn next to the house, harmonising it with old elements and forms.[8]

Ribeiro Telles (1922–2020), one of Cabral's first students, also made interventions in historic estates, such as the 1960 project framing the main façade of Mateus

Palace, in Vila Real, where, since the 19th century, there had been a picturesque garden. Here he begins by sketching a modernist composition, with simple forms and a small tank. However, the implemented proposal was a large baroque style basin, in accordance with the "spirit of the place", reinforcing it. At the Quinta do Marquês de Pombal in Oeiras, Ribeiro Telles, in the late 1950s, combines historical inspirations with modern designs.

Ilídio de Araújo (1925–2015), also a pupil of Cabral, developed projects that reveal his ability to understand the historical compositions and layouts of different types of gardens and their relationship with the wider landscape. That is the case of Quinta de Azevedo, in Barcelos, an intervention dating from 1983, and which was also not intended to restore a pre-existing layout (actually, there will not have been a previous ornamental garden in the intervention area). In the project's report[9], Araújo states:

> How to make the garden a useful space with a complementary function to that of the building, without affecting the ancestral character that those who walk through the garden gates expect to find there? That is one of the aspirations of this project.

And he describes the proposed design:

> The Renaissance tradition was used to divide the garden into three terraces arranged in a cascade—with an open perspective to the landscape—communicating with each other by flights of steps. The treatment given to the upper terrace does not differ in essence (neither in form nor in content) from what was usual in Renaissance gardens: beds bordered with boxwood with a fountain in the centre.

However, in this upper garden, he designed a large, paved area in granite slabs, where a fountain was placed in order to enable its use by a significant number of people (one of the premises of the programme); the middle terrace was designed to accommodate a swimming pool and the lower terrace a tennis court. Ilídio de Araújo thus harmoniously brought together the recreation of an old garden with elements of a modern lifestyle, designed in accordance with modernist forms, using new materials, such as concrete, and harmonising them with vernacular materials.

Two decades earlier, in 1962, Ilídio de Araújo had published the celebrated book "Arte Paisagista e Arte dos Jardins em Portugal" (Landscape Art and Garden Art in Portugal), whose content reveals what Cabral had considered central 20 years earlier. The book turned out to be a mandatory reference for all those who wanted to know and intervene in landscape heritage, and Ilídio de Araújo became an essential figure for young landscape architects who started their activity with the 1988 programme. If this programme was guided by the principles of intervention conveyed in the Florence Charter—in particular, preference for conservation actions such as 'maintenance', 'repair', 'adaptation', or 'conjectural detailing'; restrain in the use of new materials or designs; avoid the so-called 'restoration in spirit' and 're-creation'; adoption of 'reconstruction' only in cases where historical research supplied accurate and authentic information for the re-establishment of an earlier known layout—it is certain that the knowledge and sensitivity of the senior tutors allowed for an attentive and very careful orientation, directed towards the cultural significance of the gardens, namely by preserving and restoring their fundamental elements and characteristics.

It is noteworthy that the projects of the 1988 programme were concerned with gardens belonging to and managed by the State (through the IPPC) and not with private gardens, as were most of those executed by the first generation of landscape architects whose interventions were intended to conserve and improve the gardens, by emphasising their ancient and vernacular character and to adapt them to new uses required by owners.

Until 1988, the two existing degree courses in Landscape Architecture—Technical University of Lisbon (Higher Institute of Agronomy–ISA) and University of Évora (UE)—did not include any topic on the restoration of historic gardens in their curricula. From the IPPC–APAP programme and the research and projects it generated, the number of studies and academic publications on the history of gardens, their layouts and composition, and their significance and contribution to the national cultural heritage multiplied. The inclusion of subjects such as the history of Landscape Architecture, the restoration of historic gardens/cultural landscapes/landscape heritage, and landscape theory and critique has contributed to the strengthening of this knowledge and has aroused the attention of new professionals and researchers. Currently, the courses in Landscape Architecture in the five Portuguese universities (Lisbon–1942, Évora–1981, Algarve–1998, Trás-os-Montes e Alto Douro–1998, Porto–2001) are all teaching these topics.

An ongoing survey, coordinated by the FCUP Landscape Observatory, on the protection and conservation actions undertaken in Portugal since 1988, concerning national legislation, inventories, academic studies, publications, interventions in gardens, and the organisation of conferences, courses, among others, reveals that there has been a significant increase that follows the international production and the attention given to this subject by organizations worldwide. Since it is not possible to address all the topics here, we will present some of the most significant events to illustrate this evolution.

Until 1988, there was no inventory of historic gardens prepared by any private or public entity. The aforementioned book "Arte Paisagista e Arte dos Jardins em Portugal" was a major reference for scholars and professionals, but it was not intended to be a register of all the gardens and designed landscapes of historic interest. Similarly, the work of Aurora Carapinha's "Inventory of Landscape Art in Portugal", UE (1986), did not aim to be a thorough survey but became the basis for the protocol established between the Directorate-General of Buildings and National Monuments (DGEMN) and the UE Centre for the History of Art for the identification and recording of the historical gardens and sites of Portugal. Around 300 gardens and parks, monastic enclosures and hunting parks have been registered and described. This dataset was integrated in SIPA—Architectural Heritage Information System—under the responsibility of DGEMN since 1992 and DGPC since 2015. This is the official information and documentation system on the architectural and landscape heritage of Portugal or of Portuguese origin. In 1998, and by commission of the Tourism Fund, the "Survey of Historic Gardens of Mainland Portugal" was carried out, coordinated by Cristina Castel-Branco, arriving at a set of just over 120 gardens with tourist potential. Between 2017 and 2019, the AJH, under the coordination of Teresa Andresen, has drawn up an inventory with the aim of devising 'routes of historic gardens of Portugal'.[10] The activity of AJH, a non-profit cultural association founded in 2003 with the aim of contributing to the study, protection, preservation, and promotion of historic gardens is now very mature, playing a key role in the enhancement and dissemination of gardens and in the technical support for their conservation.

In addition to these inventories on a national scale, local and regional inventories are being carried out, particularly in the context of research projects: "LxGardens–Lisbon Historic Gardens and Parks"[11] (2010–14) and "Landscape Art in the North of Portugal"[12] (2005–09), the latter covering the north-east of mainland Portugal. As a result of these surveys, a greater knowledge of the typologies and distribution of the heritage of gardens and other designed landscapes was obtained.

Master's theses also contribute to broadening knowledge on our landscape heritage. Their production has increased with the growth in the number of higher education courses and their geographical distribution in the country, broadening the diversity of subjects within the theme of historic gardens and landscapes. Topics cover the history of gardens, designers, and their works; surveys of layouts and vegetation; interventions in the historical fabric; and conservation and management actions, among others.[13]

Academic maturity has also contributed to the increase in the number of doctoral theses, with emphasis on studies on Portuguese gardens, 'Portuguese modernism' and its immediate antecedents, on the origin of higher education in Landscape Architecture in Portugal, on cultural landscapes and their conservation, and on ancient water management. The number of books, book chapters, and scientific articles published in this field is already very significant, allowing access to essential sources of information for theoretical and practical teaching and for new interventions in these sites or in sites with identical characteristics.

Presently, conservation work is being carried out all over the country mostly under the coordination of landscape architects, and strategies for managing the conservation of public gardens and designed landscapes have been tested and proved to be successful.[14] They offer key material for teaching and learning in an academic context, contributing to the soundness of the professional training.

Investment has also been made in the research and dissemination of the work of landscape architects and gardens in Portugal, particularly through exhibitions aimed both to specialists and the general public, making them aware of the cultural significance of gardens. We highlight the exhibitions "From the National Stadium to the Calouste Gulbenkian Foundation Garden. Francisco Caldeira Cabral and the First Generation of Landscape Architects, 1940–1970" (Calouste Gulbenkian Foundation, 2003); "Historic Gardens of Portugal. Memory and Future" (National Library, 2020–2021); and the itinerant exhibition "80 Years of Landscape

Architecture Teaching in Portugal (1942–2022): Art and Ecology" (ISA, 2022; other Portuguese universities, 2022 and 2023).

We believe we are in a position to claim that significant steps have been taken in pursuing and strengthening the objectives set out in the document "Historic Gardens in Portugal and the Berlin Call for Action" (2018). The inclusion of the category "Cultural Landscapes" in Portuguese legislation, on the contrary, has not been well received by the relevant public authorities, but discussions are underway to find alternative paths to address and distinguish gardens and other designed landscapes from other cultural assets of a different nature, such as buildings and groups of buildings.

Notes

1 The forerunner of the present Directorate General of Cultural Heritage (DGPC).
2 www.fc.up.pt/ap/observatorio/.
3 The Berlin Call, "Cultural Heritage for the Future of Europe", "builds on the momentum created by the European Year of Cultural Heritage 2018 and seeks to engage the many organizations, institutions, governments, civil society actors and individuals in recognizing the positive and cohesive power of Europe's shared cultural heritage". It was presented at the European Cultural Heritage Summit in June 2018, Berlin.
4 www.fc.up.pt/ap/wp-content/uploads/2023/04/OS-JARDINS-HISTORICOS-EM-PORTUGAL-E-O-APELO-A-ACAO-DE-BERLIM_ENG_final.pdf.
5 In 2006, as part of the National Congress of APAP, a paper entitled "From Re-creation to Conservation—the Historic Gardens in the History of Heritage Conservation" (Portela Marques, 2006) was presented and published. In the first part, "The 1960s to the 1980s—origins of contemporary concepts and practice on the conservation of historic gardens", examples of intervention are presented, namely some projects carried out by Cabral, and in the second part, "The consolidation of ideas and practices and the systematic conservation of historic gardens", the more contemporary principles widely disseminated after the adoption of the Florence Charter in 1982, which supported the already mentioned "Conservation Programme for Historic Gardens", were discussed.
6 Andresen, T., Câmara, T.B., & Carvalho, L.G. (2003). Lugares da Arquitectura Paisagista Portuguesa. In Andresen, T. (Coord.), *Do Estádio Nacional ao Jardim Gulbenkian. Francisco Caldeira Cabral e a primeira geração de arquitectos paisagistas. 1940–1970.* Lisboa: Fundação Calouste Gulbenkian, pp. 144–314.
7 This study had been initiated in 1940 by Cabral with the support of the Portuguese Instituto para a Alta Cultura. See Antunes, A.C., Portela Marques, T., & Andresen, T. (2022). La genèse de la formation en architecture de paysage au Portugal. *Projets de paysage. Revue scientifique sur la conception et l'aménagement de l'espace*, (Hors-série).
8 Andresen, T., Câmara, T.B., & Carvalho, L.G. (2003). Lugares da Arquitectura Paisagista Portuguesa. In Andresen, T. (Coord.), *Do Estádio Nacional ao Jardim Gulbenkian. Francisco Caldeira Cabral e a primeira geração de arquitectos paisagistas. 1940–1970.* Lisboa: Fundação Calouste Gulbenkian, pp. 144–314.
9 Araújo, I. (1983). Memória Descritiva e Justificativa do Projecto de Reordenamento do Horto da Casa de Azevedo. [Unpublished].
10 Financed through the "Valorizar" Programme. Initially aimed at the inland regions of Portugal, it was later extended to the whole of mainland Portugal and the islands.
11 Centre for Applied Ecology Professor Baeta Neves (CEABN/ISA/UL) coordinated by Ana Luísa Soares.
12 Centre for Technological Studies on Environment and Life (CETAV/UTAD) coordinated by Teresa Portela Marques.

13 Approximately 150 theses in the field of historic gardens have already been identified.

14 This is the case of the publicly owned company Parques de Sintra-Monte da Lua, S.A. created after the inclusion of the 'Cultural Landscape of Sintra' in the World Heritage List (1995), that brings together the institutions with responsibility for the conservation of this cultural landscape. The state entrusted the company with the management of its main properties in the Sintra Hill. The conservation is ensured by revenues from ticket offices, shops, cafeterias, and the rental of spaces for events. In 2018, the restoration project of the Botanical Garden of Queluz Palace received the Europa Nostra Award in the conservation category.

2 Gonçalo Ribeiro Telles

Landscape architect, ecologist and politician in Portugal

Domingas Vasconcelos, Paulo Farinha-Marques and Francisco Castro Rego

Introduction

The man of the future, who is being born everywhere, will unite the city and the countryside.

Gonçalo Ribeiro Telles (Saldanha, 2010)

Purpose

Gonçalo Ribeiro Telles (1922–2020) was someone we were used to seeing as a giant of Landscape Architecture and ecology in Portugal. Based on the collection of facts on his life and work, and on the analysis of memories published at the final moment of his life, this chapter is a contribution to the history of Landscape Architecture and ecology in Portugal.

Methodology

For this work, a sample of 140 written news and opinion texts was considered (Table 2.1). It was obtained using a "Google" generic search engine. The search was carried out by combining name and date of death. All texts published from 11th to 26th November 2020 were considered. The analysis focused on identifying words or groups of words used to characterize Ribeiro Telles and how he was remembered. The specific concepts were serialized and integrated into broader components.

Concurrently, a graphic representation was built in the form of a biographical tree to identify the most important works and moments in his life. This tree contains references to Francisco Caldeira Cabral (1908–1992), the founder of Portuguese Landscape Architecture, as well as to his first disciples, who together with Ribeiro Telles constituted the first generation of Portuguese landscape architects. The tree also includes relevant facts for the disciplinary area of landscape architecture.

DOI: 10.4324/9781003454007-5

Table 2.1 Sample for analysis.

ref. Unit of Analysis *ref. Unidade de Análise*	***Text ref.***
11.01	Coentrão, Abel; Salema, I. (2020) Gonçalo Ribeiro Telles, o cultivador de utopias. *Público*. https://www.publico.pt/2020/11/11/culturaipsilon/noticia/goncalo-ribeiro-telles-19222020-cultivador-utopias-1938817
11.02	Fotogaleria. Da Gulbenkian à Mata dos Medos: imagens da paisagem que Gonçalo Ribeiro Telles mudou (2020) *Público*. https://www.publico.pt/2020/05/25/culturaipsilon/fotogaleria/obras-goncalo-ribeiro-telles-401342
11.03	Ralha, L. (2020) Morreu Gonçalo Ribeiro Telles, pioneiro do ecologismo na política portuguesa. *Jornal Económico*. https://jornaleconomico.sapo.pt/noticias/__trashed-97-592632
11.04	Morreu o arquiteto paisagista Gonçalo Ribeiro Telles aos 98 anos (2020) *Jornal de Notícias*. https://www.jn.pt/nacional/morreu-o-arquitecto-paisagista-goncalo-ribeiro-telles-13024960.html
11.05	Morreu Gonçalo Ribeiro Telles (2020) *SIC Notícias*. https://sicnoticias.pt/pais/2020-11-11-Morreu-Goncalo-Ribeiro-Telles
11.06	Morreu o arquiteto Gonçalo Ribeiro Telles aos 98 anos (2020) *Expresso*. https://expresso.pt/politica/2020-11-11-Morreu-o-arquiteto-Goncalo-Ribeiro-Telles-aos-98-anos
11.07	Barreto, D. (2020) Morreu o arquiteto paisagista Gonçalo Ribeiro Telles. *Sábado*. https://www.sabado.pt/vida/detalhe/morreu-o-arquiteto-goncalo-ribeiro-telles
11.08	Rosa, V. (2020) Morreu Gonçalo Ribeiro Telles, o arquiteto paisagista com "um dom especial." *Observador*. https://observador.pt/especiais/morreu-goncalo-ribeiro-telles-o-arquiteto-paisagista-com-um-dom-especial/
11.09	Morreu Gonçalo Ribeiro Telles (2020) *PTJornal*. https://ptjornal.com/morreu-goncalo-ribeiro-telles-535108
11.10	Morreu o arquiteto paisagista e político Gonçalo Ribeiro Telles (2020) *Agroportal*. https://www.agroportal.pt/morreu-o-arquiteto-paisagista-e-politico-goncalo-ribeiro-telles/
11.11	Morreu o arquiteto paisagista Gonçalo Ribeiro Telles. Tinha 98 anos (2020) *Sapo*. https://24.sapo.pt/atualidade/artigos/morreu-o-arquitecto-paisagista-goncalo-ribeiro-telles
11.12	Morreu o arquiteto paisagista Gonçalo Ribeiro Telles (2020) *Correio da Manhã*. https://www.cmjornal.pt/politica/detalhe/morreu-o-arquiteto-paisagista-goncalo-ribeiro-telles-avanca-o-publico
11.13	Gonçalves, P. Z. (2020) Morreu o arquiteto paisagista Gonçalo Ribeiro Telles. *CM-TV*. https://www.cm-tv.pt/atualidade/detalhe/20201111-1703-morreu-o-arquiteto-paisagista-goncalo-ribeiro-telles-avanca-o-publico
11.14	Morreu o arquiteto paisagista Gonçalo Ribeiro Telles aos 98 anos (2020) *Porto Canal*. http://portocanal.sapo.pt/noticia/243108/
11.15	Morreu o arquiteto paisagista Gonçalo Ribeiro Telles (2020) *Jornal de Negócios*. https://www.jornaldenegocios.pt/economia/detalhe/morreu-o-arquiteto-paisagista-goncalo-ribeiro-telles
11.16	Morreu o arquitecto paisagista Gonçalo Ribeiro Telles (2020) *JM-Madeira*. https://www.jm-madeira.pt/nacional/ver/109648/Morreu_o_arquitecto_paisagista_Goncalo_Ribeiro_Telles

(*Continued*)

Table 2.1 (Continued)

ref. Unit of Analysis *ref. Unidade de Análise*	***Text ref.***
11.17	Morreu o arquiteto paisagista Gonçalo Ribeiro Telles (2020) *BIG Banco de Investimento Global.* https://www.big.pt/InformacaoMercados/Noticias/Index/253266
11.18	Tomé, J. (2020) Morreu o arquiteto paisagista Gonçalo Ribeiro Telles aos 98 anos. *Dinheiro Vivo.* https://www.dinheirovivo.pt/geral/morreu-o-arquiteto-paisagista-goncalo-ribeiro-telles-aos-98-anos-13025047.html
11.19	Morreu o arquiteto paisagista Gonçalo Ribeiro Telles aos 98 anos (2020) *Visão.* https://visao.sapo.pt/atualidade/sociedade/2020-11-11-morreu-o-arquiteto-paisagista-goncalo-ribeiro-telles-aos-98-anos/
11.20	Morreu Gonçalo Ribeiro Telles (2020) *Record.* https://www.record.pt/fora-de-campo/detalhe/morreu-goncalo-ribeiro-telles
11.21	Morreu o arquitecto Gonçalo Ribeiro Telles (2020) *Notícias do Sorraia.* https://noticiasdosorraia.pt/morreu-o-arquitecto-goncalo-ribeiro-telles/
11.22	Morreu Gonçalo Ribeiro Telles, arquiteto paisagista e figura da ecologia (2020) *TSF.* https://www.tsf.pt/portugal/sociedade/morreu-goncalo-ribeiro-telles-arquiteto-paisagista-e-figura-da-ecologia-13025008.html
11.23	Morreu Gonçalo Ribeiro Telles (2020) *Bancada.* https://bancada.pt/artigos/fora-da-bancada/morreu-goncalo-ribeiro-telles
11.24	Governo decreta um dia de luto nacional na quinta-feira pela morte de Gonçalo Ribeiro Telles (2020) *Sapo.* https://24.sapo.pt/atualidade/artigos/governo-decreta-um-dia-de-luto-nacional-na-quinta-feira-pela-morte-de-goncalo-ribeiro-telles
11.25	Ribeiro Telles: Governo decreta um dia de luto nacional na quinta-feira (2020) *Visão.* https://visao.sapo.pt/atualidade/politica/2020-11-11-obito-ribeiro-telles-governo-decreta-um-dia-de-luto-nacional-na-quinta-feira/
11.26	Gonçalo Ribeiro Telles. A vida em imagens (2020) *Público.* https://www.publico.pt/2020/11/11/fotogaleria/goncalo-ribeiro-telles-403511
11.27	O arquiteto paisagista que pensou as cidades para as pessoas (2020) *Destak.* http://www.destak.pt/artigo/456610-o-arquiteto-paisagista-que-pensou-as-cidades-para-as-pessoas?switcher=desktop
11.28	Morreu o arquitecto Gonçalo Ribeiro Telles (2020) *Funchal Notícias.* https://funchalnoticias.net/2020/11/11/morreu-o-arquitecto-goncalo-ribeiro-telles/
11.29	Morreu Gonçalo Ribeiro Telles (2020) *Touroe Ouro.* http://www.touroeouro.com/article/view/21975/morreu-goncalo-ribeiro-telles
11.30	Coruche na obra do arquitecto Gonçalo Ribeiro Telles (2020) *Notícias do Sorraia.* https://noticiasdosorraia.pt/coruche-na-obra-do-arquitecto-goncalo-ribeiro-telles/
11.31	Ministro, P. (2020) Dia de Luto Nacional em homenagem a Gonçalo Ribeiro Telles. *Primeiro Ministro.* https://www.portugal.gov.pt/pt/gc22/comunicacao/comunicado?i=dia-de-luto-nacional-em-homenagem-ao-professor-goncalo-ribeiro-telles
11.32	Ambiente e Ação Climática, M. do (2020) Ministro do Ambiente e da Ação Climática manifesta pesar pelo falecimento do Arquiteto Gonçalo Ribeiro Telles. *Ministro do Ambiente e da Ação Climática.* https://www.portugal.gov.pt/pt/gc22/comunicacao/comunicado?i=ministro-do-ambiente-e-da-acao-climatica-manifesta-pesar-pelo-falecimento-do-arquiteto-goncalo-ribeiro-telles

(*Continued*)

Table 2.1 (Continued)

ref. Unit of Analysis *ref. Unidade de Análise*	***Text ref.***
11.33	Um dia de luto nacional e um parque: as reacções à morte de Gonçalo Ribeiro Telles (2020) *Público*. https://www.publico.pt/2020/11/11/culturaipsilon/noticia/governo-decreta-dia-luto-nacional-goncalo-ribeiro-telles-perda-inestimavel-1938836
11.34	Gonçalo Ribeiro Telles: o arquiteto paisagista que pensou as cidades para as pessoas (2020) *TSF*. https://www.tsf.pt/portugal/sociedade/goncalo-ribeiro-telles-o-arquiteto-paisagista-que-pensou-as-cidades-para-as-pessoas-13025270.html
11.35	Cultura, M. da (2020). Ministra da Cultura lamenta morte de Gonçalo Ribeiro Telles. *Ministra da Cultura*. https://www.portugal.gov.pt/pt/gc22/comunicacao/comunicado?i=ministra-da-cultura-lamenta-morte-de-goncalo-ribeiro-telles
11.36	Ribeiro, Luís; Carvalho, J. C. (2020) Gonçalo Ribeiro Telles, o guerreiro incansável. *Visão*. https://visao.sapo.pt/atualidade/ambiente/2020-11-11-goncalo-ribeiro-telles-o-guerreiro-incansavel/
11.37	"Um homem à frente do seu tempo". As reações à morte de Gonçalo Ribeiro Telles (2020) *Observador*. https://observador.pt/2020/11/11/um-homem-a-frente-do-seu-tempo-as-reacoes-a-morte-de-goncalo-ribeiro-telles/
11.38	Morreu Gonçalo Ribeiro Telles (2020) *Esquerda*. https://www.esquerda.net/artigo/morreu-goncalo-ribeiro-telles/71205
11.39	Gonçalo Ribeiro Telles. Um visionário que deixa um valioso legado, segundo a associação de arquitetos paisagistas (2020) *Expresso*. https://expresso.pt/politica/2020-11-11-Goncalo-Ribeiro-Telles.-Um-visionario-que-deixa-um-valioso-legado-segundo-a-associacao-de-arquitetos-paisagistas?platform=hootsuite&utm_campaign=HSCampaign
11.40	Morreu o arquiteto paisagista Gonçalo Ribeiro Telles aos 98 anos (2020) *Jornal do Luxemburgo*. https://jornaldoluxemburgo.com/2020/11/morreu-o-arquiteto-paisagista-goncalo-ribeiro-telles-aos-98-anos/
11.41	Parque Urbano da Praça de Espanha vai chamar-se Gonçalo Ribeiro Telles (2020) *Notícias ao Minuto*. https://www.noticiasaominuto.com/pais/1625233/parque-urbano-da-praca-de-espanha-vai-chamar-se-goncalo-ribeiro-telles
11.42	Soromenho, A. (2020) "Se podemos ser um exemplo, sem andar a chatear ninguém, ótimo": Gonçalo Ribeiro Telles 1922-2020. *Expresso*. https://expresso.pt/arquivos-expresso/2020-11-11-Se-podemos-ser-um-exemplo-sem-andar-a-chatear-ninguem-otimo-Goncalo-Ribeiro-Telles-1922-2020
11.43	Gomes, G. D. (2020) Gonçalo Ribeiro Telles ou a ousadia de sonhar. *Sul Informação*. https://www.sulinformacao.pt/2020/11/goncalo-ribeiro-telles-ou-a-ousadia-de-sonhar/?fbclid=IwAR04hhomKCw09XPq0VK1p13_XllHhptU7iE1c5yFP6XbX7XLd5FHtGv3hig&platform=hootsuite&utm_campaign=HSCampaign
11.44	Morreu o arquiteto paisagista Gonçalo Ribeiro Telles (2020) *TV Europa*. https://www.tveuropa.pt/noticias/morreu-o-arquiteto-paisagista-goncalo-ribeiro-telles/
11.45	Morreu Gonçalo Ribeiro Teles (2020) *Sintra Notícias*. https://sintranoticias.pt/2020/11/11/morreu-goncalo-ribeiro-teles/
11.46	Salema, I. (2020) Gonçalo Ribeiro Telles, o arquitecto que levou a paisagem para a política. *Público*. https://www.publico.pt/2020/11/11/culturaipsilon/noticia/morreu-goncalo-ribeiro-telles-arquitecto-levou-paisagem-politica-1938857

(*Continued*)

Table 2.1 (Continued)

ref. Unit of Analysis *ref. Unidade de Análise*	***Text ref.***
11.47	Morreu o arquitecto paisagista Gonçalo Ribeiro Telles (2020) *Espaço de Arquitetura*. https://espacodearquitetura.com/noticias/morreu-o-arquitecto-paisagista-goncalo-ribeiro-telles/
11.48	Queirós, L. M. (2020) "Se há hoje em Portugal ordenamento e planeamento do território, a ele o devemos." *Público*. https://www.publico.pt/2020/11/11/culturaipsilon/noticia/ha-hoje-portugal-ordenamento-planeamento-territorio-devemos-1938865
11.49	"O país é o meu sonho permanente, estou sempre a sonhá-lo". O legado de Ribeiro Telles (2020) *SIC Notícias*. https://sicnoticias.pt/pais/2020-11-11-O-pais-e-o-meu-sonho-permanente-estou-sempre-a-sonha-lo.-O-legado-de-Ribeiro-Telles
11.50	Távora, J. (2020) Hoje partiu um dos nossos maiores. *Real Associação de Lisboa*. https://monarquia-lisboa.blogs.sapo.pt/tag/gonçalo+ribeiro+telles
11.51	Ribeiro, A. M. (2020) Gonçalo Ribeiro Telles. *Jornal de Negócios*. https://anabelamotaribeiro.pt/24603.html
11.52	Morreu o arquiteto paisagista Gonçalo Ribeiro Telles (2020) *ZAP Aeiou*. https://zap.aeiou.pt/morreu-arquiteto-paisagista-goncalo-ribeiro-telles-358968
11.53	Marmé, P. (2020) O adeus do arquiteto que inspirou gerações de paisagistas. *Wattson*. https://www.wattson.pt/2020/11/11/o-adeus-do-arquiteto-que-inspirou-geracoes-de-paisagistas/https://www.publico.pt/2020/05/25/opiniao/opiniao/goncalo-ribeiro-telles-1917209
11.54	Morreu Gonçalo Ribeiro Telles, fundador da licenciatura de Arquitetura Paisagista na UÉ (2020) *Universidade de Évora*. https://www.uevora.pt/ue-media/noticias?item=30823
11.55	Morreu Gonçalo Ribeiro Telles (2020) *União das Cidades Capitais de Língua Portuguesas (UCCLA)*. https://www.uccla.pt/noticias/morreu-goncalo-ribeiro-telles-0
11.56	Gonçalo Ribeiro Telles "foi o primeiro grande ambientalista." (2020) TViI24. https://www.msn.com/pt-pt/video/assistir/gonçalo-ribeiro-telles-"foi-o-primeiro-grande-ambientalista"/vi-BB1aV5tF
11.57	Morreu Gonçalo Ribeiro Telles (2020) SIC Famashow. https://famashow.pt/famosos/2020-11-11-Morreu-Goncalo-Ribeiro-Telles
11.58	Nunes, E. (2020) Morreu Gonçalo Ribeiro Telles. Record Fm. https://www.recordfm.pt/morreu-goncalo-ribeiro-telles/
11.59	Nota de pesar pelo falecimento de Gonçalo Ribeiro Telles (2020) *Partido Social Democrata*. https://www.psd.pt/pt/noticias/nota-de-pesar-pelo-falecimento-de-goncalo-ribeiro-tellesde Gonçalo Ribeir. (2020). Partido Social Democrata. https://www.psd.pt/pt/noticias/nota-de-pesar-pelo-falecimento-de-goncalo-ribeiro-telles
11.60	Nota de pesar pelo falecimento de Gonçalo Ribeiro Telles (2020) *Partido Socialista*. https://ps.pt/index.php/2020/11/11/nota-de-pesar-pelo-falecimento-de-goncalo-ribeiro-telles/
11.61	Morreu o arquiteto Gonçalo Ribeiro Telles (2020) *Press Net*. https://pressnet.pt/2020/11/11/morreu-o-arquiteto-goncalo-ribeiro-telles/

(*Continued*)

Table 2.1 (Continued)

ref. Unit of Analysis *ref. Unidade de Análise*	***Text ref.***
11.62	Sousa, M. R. de (2020) Presidente da República assina Luto Nacional em honra de Gonçalo Ribeiro Telles. *Presidência da República*. http://www.presidencia.pt/?idc=10&idi=180600
11.63	Sousa, M. R. de (2020) Presidente da República honra a memória de Gonçalo Ribeiro Telles. *Presidência da República*. http://www.presidencia.pt/?idc=18&idi=180577
11.64	Santos, A. D. dos (2020) Morreu o arquitecto Gonçalo Ribeiro Telles. *Portal de Angola*. https://www.portaldeangola.com/2020/11/11/morreu-o-arquitecto-goncalo-ribeiro-telles/
11.65	Coelho, R. (2020) Morreu Gonçalo Ribeiro Telles, arquiteto paisagista e figura da ecologia. *Plataforma Cascais*. https://plataformacascais.com/artigos/portugal/192335-morreu-goncalo-ribeiro-telles-arquiteto-paisagista-e-figura-da-ecologia.html
11.66	Presidente da Assembleia da República homenageia Gonçalo Ribeiro Telles (2020) *Assembleia da República*. https://www.parlamento.pt/Paginas/2020/novembro/PAR-lamenta-falecimento-Goncaço-Ribeiro-Telles.asp
11.67	Morreu o arquiteto paisagista Gonçalo Ribeiro Telles aos 98 anos (2020) *O Minho*. https://ominho.pt/morreu-o-arquiteto-paisagista-goncalo-ribeiro-telles-aos-98-anos/
11.68	Morreu Gonçalo Ribeiro Telles. Tinha 98 anos (2020) *O Jogo*. https://www.ojogo.pt/extra/noticias/morreu-goncalo-ribeiro-telles-tinha-98-anos-13025143.html
11.69	Morreu Gonçalo Ribeiro Telles: o homem que lutou pela defesa do Ambiente e valorização do Mundo Rural (2020) *O Instalador*. https://www.oinstalador.com/Artigos/318657-Morreu-Goncalo-Ribeiro-Telles-homem-lutou-na-defesa-do-Ambiente-na-valorizacao-do-Mundo.html
11.70	Faleceu Gonçalo Ribeiro Telles (2020) *Ordem dos Biólogos*. http://ordembiologos.pt/2020/11/11/faleceu-goncalo-ribeiro-telles/
11.71	A Ordem dos Arquitectos endereça sentidas condolências à família e amigos do Arquitecto Paisagista Gonçalo Ribeiro Telles (2020) *Ordem dos Arquitetos*. https://www.arquitectos.pt/?no=2020497343,154
11.72	Morreu o arquiteto paisagista Gonçalo Ribeiro Telles aos 98 anos (2020) *Notícias de Coimbra*. https://www.noticiasdecoimbra.pt/morreu-o-arquiteto-paisagista-goncalo-ribeiro-telles-aos-98-anos/
11.73	Gonçalo Ribeiro Telles, 1922-2020 (2020) *Monumentos*. http://www.monumentos.gov.pt/Site/APP_PagesUser/News.aspx?id=a3abb356-02f9-46a2-a669-d9dc96d85601&rid=93073076-c788-41cb-99ea-01b609881c92
11.74	Gonçalo Ribeiro Telles: "Sou fascinado pelas paisagens que tenham relação com a história." (2020) *Revista Jardins*. https://revistajardins.pt/goncalo-ribeiro-telles-fascinado-pelas-paisagens-tenham-relacao-historia/
11.75	Magalhães, M. M. R. de (2020) Professor Gonçalo Pereira Ribeiro Telles (1922-). *Instituto Superior de Agronomia (ISA)*. https://www.isa.ulisboa.pt/files/id/g-ribeiro-telles/Goncalo_Ribeiro_Telles-Prof_M_Raposo_Magalhaes.pdf
11.76	Castel-Branco, C. (2020) Tributo a Gonçalo Ribeiro Telles. *Instituto Superior de Agronomia (ISA)*. Tributo a Gonçalo Ribeiro Telles, por Professora Cristina Castel-Branco, 11 Nov. 2020

(*Continued*)

Table 2.1 (Continued)

ref. Unit of Analysis *ref. Unidade de Análise*	***Text ref.***
11.77	Falecimento do Professor Gonçalo Ribeiro Telles (2020) *Instituto Superior de Agronomia (ISA)*. http://www.isa.ulisboa.pt/vida-no-isa/destaques/noticias/20201112-falecimento-do-professor-goncalo-ribeiro-telles
11.78	Morreu o arquiteto paisagista Gonçalo Ribeiro Telles aos 98 anos (2020) *Impala News*. https://www.impala.pt/noticias/atualidade/morreu-o-arquiteto-paisagista-goncalo-ribeiro-telles-aos-98-anos/
11.79	Conselho de Administração FCG (2020) Gonçalo Ribeiro Telles (1922-2020). *Fundação Calouste Gulbenkian (FCG)*. https://gulbenkian.pt/noticias/goncalo-ribeiro-telles-1922-2020/
11.80	Morreu o arquiteto paisagista Gonçalo Ribeiro Telles (2020) *Executive Digest*. https://executivedigest.sapo.pt/morreu-o-arquiteto-paisagista-goncalo-ribeiro-telles/
11.81	Divisão de Comunicação (2020) Gonçalo Ribeiro Telles foi um pioneiro que esteve à frente do seu tempo. *Ensino Magazine*. https://www.ensino.eu/ensino-magazine/universidade/2020/morteribeiroteles/#
11.82	Morreu o arquitecto paisagista Gonçalo Ribeiro Telles aos 98 anos (2020) *Diário Imobiliário-Lusa*. https://www.diarioimobiliario.pt/Arquitectura/Morreu-o-arquitecto-paisagista-Goncalo-Ribeiro-Telles-aos-98-anos
11.83	Luto nacional por Gonçalo Ribeiro Telles (2020) *Construir*. https://construir.pt/2020/11/11/luto-nacional-por-goncalo-ribeiro-telles/
11.84	Martins, G. d'Oliveira (2020) Gonçalo Ribeiro Telles (1922-2020). *Centro Nacional de Cultura (CNC)*. https://www.cnc.pt/goncalo-ribeiro-telles-1922-2020/
11.85	Morreu Gonçalo Ribeiro Telles (2020) *Câmara Municipal de Lisboa*. https://www.lisboa.pt/atualidade/noticias/detalhe/morreu-goncalo-ribeiro-telles
11.86	Morreu o Arquiteto Paisagista Gonçalo Ribeiro Telles aos 98 anos (2020) *Correio da Manhã*. https://www.correiodamanhacanada.com/morreu-o-arquiteto-paisagista-goncalo-ribeiro-telles-aos-98-anos/
11.87	Gonçalo Ribeiro Telles (2020) *Blasfémias*. https://blasfemias.net/2020/11/11/goncalo-ribeiro-telles/
11.88	Lagiosa, J. (2020) "Livre" expressa pesar pela morte de Ribeiro Telles. *Beira News*. https://beiranews.pt/2020/11/11/livre-expressa-pesar-pela-morte-de-ribeiro-telles/
11.89	Lagiosa, J. (2020) Morreu o arquiteto paisagista Gonçalo Ribeiro Telles aos 98 anos. *Beira News*. https://beiranews.pt/2020/11/11/morreu-o-arquiteto-paisagista-goncalo-ribeiro-telles-aos-98-anos/
11.90	Direção da Associação Portuguesa dos Arquitectos Paisagistas (2020) Comunicado: Professor Gonçalo Ribeiro Telles 1922-2020. *Associação Portuguesa dos Arquitectos Paisagistas*. https://m.facebook.com/story.php?story_fbid=3094484833991506&substory_index=0&id=299345743505443
11.91	Gonçalo Ribeiro Telles, 1922-2020 (2020) *Antifascistas da Resistência*. https://www.facebook.com/FascismoNuncaMais/photos/gonçalo-ribeiro-telles-n1922arquitecto-paisagista-esteve-presente-na-resistência/733696353406413
11.92	Morreu o arquiteto paisagista Gonçalo Ribeiro Telles aos 98 anos (2020) *Antena Minho*. https://www.antenaminho.pt/noticias/morreu-o-arquiteto-paisagista-goncalo-ribeiro-telles-aos-98-anos/11398

(*Continued*)

Table 2.1 (Continued)

ref. Unit of Analysis *ref. Unidade de Análise*	***Text ref.***
11.93	Morreu Gonçalo Ribeiro Telles: o homem que lutou pela defesa do Ambiente e valorização do Mundo Rural (2020) *Agriterra*. https://agriterra.pt/Artigos/318657-Morreu-Goncalo-Ribeiro-Telles-homem-lutou-na-defesa-do-Ambiente-na-valorizacao-do-Mundo.html
11.94	Dia de Luto Nacional em homenagem a Gonçalo Ribeiro Telles (2020) *Mais Ribatejo*. https://maisribatejo.pt/2020/11/11/dia-de-luto-nacional-em-homenagem-a-goncalo-ribeiro-telles/
11.95	Conselho de Ministros (2020) Decreto n.º 8-A/2020, de 11 de novembro. Diário da República—I Série, 220, 19-(2). *Presidência do Conselho de Ministros*. https://files.dre.pt/1s/2020/11/22001/0000200002.pdf
12.01	Reis, B. (2020) Gonçalo Ribeiro Telles, o homem que rouba ideias aos pássaros. *Público*. https://www.publico.pt/2020/11/12/culturaipsilon/entrevista/goncalo-ribeiro-telles-homem-rouba-ideias-passaros-1938878
12.02	Covas, A. (2020) A cidade do futuro, metabolismo urbano e rede de bens comuns. *Público*. https://www.publico.pt/2020/11/12/opiniao/opiniao/cidade-futuro-metabolismo-urbano-rede-bens-comuns-1938874?platform=hootsuite&utm_campaign=HSCampaign
12.03	Cerimónia fúnebre de Gonçalo Ribeiro Telles hoje na Igreja dos Jerónimos em Lisboa (2020) *24 Sapo*. https://24.sapo.pt/atualidade/artigos/cerimonia-funebre-de-goncalo-ribeiro-telles-hoje-na-igreja-dos-jeronimos-em-lisboa
12.04	Fernandes, J. S. (2020) Gonçalo Ribeiro Telles, Mester da Paisagem. *Expresso*. https://expresso.pt/opiniao/2020-11-12-Goncalo-Ribeiro-Telles-Mester-da-Paisagem?platform=hootsuite&utm_campaign=HSCampaign
12.05	Cerimónia fúnebre de Ribeiro Telles é esta quinta-feira na Igreja dos Jerónimos (2020) *Público*. https://www.publico.pt/2020/11/12/culturaipsilon/noticia/cerimonia-funebre-ribeiro-telles-quintafeira-igreja-jeronimos-1938888
12.06	Reis, T. (2020) O que nos deu Gonçalo Ribeiro Telles? *Público*. https://www.publico.pt/2020/11/12/p3/cronica/deu-goncalo-ribeiro-telles-1938900
12.07	Portugal vive dia de luto por Gonçalo Ribeiro Telles, um «jardineiro de Deus» (2020) *Agência Ecclesia*. https://agencia.ecclesia.pt/portal/sociedade-portugal-vive-dia-de-luto-por-goncalo-ribeiro-telles-um-jardineiro-de-deus/
12.08	Fotogaleria: o legado de Gonçalo Ribeiro Telles em 23 fotografias (2020) *Expresso*. https://expresso.pt/sociedade/2020-11-12-Fotogaleria-o-legado-de-Goncalo-Ribeiro-Telles-em-23-fotografias
12.09	Naves, P. (2020) Parque da Praça de Espanha vai chamar-se Parque Gonçalo Ribeiro Telles. *NiT*. https://www.nit.pt/fora-de-casa/na-cidade/parque-urbano-da-praca-de-espanha-vai-chamar-se-parque-goncalo-ribeiro-telles
12.10	Pato, H. (2020) Gonçalo Ribeiro Telles (1922—2020). *Jornal Tornado*. https://www.jornaltornado.pt/goncalo-ribeiro-telles/
12.11	Pato, H. (2020) Gonçalo Ribeiro Telles (1922—2020). *Jornal Tornado*. https://www.jornaltornado.pt/goncalo-ribeiro-telles/
12.12	10 mandamentos de Gonçalo Ribeiro Telles para a criação de um jardim que ficam para a história (2020) *Life Style. Sapo.Pt*. https://lifestyle.sapo.pt/casa-e-lazer/decoracao-e-jardins/artigos/os-10-mandamentos-do-arquiteto-goncalo-ribeiro-telles-para-a-criacao-de-um-jardim

(*Continued*)

Table 2.1 (Continued)

ref. Unit of Analysis *ref. Unidade de Análise*	***Text ref.***
12.13	Um Adeus Especial ao arquiteto paisagista Gonçalo Ribeiro Telles (2020) *Ribatejo News*. https://ribatejonews.net/2020/11/12/um-adeus-especial-ao-arquiteto-paisagista-goncalo-ribeiro-telles/
12.14	Gonçalo Ribeiro Telles: o adeus ao Homem da Paisagem (2020) *Revista Rua*. https://www.revistarua.pt/goncalo-ribeiro-telles-o-adeus-ao-homem-da-paisagem/
12.15	Távora, J. (2020) Um dos nossos maiores. *Ponto SJ*. https://pontosj.pt/especial/ribeiro-telles-um-dos-nossos-maiores/
12.16	Baptista-Martins, L. (2020) Perdemos Gonçalo Ribeiro Telles. *O Interior*. https://ointerior.pt/opiniao/perdemos-goncalo-ribeiro-telle/
12.17	Gonçalo Ribeiro Telles, 1922-2020 (2020) *Minha Terra—Federação Portuguesa das Associações de DesenvolvimentoLocal*. https://www.minhaterra.pt/goncalo-ribeiro-telles-1922-2020.T13572.php
12.18	Saraiva, J. C. (2020) Gonçalo Ribeiro Telles: o éden ao alcance de todos. *Jornal I*. https://ionline.sapo.pt/artigo/714863/goncalo-ribeiro-telles-o-eden-ao-alcance-de-todos?seccao=Opinião_i
12.19	Justino, J. (2020) Gonçalo Ribeiro Telles. *Estrategizando*. https://estrategizando.pt/2020/11/12/goncalo-ribeiro-telles/
12.20	Roque, M. I. (2020). Gonçalo Ribeiro Telles, um homem de lutas e sonhos. *Hypotheses-A.Muse.Arte*. https://amusearte.hypotheses.org/6922
12.21	Marujo, A. (2020) Gonçalo Ribeiro Telles (1922-2020): Portugal de luto pelo ambientalista, católico antifascista e pensador do território. *Sete Margens*. https://setemargens.com/goncalo-ribeiro-telles-1922-2020-portugal-de-luto-pelo-ambientalista-catolico-antifascista-e-pensador-do-territorio/
12.22	"Fui incómodo em muitas situações mas se voltasse atrás ainda teria sido mais incómodo." (2020) *O Mirante*. https://omirante.pt/textos-que-fizeram-historia/2020-11-12-Fui-incomodo-em-muitas-situacoes-mas-se-voltasse-atras-ainda-teria-sido-mais-incomodo?platform=hootsuite&utm_campaign=HSCampaign
12.23	Gonçalo Ribeiro Telles, 1922-2020 (2020). *Ordem dos Arquitetos-Secção Regional Sul*. https://www.oasrs.org/noticias/816/
12.24	Município de Coruche aprova voto de pesar pela morte de Gonçalo Ribeiro Telles. (2020). *Mais Ribatejo*. https://maisribatejo.pt/2020/11/12/municipio-de-coruche-aprova-voto-de-pesar-pela-morte-de-goncalo-ribeiro-telles/
12.25	Ribeiro Telles : o paisagista que pensou as cidades para as pessoas (2020) *Jornal de Negócios*. https://www.pressreader.com/portugal/jornal-de-negocios/20201112/page/36/textview
12.26	Cardoso, M. E. (2020) O Gonçalo que ficou. *Público*. https://www.publico.pt/2020/11/13/opiniao/cronica/goncalo-ficou-1938977
12.27	Tomé, J. (2020) Morreu o arquiteto paisagista Gonçalo Ribeiro Telles aos 98 anos. *Dinheiro Vivo*. https://www.dinheirovivo.pt/geral/morreu-o-arquiteto-paisagista-goncalo-ribeiro-telles-aos-98-anos-13025047.html
12.28	Coletivos aveirenses recordam Gonçalo Ribeiro Telles e dizem que mensagem continua atual. (2020). *Terra Nova*. https://www.terranova.pt/noticia/sociedade/coletivos-aveirenses-recordam-goncalo-ribeiro-telles-e-dizem-que-mensagem-continua
12.29	Gomes, G. D. (2020) O Ribeiro Telles que fica em nós. *LugarAo Sul*. http://www.lugaraosul.pt/home/o-ribeiro-telles-que-fica-em-nos?fbclid=IwAR11CaUslZgL0FoVSuuyGPc3zi8C5L5PrsZpET1l9kznlFof3Vtb9Pj7oH4&platform=hootsuite&utm_campaign=HSCampaign

(*Continued*)

Table 2.1 (Continued)

ref. Unit of Analysis *ref. Unidade de Análise*	***Text ref.***
12.30	"A visionary who leaves a valuable legacy"—Gonçalo Ribeiro Telles 1922-2020 (2020) *International Federation of Landscape Architects(IFLA)*. https://iflaeurope.eu/index.php/site/news-single/a-visionary-who-leaves-a-valuable-legacy-goncalo-ribeiro-telles-1922-2020?platform=hootsuite&utm_campaign=HSCampaign
12.31	Fernandes, F. (2020) Os ulmeiros de Ribeiro Telles e outras árvores. *Público*. https://www.publico.pt/2020/11/15/opiniao/opiniao/ulmeiros-ribeiro-telles-arvores-1939152
12.32	Reis, T. (2020) Como projectar o legado de Gonçalo Ribeiro Telles no século XXI. *Observador*. https://observador.pt/opiniao/como-projectar-o-legado-de-goncalo-ribeiro-telles-no-seculo-xxi/?platform=hootsuite&utm_campaign=HSCampaign
12.33	Gonçalo Ribeiro Telles: O monárquico que serviu a República (2020) *Renascença*. https://www.youtube.com/watch?v=XmJ632ma_MA
12.34	Leitão, J. (2020) Voto 124/05 (Mesa)—Pesar pelo falecimento de Gonçalo Ribeiro Telles. *Assembleia Municipal de Lisboa*. https://www.am-lisboa.pt/303000/1/015017,000428/index.htm
12.35	Movimento o Partido da Terra-MPT (2020) Voto 124/01 (MPT)—Pesar pelo falecimento de Gonçalo Ribeiro Telles. *Assembleia Municipal de Lisboa*. https://www.am-lisboa.pt/303000/1/015012,000428/index.htm
12.36	Spranger, D. (2020) Pesar por Gonçalo Ribeiro Telles. *JM-Madeira*. https://www.jm-madeira.pt/regiao/ver/110208/Pesar_por_Goncalo_Ribeiro_Telles
12.37	Continuamos contigo, Gonçalo (2020) *Público*. https://www.publico.pt/2020/11/19/opiniao/opiniao/continuamos-contigo-goncalo-1939747
12.38	Covas, A. (2020) Mais campo na cidade e mais cidade no campo (em memória do Arquiteto Gonçalo Ribeiro Telles). *Sul Informação*. https://www.sulinformacao.pt/2020/11/mais-campo-na-cidade-e-mais-cidade-no-campo-em-memoria-do-arquiteto-goncalo-ribeiro-telles/?platform=hootsuite&utm_campaign=HSCampaign
12.39	Schmidt, L. (2020) Tudo na sua obra vai dar à saúde humana, ao sustento e à segurança coletiva. *Expresso*. https://expresso.pt/opiniao/2020-11-20-Continuar-Ribeiro-Telles
12.40	Guedes, G. M. (2020) Gonçalo Ribeiro Telles e o jardim. *Expresso*. https://leitor.expresso.pt/semanario/semanario2508/html/revista-e/vicios/design/goncalo-ribeiro-telles-e-o-jardim
12.41	Saraiva, J. C. (2020) Gonçalo Ribeiro Telles: o éden ao alcance de todos. *Jornal I*. https://ionline.sapo.pt/artigo/714863/goncalo-ribeiro-telles-o-eden-ao-alcance-de-todos?seccao=Opinião_i
12.42	Parlamento unânime no pesar pela morte do arquiteto, político e ecologista Ribeiro Telles (2020) *Observador*. https://observador.pt/2020/11/26/parlamento-unanime-no-pesar-pela-morte-do-arquiteto-politico-e-ecologista-ribeiro-telles/
05.01	Franco, L. (2020) Memórias do Mestre: a obra e carreira de Gonçalo Ribeiro Telles. *Ponto SJ*. https://pontosj.pt/especial/memorias-do-mestre-a-obra-e-carreira-de-goncalo-ribeiro-telles/
05.02	Pessoa, F. S. (2020) Gonçalo Ribeiro Telles. *Público*. https://www.publico.pt/2020/05/25/opiniao/opiniao/goncalo-ribeiro-telles-1917208
05.03	Sousa, M. R. de (2020) Presidente da República saúda Gonçalo Ribeiro Telles pelo seu 98.o aniversário. *Presidência da República*. http://www.presidencia.pt/?idc=18&idi=177534

Gonçalo Ribeiro Telles

When considering landscape architecture and ecology in Portugal, we always meet Ribeiro Telles. Most of his biography is public and widely known. Certainly, there are many aspects that deserve to be examined and shown in more detail. However, here we will focus only on facts of his life and work recognized as relevant.

Ribeiro Telles was born on 25th May 1922. He grew up between Lisbon and Coruche. These two worlds' experiences made him sensitive to the importance of respecting nature and the need for a connection between the city and countryside.

He graduated in Agronomic Engineering and completed the 'Curso Livre' of Landscape Architecture (1952) from the Technical University of Lisbon (Higher Institute of Agronomy–ISA), among those who would come to be known as the first generation of landscape architects in Portugal. 'Curso Livre' was the initiative of Cabral to introduce the study of Landscape Architecture at the university level, in 1941, as an experimental course, and a year later as complimentary education to agronomy or forestry.

His career spanned over six decades, as a designer, a teacher (ISA and University of Évora) and in public administration (Lisbon City Council and Fundo de Fomento da Habitação). He also served as a member of several governments of Portugal. The promotion of Landscape Architecture has always been one of the main objectives of his career. He was a key figure in the development of Landscape Architecture training in Portugal. He taught in the first course at ISA and was the founder of the degree in landscape architecture at the University of Évora, where he was also professor.

The biographical tree representation

The biographical tree presented in Figure 2.1 is a graphic representation of Ribeiro Telles' life and work, including some facts of the context in which he progressed. It helps to find, in an easy way, relations between his life and the field of landscape architecture.

On the right side of the tree, facts and works from Ribeiro Telles' life are presented; on the left side, the events are related to the scope of Landscape Architecture in Portugal. The two horizontal bars at the bottom of the graph represent the two world wars. The first associations of landscape architects were created in the United States of America and in Germany before the First World War. The beginning of the Second World War coincided with Cabral's return to Lisbon from Berlin, where he graduated in Landscape Architecture. In 1940, Cabral began teaching at the ISA. The following year he laid the foundations for the 'Curso Livre' of Landscape Architecture. Ribeiro Telles completed his academic training with this course. In the centre, perpendicular and crossing the wars' lines, we have the lifelines. The first one on the right corresponds to Ribeiro Telles. To the immediate left is Cabral's lifeline. There are lifelines to each one of the first generation

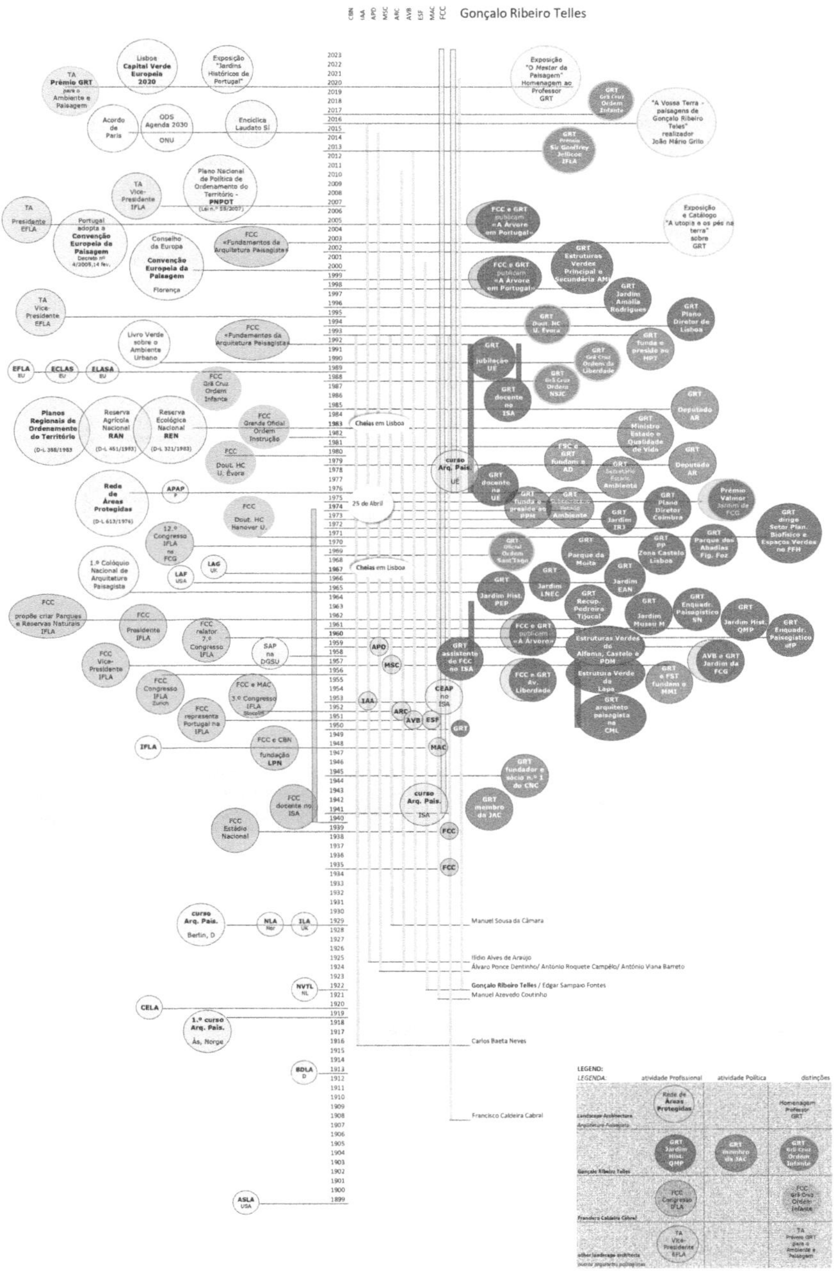

Figure 2.1 Gonçalo Ribeiro Telles biographical tree and Landscape Architecture context.

List of abbreviations and acronyms for Gonçalo Ribeiro Telles biographical tree.

List of Abbreviations and Acronyms / *Lista de Abreviaturas e Acrónimos*

Democratic Alliance	**AD**	*Aliança Democrática*
Lisbon Metropolitan Area	**AML**	*Área Metropolitana de Lisboa*
Portuguese Association of Landscape Architects	**APAP**	*Asociação Portuguesa dos Arquitetos Paisagistas*
Álvaro Ponce Dentinho	**APD**	*Álvaro Ponce Dentinho*
Portuguese Parliament	**AR**	*Assembleia da República*
António Roquete Campêlo	**ARC**	*António Roquete Campêlo*
Landscape Architecture	**Arq. Pais.**	*Arquitetura Paisagista*
American Society of Landscape Architects	**ASLA**	*Sociedade Americana de Arquitetos Paisagistas*
António Viana Barreto	**AVB**	*António Viana Barreto*
Bund DeutscherLandschaftsarchitekten	**BDLA**	*Federação Alemã dos Arquitetos Paisagistas*
Carlos Baeta Neves	**CBN**	*Carlos Baeta Neves*
Centre for Landscape Architecture Studies	**CEAP**	*Centro de Estudos de Arquitectura Paisagista*
Council of Educators in Landscape Architecture	**CELA**	*Concelho dos Educadores em Arquitetura Paisagista*
Lisbon Municipality	**CML**	*Câmara Municipal de Lisboa*
National Centre for Culture	**CNC**	*Centro Nacional de Cultura*
Germany	**D**	*Alemanha*
Department of Urbanisation Services	**DGSU**	*Direção-Geral dos Serviços de Urbanização*
Decree-Law	**D-L**	*Decreto-Lei*
National Agronomy Station	**EAN**	*Estação Agronómica Nacional*
European Council of Landscape Architecture Schools	**ECLAS**	*Conselho Europeu de Escolas de Arquitetura Paisagista*
European Foundation for Landscape Architecture	**EFLA**	*Federação Europeia para a Arquitetura Paisagista*
European Landscape Architecture Students Association	**ELASA**	*Associação Europeia de Estudantes de Arq. Paisagista*
Framework	**Enquadr.**	*Enquadramento*
Edgar Sampaio Fontes	**ESF**	*Edgar Sampaio Fontes*
European Union	**EU**	*União Europeia*
Francisco Caldeira Cabral	**FCC**	*Francisco Caldeira Cabral*
Calouste Gulbenkian Foundation	**FCG**	*Fundação Calouste Gulbenkian*
Housing Development Fund	**FFH**	*Fundo de Fomento da Habitação*
Figueira da Foz	**Fig. Foz**	*Figueira da Foz*
Francisco Sá Carneiro	**FSC**	*Francisco Sá Carneiro*
Gonçalo Ribeiro Telles	**GRT**	*Gonçalo Ribeiro Telles*
Ilídio Alves de Araújo	**IAA**	*Ilídio Alves de Araújo*
International Federation of Landscape Architects	**IFLA**	*Federação Internacional dos Arquitetos Paisagistas*
Institute of Landscape Architects	**ILA**	*Instituto dos Arquitetos Paisagistas*
Ricardo Jorge Institute	**IRJ**	*Instituto Ricardo Jorge*
Higher Institute of Agronomy	**ISA**	*Instituto Superior de Agronomia*
Catholic Agrarian Youth	**JAC**	*Juventude Agrária Católica*
Historic Garden	**Jardim Hist.**	*Jardim Histórico*
Landscape Architecture Foundation	**LAF**	*Fundação da Arquitetura Paisagista*
National Civil Engineering Laboratory	**LNEC**	*Laboratório Nacional de Engenharia Civil*
Nature Protection League	**LPN**	*Liga de Proteção da Natureza*
Manuel Azevedo Coutinho	**MAC**	*Manuel Azevedo Coutinho*
Movement of Independent Monarchists	**MMI**	*Movimento dos Monárquicos Independentes*
Land Party Movement	**MPT**	*Movimento Partido da Terra*
Manuel Sousa da Câmara	**MSC**	*Manuel Sousa da Câmara*
Marine Museum	**Museu M**	*Museu de Marinha*
Netherlands	**NL**	*Países Baixos*
Norske landskapsarkitekters forening	**NLA**	*Associação Norueguesa dos Arquitetos Paisagistas*
Norway	**Nor**	*Noruega*
Our Lord Jesus Christ	**NSJC**	*Nosso Senhor Jesus Cristo*
Nederlandse Vereniging voor Tuin- en landschapsarchitectuur	**NVTL**	*Associação Holandesa dos Arquitetos Paisagistas*
Sustainable Development Goals	**ODS**	*Objetivos de Desenvolvimento Sustentável*
United Nations Organisation	**ONU**	*Organização das Nações Unidas*
Portugal	**P**	*Portugal*
Episcopal Palace of Portalegre	**PEP**	*Paço Episcopal de Portalegre*
Popular Monarchist Party	**PPM**	*Partido Popular Monárquico*
Marqueses de Pombal Estate	**QMP**	*Quinta dos Marqueses de Pombal*
Recovering	**Recup.**	*Recuperação*
Landscape Architecture Service	**SAP**	*Serviço de Arquitetura Paisagista*
National Steel Plant	**SN**	*Siderurgia Nacional*
Teresa Andresen	**TA**	*Teresa Andresen*
University	**U.**	*Universidade*
University of Évora	**UÉ**	*Universidade de Évora*
Petrochemical plant	**ufP**	*unidade fabril da Petroquímica*
United Kingdom	**UK**	*Reino Unido*
United States of America	**USA**	*Estados Unidos da América*

Figure 2.1 (Continued)

landscape architects: Manuel Azevedo Coutinho, Edgar Sampaio Fontes, António Viana Barreto, António Roquete Campelo, Manuel de Sousa da Câmara, Álvaro Ponce Dentinho and Ilídio Alves de Araújo. The years when each completed the course are highlighted. Two floods that had a devastating impact in Lisbon (1967 and 1983) have also been identified, as well as the 25th of April 1974 ("The Carnation Revolution"). Following these events, Ribeiro Telles had relevant public interventions.

The 1974 Revolution brought him to the forefront of the country's political life: in 1975, as secretary of state for the environment and from 1981 to 1983, as state minister and quality of life minister. He then established the legal foundations for the current framework laws on nature conservation, spatial planning and environmental quality. This legislative regulation made Portugal a pioneer country that would come to influence others in these matters. It is important to note, on the left side of the tree, the references to the publication of the seminal legal diplomas of the Land-planning Policy: the National Agricultural Reserve (RAN), the National Ecological Reserve (REN) and the Protected Areas Network. In 1976, the Portuguese Association of Landscape Architects (APAP) was created. In all these initiatives, Ribeiro Telles was actively involved.

In the following decade, as a Lisbon City councillor, he fought for the protection and enhancement of the city's ecological structure. His action led to the 1994 approval of the Lisbon Green Plan, embedded in the Municipal Master Plan.

Alongside the central trunk, the darker bars represent the landscape architecture courses at ISA and UÉ. Having been a student and lecturer at the former, Ribeiro Telles was a founder of the latter. In Évora, he was a professor until his 1992 retirement. The UÉ awarded him an honorary doctorate. In Portugal, these were the first higher education courses in Landscape Architecture. Currently, the universities of Porto, Algarve and Trás-os-Montes e Alto Douro also provide this degree.

As another reading example, in 2020, Lisbon was the European Green Capital (Figure 2.2). This coincidence may be considered as a corollary of Ribeiro Telles' lifetime work.

It is also noted that in 2013, the International Federation of Landscape Architects (IFLA) conferred him the Sir Geoffrey Jellicoe Award, the highest international distinction to a landscape architect.

A few years earlier, in 1999, the second edition of the book "The Tree" was published, under the title "The Tree in Portugal". The first edition, in 1960, resulted from the work of Cabral and Ribeiro Telles (Table 2.2).

His most famous project, the gardens of the Calouste Gulbenkian Foundation, which became a reference in Lisbon, was designed in 1962–1963 with Viana Barreto. In 1975, this work was awarded the Valmor Prize, the most prestigious architecture prize in Lisbon, awarded since 1902.

From this biographical tree, it is possible to make other readings, as well as filter or add facts. It may be included, for example, the "Gonçalo Ribeiro Telles Award for Environment and Landscape", first granted in 2020.

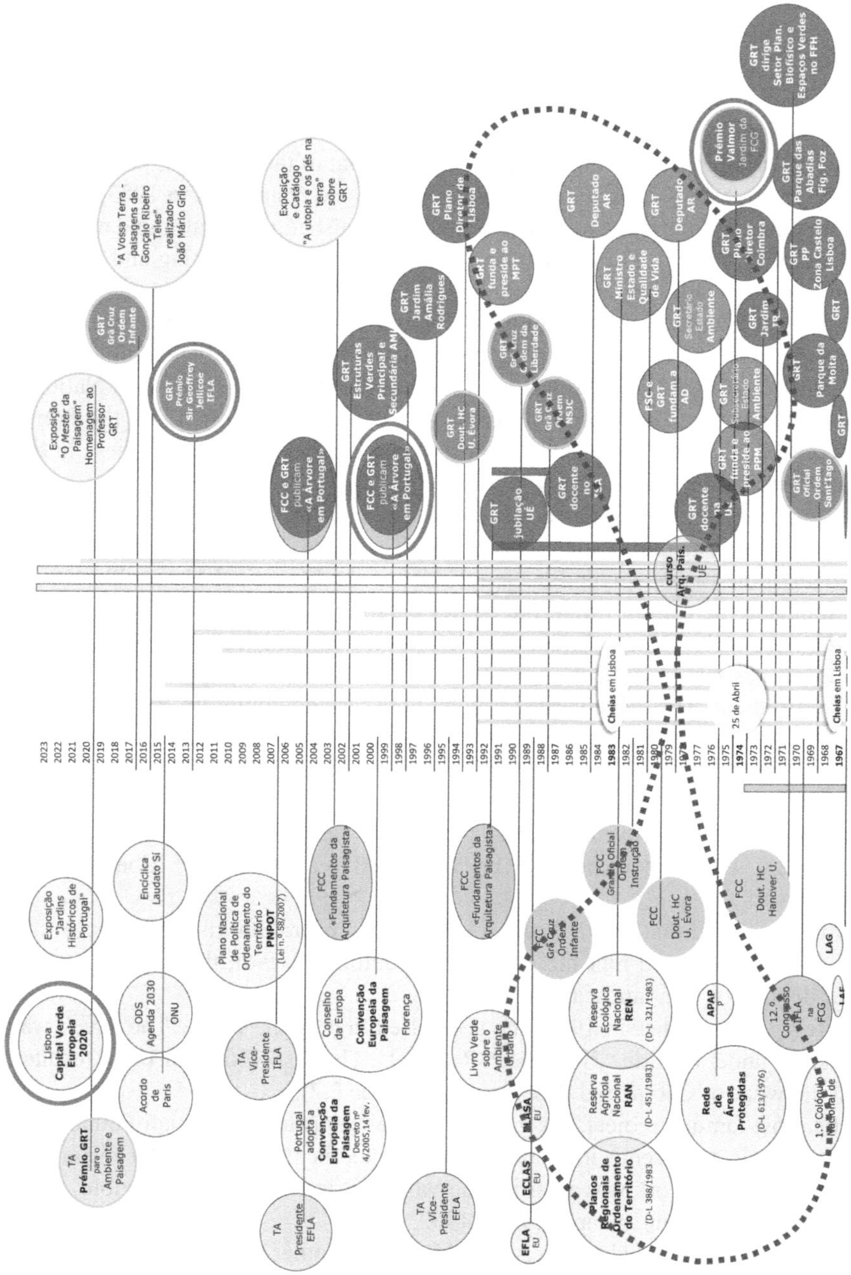

Figure 2.2 For a reading: extract from Gonçalo Ribeiro Telles' biographical tree.

Table 2.2 Publications by Gonçalo Ribeiro Telles, with his involvement or about him.

	Author	***Year***	***Title***	***Local***	***Publisher***	***Book, journal or ISBN***	***n.º***	***p.***
1. GRT_Author								
	Cabral, F. C.; Telles, G. R.	1960	***A Árvore****: o seu valor na paisagem urbana e rural, a sua cultura, a poda*	Lisboa	Centro de Est. de Urbanismo			
	Telles, G. P. R.; Fontes, E. S. F.; Campelo, A. J. S. R. de	1964	*Um caso concreto de ordenamento paisagístico*	Lisboa		Sep. Agros	45	
	Telles, G. R. (col.)	1969	*A Evolução dos Espaços Verdes de Lisboa*	Lisboa		*Arquitectura*	108	44–51
	Telles, G. R. (org.)	1971	*Convergência monárquica: 1.º ano de acção*	Lisboa				
	Telles, G. R.	1973	*Burle Marx: arquitecto paisagista*	Lisboa	Fundação Calouste Gulbenkian	*Colóquio/Artes*		
	Telles, G. R.	1975	*Da formação do solo à socialização da paisagem*	Lisboa	Seara Nova	*Uns comem os figos*		55–150
	Telles, G. R.	1976	*Reforma agrária: o homem e a terra*	Lisboa	PPM			
	Telles, G. R.	1980	*Administração pública e ordenamento territorial*	Lisboa	Inst. Democracia e Liberdade	Sep. Democracia e Liberdade	13	
	Telles, G. R.	1981	*Região, desenvolvimento e ordenamento do território*	Lisboa	Inst. Democracia e Liberdade-Inst. Amaro da Costa	Sep. Democracia e Liberdade	19	
	Telles, G. R.	1985	*Para além da Revolução*	Lisboa	Edições Salamandra			

(*Continued*)

Table 2.2 (Continued)

Author	Year	Title	Local	Publisher	Book, journal or ISBN	n.º	p.
Telles, G. R.	1987	*Lisboa: entre o passado e o futuro*	Lisboa	Universidade Nova de Lisboa—Faculdade de Ciências e Tecnologia			
Telles, G. R.	1992	*Cinquenta anos de Ensino da Arquitectura Paisagista do Instituto Superior de Agronomia 1942–1992*			*Agros*: Revista Ass. Estud. ISA (AEISA)	74: 1	42–44
Telles, G. R.	1994	*Paisagem Global, um Conceito para o Futuro*			*Iniciativa*	Espec.	28–33
Telles, G. R.	1995	*A propósito do início da política de ambiente em Portugal*			*Revista das Autarquias*	2	73–80
Telles, G. R.	1996	*Um novo conceito de cidade: a paisagem global*	Matosinhos	Contemporânea Editora; Câmara Municipal de Matosinhos			
Telles, G. R.; Pessoa, F. S.	1996	*Portugal, paisagens e espaços naturais*	Alfragide	Clube Internacional do Livro	ISBN: 9729700303		
Telles, G. R. (coor.)	1997	*Plano verde de Lisboa. Componente do Plano Director Municipal de Lisboa*	Lisboa	Edições Colibri			
Cabral, F. C.; Telles, G. R.	1999	***A árvore** em Portugal*	Lisboa	Assírio & Alvim	ISBN: 9789728288747		
Telles, G. R.; Bacalhau, D. (conceção)	2003	*Já pensaste na paisagem?*	Évora	Museu de Évora	ISBN: 9789723705386		
Cabral, F. C.; Telles, G. R.	2005	***A árvore** em Portugal*	Lisboa	Assírio & Alvim	2.ª ed.		
Fernandes, J. M. (coor.)	2005	*Património arquitectónico moderno: do contexto internacional ao concelho de Oeiras*	Oeiras	Câmara Municpal de Oeiras			

(*Continued*)

Table 2.2 (Continued)

	Author	*Year*	*Title*	*Local*	*Publisher*	*Book, journal or ISBN*	*n.º*	*p.*
	Telles, G. R.	2009	[Francisco Caldeira Cabral]	Lisboa	Bertrand Editora	*Memórias do mestre no centenário do seu nascimento: Francisco Caldeira Cabral*		14–16
	Telles, G. R.	2012	*Desenvolver o país*	Lisboa	Bertrand Editora	Plano C: o combate da cidadania.		
	Telles, G. R.; Pessoa, F. S. (seleção)	2016	*Gonçalo Ribeiro Telles: textos escolhidos*	Lisboa	Argumentum	ISBN: 9789728479985		
	Telles, G. R.	2017	*Por que sou monárquico*	Lisboa	Real Associação de Lisboa	ISBN: 9789896916473		
2. GRT_Foreword |Posface|Thesis or Dissertation Orientation								
foreword	Simonet, D.	1983	*O que é a ecologia?*	Lisboa	Notícias	Biblioteca de conhecimentos básicos	1	
presentation	Martins, F. E. de O.	1991	*Arquitectura popular do Ramo Grande: uma arquitectura entre dois terramotos (1841–1980)*	Praia da Vitória	Câmara Municipal da Praia da Vitória			
foreword	Eloy, A.	1992	*Planetas vivos são difíceis de encontrar: defendendo a vida na Terra*	Lisboa	Amigos da Terra			
foreword	Cabral, F. C.	1993	*Fundamentos da arquitectura paisagista*	Lisboa	Inst. Conservação da Natureza			
orientation	Carapinha, A. da C. P.	1995	*Da essência do jardim português*	Évora				
orientation	Pacheco, P. de A.	2001	*Estação de Barca de Alva [Texto policopiado]: memória e renascimento*	Lisboa				

(*Continued*)

Table 2.2 (Continued)

	Author	***Year***	***Title***	***Local***	***Publisher***	***Book, journal or ISBN***	***n.º***	***p.***
foreword	Barata, F.T.	2002	*Peservando a memória do território: o Parque Cultural Tourega-Valverde = Peserving the Land's Memories: The Tourega-Valverde Park*	Évora	CEEM-Univ	Cadernos do Centro Est. Ecossistemas Mediterrânicos		
orientation	Branco, O. R.	2004	*Subjectividades construídas pelos habitantes tradicionais e nobilitadores do Bairro de Alfama [Texto policopiado]: para uma teoria ecológica do self*	Évora				
foreword	Saraiva, A. M. de P.	2005	*Princípios de arquitectura paisagista e de planeamento do território*	Mirandela	João Azevedo			
orientation	Silva, M. L. P. da	2005	*A Ribeira Grande em Monforte, Fronteira e Avis [Texto policopiado]: bases para uma proposta metodológica de recuperação e valorização da paisagem*	Évora				
posface	Henriques, M. C.	2006	*Dom Duarte e a democracia: uma biografia portuguesa*	Lisboa	Bertrand			
orientation	Matos, I. N. de	2006	*Convento de Santa Cruz de Sintra [Texto policopiado]: bases para uma proposta metodológica de recuperação, manutenção e valorização*	Évora				
orientation	Querido, C. J. V.	2006	*As estradas e o seu património [Texto policopiado]: estudo de uma abordagem para a sua revitalização*	Évora				

(*Continued*)

Table 2.2 (Continued)

	Author	*Year*	*Title*	*Local*	*Publisher*	*Book, journal or ISBN*	*n.º*	*p.*
foreword	Mendonça, N. J. de N.	2006	*Rio Côa: a arte da água e da pedra*	Évora	Casa do Sul. Centro de História da Arte da Universidade de Évora			
foreword	Saraiva, A. M. de P.	2007	*Princípios de arquitectura paisagista e de planeamento do território 2.ª ed*	Mirandela	João Azevedo			
posface	Henriques, M. C.	2007	*Dom Duarte e a democracia: uma biografia portuguesa 2.ª ed.*	Lisboa	Bertrand			
foreword	Rodrigues, L. F.	2011	*Manual de crimes urbanísticos: exemplos práticos para compreender os negócios insustentáveis da especulação imobiliária*	Lisboa	Guerra e Paz			
foreword	Loução, P. A. et. al.	2013	*A magia das Aldeias de Montanha*	Lisboa	Ésquilo			
foreword	Rodrigues, L. F.	2017	*Manual de crimes urbanísticos: exemplos práticos para compreender os negócios insustentáveis da especulação imobiliária 2.ª ed. rev. e aument.*	Lisboa	Guerra e Paz			
3. about_GRT								
	Universidade de Évora-Sec. Ed. (coord.)	1992	*Jubilação do Professor Gonçalo Ribeiro Telles*	Évora	Unviversidade de Évora			
	Universidade de Évora	1995	*Doutoramento "Honoris Causa" de Prof. Doutor Gonçalo Ribeiro Telles*	Évora	Unviversidade de Évora			
	Pessoa, F. S.	2002	*Gonçalo Ribeiro Telles: esboço biográfico*	[s. l.]	Paulo Trancoso—Costa do Castelo Filmes			

(*Continued*)

Table 2.2 (Continued)

	Author	Year	Title	Local	Publisher	Book, journal or ISBN	n.º	p.
	Caetano, J. (coord.); Abreu, A. C. de (texto); Cunha, R. (foto.)	2003	*A utopia e os pés na terra: Gonçalo Ribeiro Telles [catálogo]*	Lisboa	Instituto Português de Museus	ISBN: 9789727761715		
	Andresen, T. (coord.)	2003	*Do Estádio Nacional ao Jardim Gulbenkian. Francisco Caldeira Cabral e a primeira geração de arquitectos paisagistas (1940–1970)*	Lisboa	Fundação Calouste Gulbenkian			
	Pessoa, F. S.	2011	*Gonçalo Ribeiro Telles: a fotobiografia*	Lisboa	Argumentum			
	Pessoa, F. S.	2016	*Gonçalo Ribeiro Telles: textos escolhidos*	Lisboa	Argumentum			
	Pessoa, F. S.	2021	*Gonçalo Ribeiro Telles: o homem e a obra*	Lisboa	Argumentum			
	d'Abreu, M. C. (coord.); Camara, T. B. da (coord.)	2021	*O mester da paisagem/Master of landscape*	Lisboa	Câmara Municipal de Lisboa	https://goncaloribeirotelles.pt/wp-content/uploads/2021/06/CATALOGO-O-MESTER-DA-PAISAGEM.pdf ISBN: 9789728543624		
	Camara, T. B. da (coord.)	2021	*Espaço público de Lisboa: plano, projeto e obra da primeira geração de arquitetos paisagistas, 1950–1970*	Lisboa	Câmara Municipal de Lisboa	ISBN: 9789895314362		

Gonçalo Ribeiro Telles remembered at the end of his life

As Ribeiro Telles reached the end of his life, we sought to identify which aspects of his character, his performance or his work emerge in the expressions of regret and appreciation published. The research universe covered the texts available online.

Sample

The analysis focused on a sample of 140 texts (Table 2.1). The sample includes a wide range of texts: brief news of decease, news with reactions from personalities or institutions, news with collection of testimonies, news with longer or shorter biographical notes, news with republishing of Ribeiro Telles interviews, opinion articles, votes of regret and the decree declaring national mourning. They were collected using "Google" generic search engine between the 11th and the 26th November 2020. This sample includes 95 texts published on 11th November, 42 released from 12th to 26th November, and three preceding his death.

The three texts prior his death are from 2020. Two were written in May, on his birthday. The other one, published in November, recommends a visit to the exhibition about his work, entitled "O *Mester* da Paisagem", held in Lisbon. They were included as they are also homage texts and were written in the last days of his life.

The most recent text considered is the vote of regret unanimously approved in the Portuguese Parliament on 26th November.

Components

The analysis focused on the words and groups of words used to characterize Ribeiro Telles that express the way it was remembered. He is often referred to not just for one, but for several aspects. The identified attributes relating to Ribeiro Telles were grouped by association with similar concepts. The four components used in the analysis were then selected and defined: 'Ecology-Environment', 'Landscape', 'Politics' and 'Teaching-Education'.

'Ecology-Environment' groups 'ecology', 'environment', 'National Agricultural and Ecological Reserves (RAN and REN)', among others.

'Landscape' includes concepts such as 'landscape', 'landscape architecture', 'landscape architect', 'designer', 'garden', 'park', for example.

'Politics' encompasses 'policy', 'founder' or 'party leader', 'ruler', 'minister', 'legislator', 'deputy', 'councillor', 'bases of environmental policy', 'Constitution of the Portuguese Republic', and others.

'Teaching-Education' ranges from 'teacher', 'professor', 'master', to 'developing ecological awareness'.

Analysis

It is observed that sets of components are repeated in several texts (Table 2.3). In the sample universe, 11 different types of combinations were identified (Figure 2.3).

Table 2.3 Component registration and cataloguing.

reference referência	**LANDSCAPE** PAISAGEM	**ECOLOGY/ENVIRONMENT** ECOLOGIA/AMBIENTE	**POLITICS** POLITICA	**TEACHING/EDUCATION** ENSINO/EDUCAÇÃO
11.01	1	1	1	1
11.02	1	0	0	0
11.03	1	1	1	0
11.04	1	1	1	0
11.05	1	1	1	0
11.06	1	0	1	0
11.07	1	1	1	0
11.08	1	0	1	0
11.09	1	1	1	1
11.10	1	1	1	0
11.11	1	1	1	0
11.12	1	1	1	1
11.13	1	1	1	1
11.14	1	1	1	0
11.15	1	1	1	0
11.16	1	1	1	1
11.17	1	1	1	0
11.18	1	1	1	0
11.19	1	1	1	0
11.20	1	1	1	0
11.21	1	1	1	1
11.22	1	1	1	0
11.23	1	1	1	0
11.24	1	0	1	0
11.25	1	0	1	0
11.26	1	1	1	0
11.27	1	0	0	0
11.28	1	0	1	1
11.29	1	0	1	0
11.30	1	0	0	0
11.31	0	1	1	1
11.32	1	1	1	1
11.33	1	1	1	0
11.34	1	1	1	1
11.35	1	1	1	1
11.36	1	1	0	0
11.37	1	1	1	1
11.38	1	1	1	0
11.39	1	1	1	0
11.40	1	1	1	0
11.41	1	0	0	0
11.42	1	1	1	1
11.43	1	1	1	1
11.44	1	1	1	1
11.45	1	1	1	0
11.46	1	1	1	1
11.47	1	1	1	1
11.48	1	1	1	1
11.49	1	0	1	0
11.50	1	1	1	0
11.51	1	0	0	0
11.52	1	1	1	0
11.53	1	1	1	0
11.54	1	0	0	1
11.55	1	1	1	0
11.56	0	1	0	0
11.57	1	1	1	0
11.58	1	1	1	0
11.59	1	1	1	0
11.60	1	1	1	0
11.61	1	1	1	0
11.62	1	1	1	0
11.63	1	1	1	0
11.64	1	1	1	0
11.65	1	1	1	1
11.66	1	1	1	0
11.67	1	1	1	0
11.68	1	1	1	0
11.69	1	1	1	1
11.70	1	1	0	0
11.71	1	1	1	1
11.72	1	1	1	0
11.73	1	1	0	0
11.74	1	1	1	1
11.75	1	1	1	1
11.76	1	1	1	1
11.77	1	1	1	1
11.78	1	1	1	0
11.79	1	1	1	1
11.80	1	1	1	0
11.81	1	0	0	1
11.82	1	1	1	1
11.83	1	1	1	1
11.84	1	1	0	1
11.85	1	1	0	0
11.86	1	1	1	0
11.87	1	1	0	0
11.88	1	1	1	0
11.89	1	1	1	0
11.90	1	1	1	1
11.91	1	1	1	1
11.92	1	1	1	0
11.93	1	1	1	1
11.94	1	1	1	1
11.95	1	1	1	1
12.01	1	1	0	1
12.02	1	1	0	0
12.03	1	1	1	0
12.04	1	1	1	1
12.05	1	1	1	0
12.06	1	1	1	0
12.07	1	1	1	0
12.08	1	0	0	0
12.09	1	1	0	0
12.10	1	1	1	0
12.11	1	1	1	1
12.12	1	0	0	0
12.13	1	1	1	0
12.14	1	1	1	0
12.15	1	1	1	0
12.16	1	1	1	0
12.17	1	0	0	0
12.18	1	0	0	0
12.19	1	0	1	0
12.20	1	1	1	0
12.21	1	1	1	0
12.22	1	1	1	1
12.23	1	1	1	1
12.24	1	0	0	0
12.25	1	0	0	0
12.26	0	1	0	0
12.27	1	1	1	1
12.28	1	1	1	0
12.29	1	1	1	1
12.30	1	1	1	1
12.31	1	1	1	1
12.32	1	1	1	1
12.33	1	1	1	0
12.34	1	1	1	1
12.35	1	1	1	1
12.36	1	0	0	0
12.37	0	1	1	0
12.38	1	1	0	0
12.39	1	1	1	0
12.40	1	0	0	0
12.41	1	0	0	0
12.42	1	1	1	1
05.01	1	1	1	1
05.02	1	1	1	1
05.03	0	1	0	0

LANDSCAPE *PAISAGEM*	ECOLOGY/ENVIRONMENT *ECOLOGIA/AMBIENTE*	POLITICS *POLÍTICA*	TEACHING/EDUCATION *ENSINO/EDUCAÇÃO*	A	45	32%
LANDSCAPE *PAISAGEM*	ECOLOGY/ENVIRONMENT *ECOLOGIA/AMBIENTE*	POLITICS *POLÍTICA*		B	56	40%
LANDSCAPE *PAISAGEM*		POLITICS *POLÍTICA*		C	7	5%
LANDSCAPE *PAISAGEM*		POLITICS *POLÍTICA*	TEACHING/EDUCATION *ENSINO/EDUCAÇÃO*	D	1	1%
	ECOLOGY/ENVIRONMENT *ECOLOGIA/AMBIENTE*	POLITICS *POLÍTICA*	TEACHING/EDUCATION *ENSINO/EDUCAÇÃO*	E	1	1%
	ECOLOGY/ENVIRONMENT *ECOLOGIA/AMBIENTE*	POLITICS *POLÍTICA*		F	1	1%
	ECOLOGY/ENVIRONMENT *ECOLOGIA/AMBIENTE*			G	3	2%
LANDSCAPE *PAISAGEM*				H	14	10%
LANDSCAPE *PAISAGEM*	ECOLOGY/ENVIRONMENT *ECOLOGIA/AMBIENTE*		TEACHING/EDUCATION *ENSINO/EDUCAÇÃO*	I	2	1%
LANDSCAPE *PAISAGEM*	ECOLOGY/ENVIRONMENT *ECOLOGIA/AMBIENTE*			J	8	6%
LANDSCAPE *PAISAGEM*			TEACHING/EDUCATION *ENSINO/EDUCAÇÃO*	L	2	1%
					140	100%

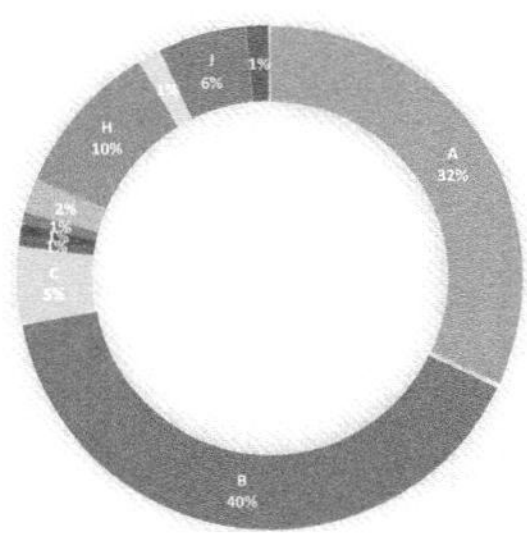

Figure 2.3 Components.

The three components combination is the most common, appearing in nearly half of the cases (61 times, 44%). The four components combination occurs 45 times, i.e. in about one third of the total sample. Two-component groups appear in 18 texts (13%). In a smaller number are the texts in which Ribeiro Telles is identified with only one component (16 times, 11%) (Figure 2.4).

Considering the number of times each component is mentioned in the sample universe, the most mentioned is 'Landscape' (135 times, 96%), then 'Ecology-Environment' (116 times; 83%), followed by 'Politics' (111 times; 79%), with the least employed being 'Teaching-Education' (51 times; 36%) (Figure 2.5).

From these data, additional information was obtained through statistical analysis using the IBM SPSS Statistics software.

In the data dimensional reduction using the principal components analysis method two main dimensions are observed (Figure 2.6). One is that which we could designate as professional intervention related to the 'Landscape' component. The other, grouping the components 'Ecology-Environment', 'Politics' and 'Teaching-Education', would be the dimension of public intervention.

By including the origin and type of text in this analysis, we can observe that the regional and/or landscape architecture/agronomy publications are those that preferentially refer to 'Landscape' and 'Teaching-Education' components (Figure 2.7). It is also observed that references to 'Teaching-Education' come mainly from institutional sources (Figure 2.8).

n.º of Components	n.º UA	
1	16	11%
2	18	13%
3	61	44%
4	45	32%
	140	100%

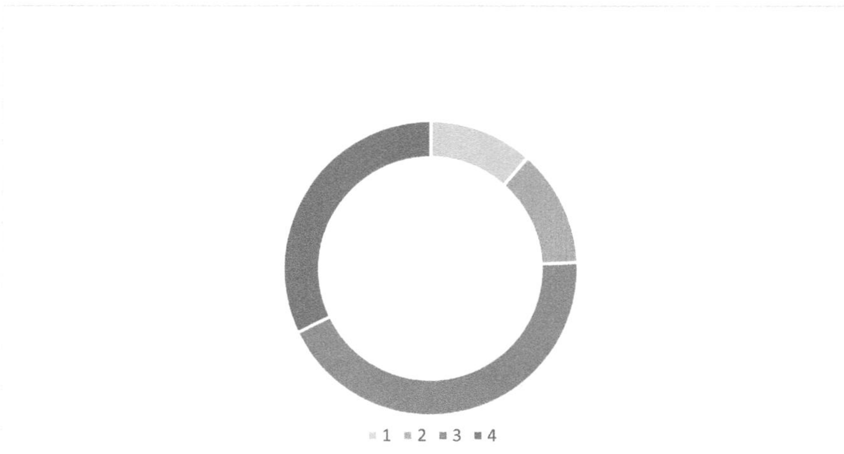

Figure 2.4 Total n.º of components per unit of analysis in the sample universe.

TEACHING/EDUCATION *ENSINO/EDUCAÇÃO*	51	36%
POLITICS *POLÍTICA*	111	79%
ECOLOGY/ENVIRONMENT *ECOLOGIA/AMBIENTE*	116	83%
LANDSCAPE *PAISAGEM*	135	96%
UA TOTAL	140	100%

Figure 2.5 Totals per component in the sample universe.

The hierarchical cluster analysis indicates that the relationship of greater proximity is established between the components 'Ecology-Environment' and 'Politics'. 'Ecology-Environment' is also very close to the 'Landscape' component (Figure 2.9). 'Teaching-Education' component, being the least referred to, has a more distant relationship with the others, appearing in less defined groups (Figure 2.10). It appears in the group where the four components are mentioned (type A). It never appears in isolation. The four other groups in which this component is found (types D, E, I and L) are formed by either one (D and E) or two (I and L) units of analysis each.

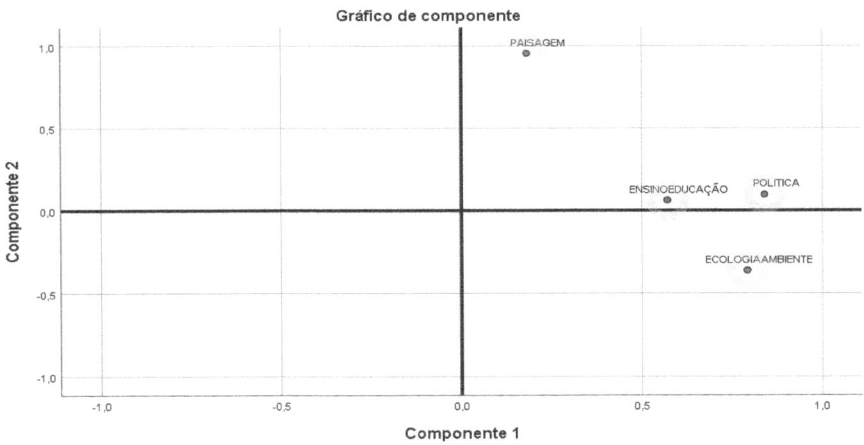

Figure 2.6 Professional and public intervention.

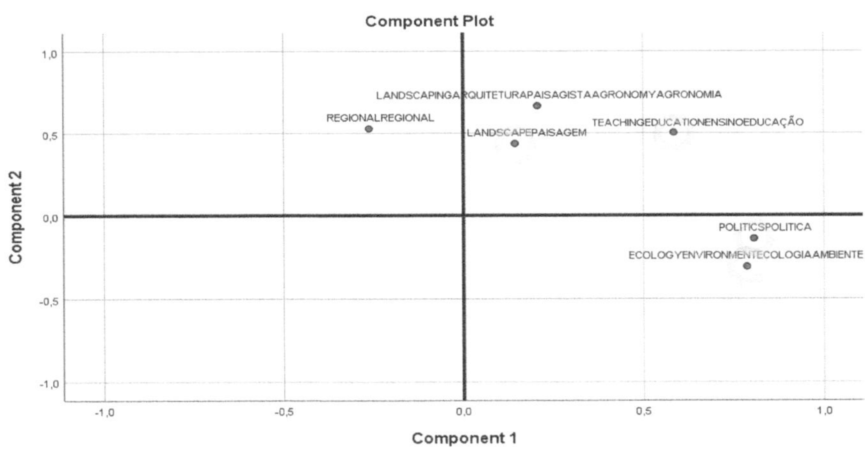

Figure 2.7 'Teaching-Education' and regional and/or landscape architecture/agronomy publications.

When including in this analysis also the type and origin of the publications, we confirm that 'Teaching-Education' component is hardly mentioned in the media. Apart from the 32 units of analysis of A01 group, only two more references appeared, both associated with 'Landscape' (I04 and D07; 34 in 113 texts, 30%). In institutional publications, on the contrary, it appears in approximately two thirds of the texts (17 references in a total of 27 units of analysis) (Figure 2.11).

Critical reflection: A cork oak that remains in the Portuguese landscape

Ribeiro Telles' activity touched many areas of society: professional, as a landscape architect; academic, as a teacher; political, and as a minister; civic, as a member

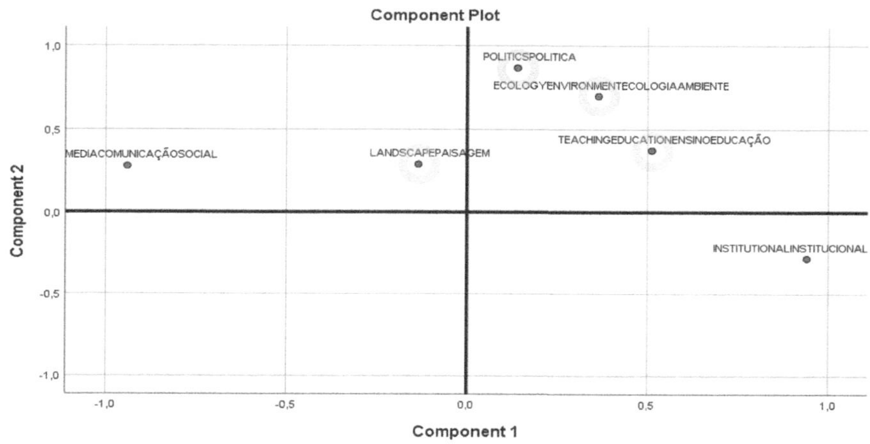

Figure 2.8 'Teaching/Education' and institutional sources.

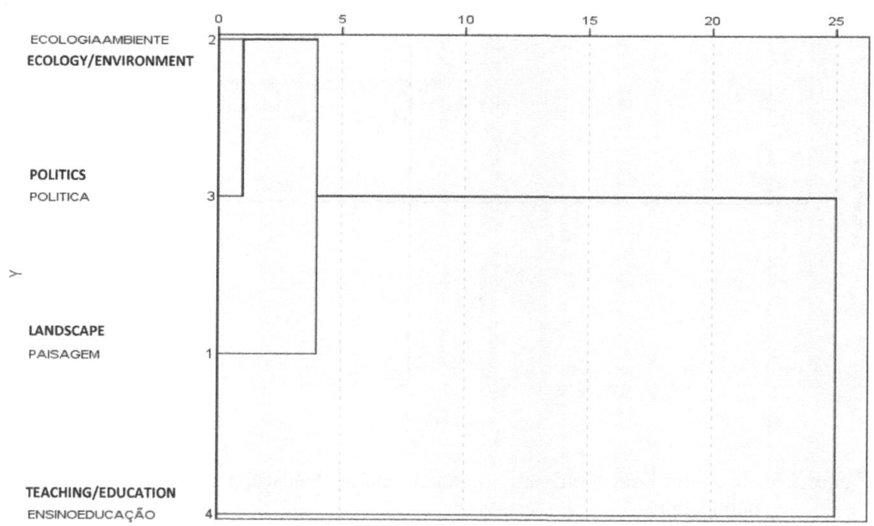

Figure 2.9 Component proximity.

or even founder of various associations or movements. The perception of his multifaceted character is confirmed in the analysis carried out by the predominance of references with three or four components.

The concept most often associated with Ribeiro Telles is 'Landscape'. The next is 'Ecology-Environment'. It happens even in texts where only one component is present, where we find 'Landscape' or, more rarely, 'Ecology-Environment'.

These concepts call different but related areas of knowledge. The search for a consensual definition for each one, even if only from the point of view of landscape architecture, could take us far. Such an undertaking is outside the purpose of this

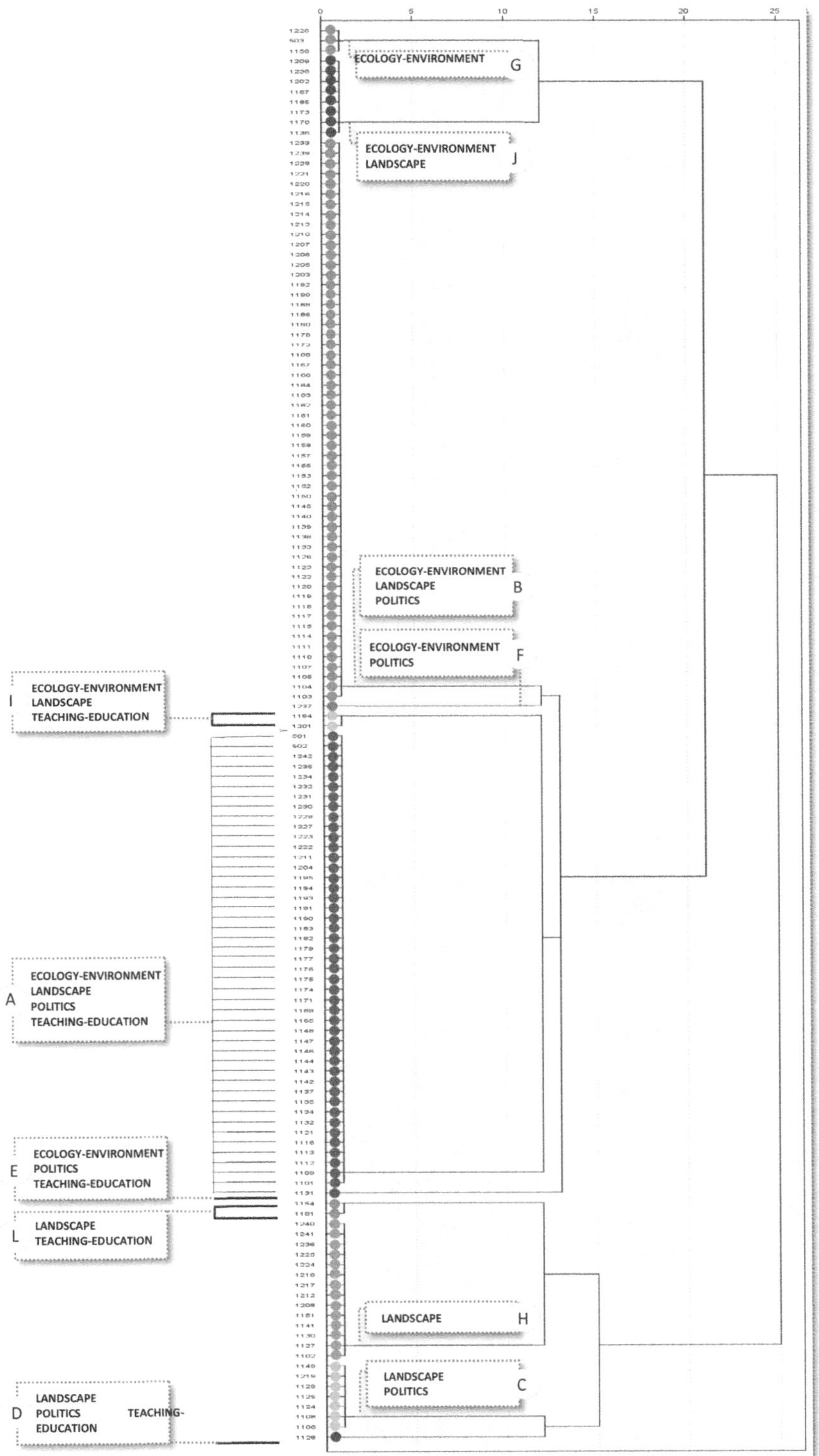

Figure 2.10 'Teaching-Education' component and small clusters.

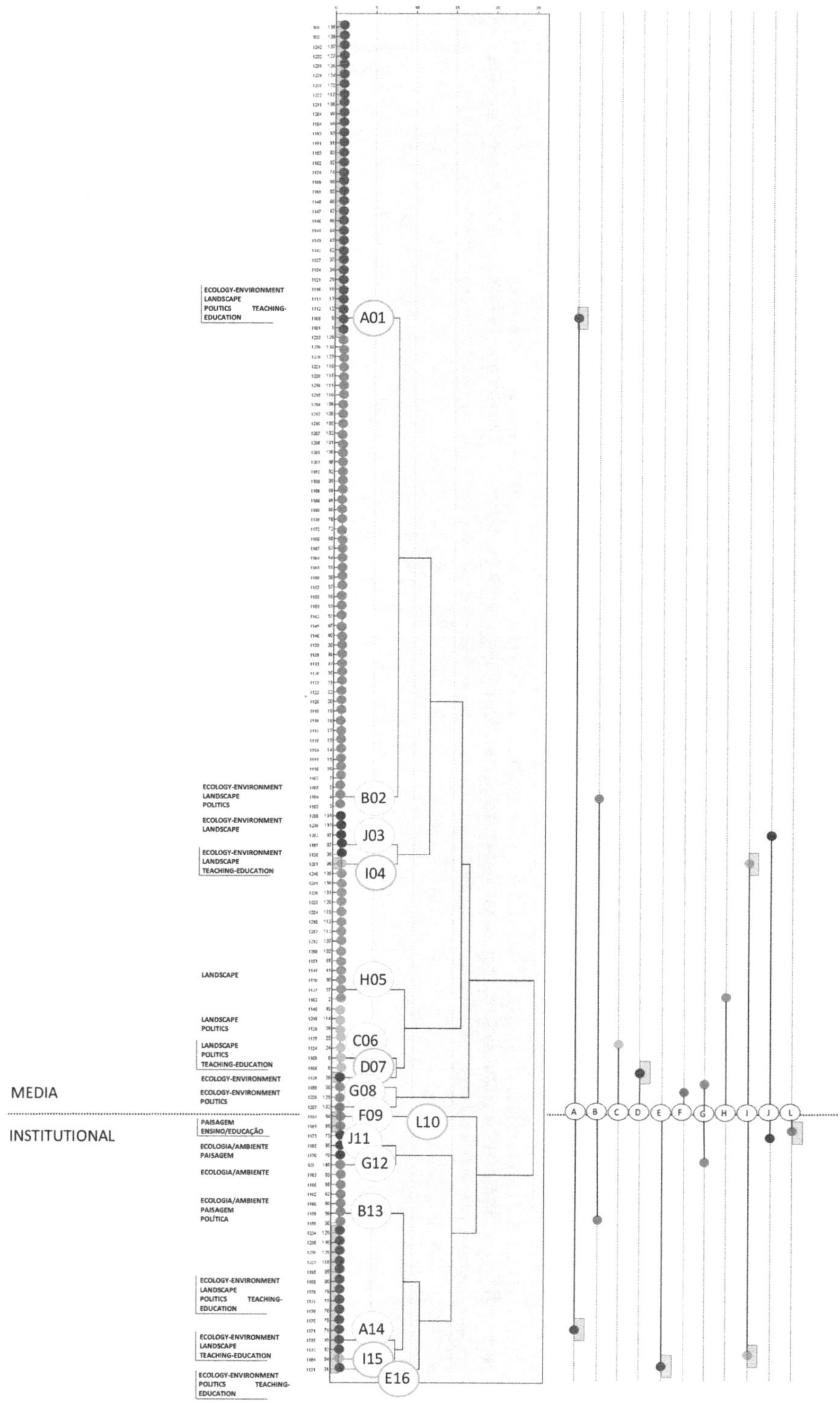

Figure 2.11 Clusters with 'Teaching/Education' component and institutional sources.

chapter; however, we will recall here the need to benchmark concepts, which is at the basis of fruitful communication and good understanding. Both Cabral and Ribeiro Telles were sensitive to this demand and contributed to the search for conceptual elucidation.

The word "ecology" derives from the Greek "oikos" (house) and "logos" (study). It was proposed by Ernst Haeckel in 1866, in "Generelle Morphologie der Organismen", to designate the scientific study of the interactions between organisms and their environment (apud Stauffer, 1957, p. 138). This study considers various levels of proximity, the broadest being the biosphere, which includes all living beings on planet Earth.

The training in Berlin provided Cabral with the opportunity to combine methodological and technical aspects of the art of gardens with theoretical and scientific knowledge, converging in the practice of design and planning from the study of the landscape.

In his words, "Landscape is the biosphere's figuration and it results from the complex action of man and all living beings—plants and animals—in equilibrium with the physical factors of the environment" (Cabral apud CEAP, 2010, p. 13). Ribeiro Telles adds, arguing that "Landscape concept is no longer exclusively contemplative, pictorial and literary, becoming also interpretative and, therefore, widened to science, thus enabling the planned intervention" (apud CEAP, 2010, p. 13).

According to Ribeiro Telles,

> landscape structure, the skeleton of form, organises space [. . .]. The overall landscape structure includes the ecological structure (water and air flow, organic matter and wildlife circulation, suitability definition and enhancement) and the built structure (paths, crossings, vegetation or built elements as a framework). [. . .] Landscape has to be considered as an organic and biological whole where every element is mutually dependent, both influencing and suffering from the presence of the other participants. Reciprocity is Nature's fundamental law.
>
> (apud Pessoa, 2021, p. 21)

He further explains that a Landscape Architecture work has "several components: the first being the sky that binds up everything. Afterward light and colour. Then the form and the figure. And then the movement. Therein lies life" (Reis, 2020).

Why did Ribeiro Telles become, so recurrently, remembered as a landscape architect at the end of his life? Landscape Architecture is not a particularly well-known profession in Portugal as, for example, medicine or engineering are. A clue to search for the answer to this question may lie in the frequency combination type B ('Landscape' + 'Ecology-Environment' + 'Politics') appears on the texts (40%). Political and civic participation, in many ways, was a constant throughout his life. If we associate the type A combination to this, which to the previous one adds the 'Teaching/Education' component, we include more than 70% of the total sample.

Actually, one of the missions he has most persistently engaged himself in was making the fundamentals of ecology understandable. He has done this in a systemic way, within the most diverse contexts, both in the most restricted or academic ones as well as in the public space. It is not surprising since landscape architecture lies very much in ecology.

The fact that Portuguese government declared a day of national mourning in his memory shows how important his life was for the country. The decree expresses recognition—which happens in similar tributes. Furthermore, it also expressly states gratitude for his legacy which so many directly enjoy in their daily lives:

> Gonçalo Ribeiro Telles was one of the pioneers of ecology in Portuguese politics. [. . .] The nation owes him an immense debt of gratitude, both for laying the foundations of environmental policy in Portugal and for developing an ecological awareness. His lasting legacy benefits all Portuguese people.
>
> (Conselho de Ministros, 2020)

Conclusions

Some years ago, when asked which tree he would choose to represent himself, Ribeiro Telles replied:

> A cork oak. I have known this tree since I was a child. When someone asks me where do I run away to, I always answer: to the *montado* [cork oak wood]. I hold this sense of existence I don't know why. It might be due to the presence of a large animal life there, which I think is shaped accordingly, and a continuous human presence. And then there are these amazing trees, the cork oaks. Are you aware of the mystery of the *montado*? The secret is that you enter, you don't know the paths and you only see trunks in front of you. Everything can appear and everything happens over and over again. Mystery always lies ahead of us in the *montado*. So is eternity.
>
> (Soromenho, 2020)

At the end of his life, Ribeiro Telles was recognized both as a landscape architect and a tireless advocate of ecological practices and a healthier environment. He is also remembered as the politician who laid the foundations of the legal framework for territorial planning oriented towards a cultural landscape, interrelating rural and urban in a system of interconnected continuities. He is moreover evoked as a Master, who contributed both to consolidate the academic training in Landscape Architecture and to implement a widespread ecological awareness.

He was truly all of this and in all he enthusiastically displayed his remarkable communication skills. Gonçalo Ribeiro Telles thus became renowned and came to be recognised as a leader in both the landscape architecture and environmental policy domains in Portugal.

Acknowledgments

The authors thank to FCT for the D.V. PhD scholarship (Ref. 2022.11757.BD).

References

CEAP. (2010). *Caracterização da arquitectura paisagista em Portugal.* Lisboa: Centro de Estudos de Arquitectura Paisagista "Prof. Caldeira Cabral", Instituto Superior de Agronomia—Universidade Técnica de Lisboa.

Conselho de Ministros. (2020). *Decreto n.º 8-A/2020 de 11 de novembro.* November 11. Available at: https://dre.pt/home/-/dre/148328004/details/maximized [Accessed 4th October 2023].

Pessoa, F.S. (2021). *Gonçalo Ribeiro Telles-o Homem e a Obra.* Lisboa: ARGUMENTUM-Edições.

Reis, B. (2020). *Gonçalo Ribeiro Telles, o homem que rouba ideias aos pássaros.* November 12. Available at: www.publico.pt/2020/11/12/culturaipsilon/entrevista/goncalo-ribeiro-telles-homem-rouba-ideias-passaros-1938878 [Accessed 4th October 2023].

Saldanha, R. (2010). *Em Nome da Terra—Gonçalo Ribeiro Telles.* Available at: www.rtp.pt/programa/tv/p26478 [Accesed 4th October 2023].

Soromenho, A. (2020). *Se podemos ser um exemplo, sem andar a chatear ninguém, ótimo.* Gonçalo Ribeiro Telles 1922–2020. November 11. Available at: https://expresso.pt/arquivos-expresso/2020-11-11-Se-podemos-ser-um-exemplo-sem-andar-a-chatear-ninguem-otimo-Goncalo-Ribeiro-Telles-1922–2020 [Accesed 4th October 2023].

Stauffer, R.C. (1957). Haeckel, Darwin, and ecology. *The Quarterly Review of Biology*, 32(2), 138–144. Available at: www.jstor.org/stable/2816117 [Accesed 4th October 2023].

3 Francisco Caldeira Cabral and the “Panorama” magazine

Introducing a landscape conscience in tourist propaganda

Gonçalo Duarte Gomes

Portugal, 1933

In 1933, following and ending a period known as Ditadura Militar (Military Dictatorship), a new constitution was approved in Portugal, by plebiscite, instating the Second Republic and the Estado Novo, an authoritarian political regime.

Drawing inspiration from other political movements, the Estado Novo will perfectly insert itself into the European zeitgeist. Within these inspirations, Adolph Hitler and the National Socialist German Workers' Party (NSGWP) regime, played a major part. The chronological coincidence between both movements—Hitler was elected as chancellor in 1933—and some parallels that could somehow be established between the situation of both countries, allied to the effectiveness of that regime's propaganda and imagery, led "part of the Portuguese elite to look upon Germany with admiration" (Ninhos, 2017, p. 14).

Among them, António Ferro.

Secretariado de Propaganda Nacional

Ferro would become the man responsible for the "magnum task of *metteur en scène*" (Victorino, 2018, p. 11), the Estado Novo, leading the Secretariado de Propaganda Nacional (SPN, the National Propaganda Secretariat), to "integrate the Portuguese into the moral thinking that must guide the Nation" (Paulo, 1994, p. 73).

Directly chosen and appointed by Salazar, Ferro designed and executed a widespread political marketing operation "with the same commitment that Goebbels" (Raimundo, 2015, p. 176), which he called "Policy of the Spirit" (Ferro, 2007, pp. 225–229).

The instruments of such policy were diverse, and in spite of an initial impulse, considered to be a "strong doctrinal bent of pro-fascist tendency" (Victorino, 2019, p. 147), the SPN would later evolve to a more moderate tone. Furthermore, it would become "a tourist promotion organism . . . not only because of the interest in sectoral promotion of this activity, but also because of the political project, perfectly symbolic, that it promoted" (Gomes, 2022, pp. 169–170).

Within such context, the "Panorama" magazine was created.

DOI: 10.4324/9781003454007-6

"Panorama" magazine

The "Panorama" magazine, first published in June of 1941, aimed to be "a place to evoke what is most alive and characteristic in the country, and therefore gives it its own physiognomy, a differentiated expression" (Ferro, 1941, p. 11).

This editorial project was a somewhat unusual instrument of propaganda. From the paper to the graphic, artistic and literary value of its contents, "Panorama" equaled the "best foreign magazines, without however losing, and rather accentuating, its very Portuguese character" (Diário da Manhã, 24–6–1941 in Victorino, 2018, p. 26).

Although an instrument of propaganda, it somehow softened the message, with formal aspects and contents that facilitated its assimilation, and integrated a broader perspective, an ambitious socio-economic model intended for the country. In fact, "Panorama" 's value as a communicational object "also consists of its action, both formative and didactic" (Victorino, 2018, p. 28), with breadth, credibility and depth to its message given through the participation of renowned authors and artists.

Cabral in the "Panorama" magazine

Cabral's collaboration with the "Panorama" magazine was brief.

He wrote only three articles, which were published between April and December of 1943. All three articles were included in a recurring section of the magazine dedicated to matters regarding the Portuguese landscape, named "Em defesa da paisagem continental" (in defence of the continental landscape, Figure 3.1).

The first article was titled "A erosão e a paisagem" (erosion and the landscape, issue 14, April 1943), and alerted for the main causes of erosion, such as intensive cereal crops, pointing out historical and contemporary examples, across the

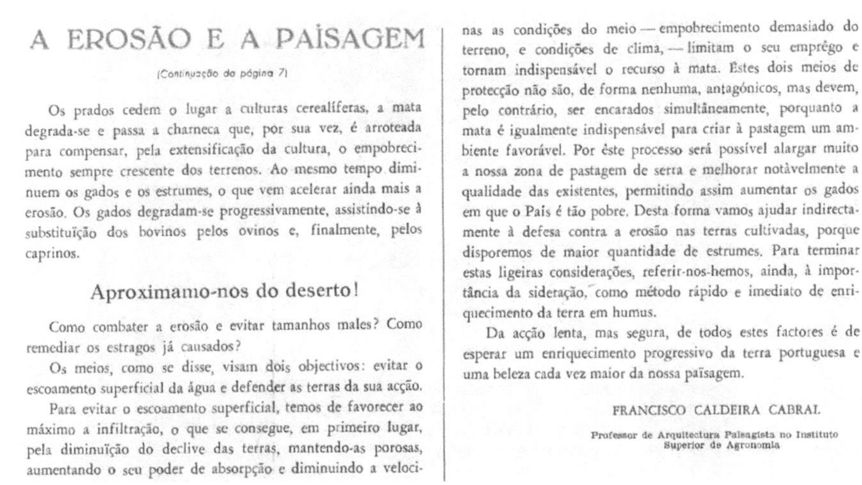

A EROSÃO E A PAÌSAGEM

(Continuação da página 7)

Os prados cedem o lugar a culturas cerealíferas, a mata degrada-se e passa a charneca que, por sua vez, é arroteada para compensar, pela extensificação da cultura, o empobrecimento sempre crescente dos terrenos. Ao mesmo tempo diminuem os gados e os estrumes, o que vem acelerar ainda mais a erosão. Os gados degradam-se progressivamente, assistindo-se à substituïção dos bovinos pelos ovinos e, finalmente, pelos caprinos.

Aproximamo-nos do deserto!

Como combater a erosão e evitar tamanhos males? Como remediar os estragos já causados?

Os meios, como se disse, visam dois objectivos: evitar o escoamento superficial da água e defender as terras da sua acção.

Para evitar o escoamento superficial, temos de favorecer ao máximo a infiltração, o que se consegue, em primeiro lugar, pela diminuïção do declive das terras, mantendo-as porosas, aumentando o seu poder de absorpção e diminuindo a velocinas as condições do meio — empobrecimento demasiado do terreno, e condições de clima, — limitam o seu emprêgo e tornam indispensável o recurso à mata. Êstes dois meios de protecção não são, de forma nenhuma, antagónicos, mas devem, pelo contrário, ser encarados simultâneamente, porquanto a mata é igualmente indispensável para criar à pastagem um ambiente favorável. Por êste processo será possível alargar muito a nossa zona de pastagem de serra e melhorar notàvelmente a qualidade das existentes, permitindo assim aumentar os gados em que o País é tão pobre. Desta forma vamos ajudar indirectamente à defesa contra a erosão nas terras cultivadas, porque disporemos de maior quantidade de estrumes. Para terminar estas ligeiras considerações, referir-nos-hemos, ainda, à importância da sideração, como método rápido e imediato de enriquecimento da terra em humus.

Da acção lenta, mas segura, de todos estes factores é de esperar um enriquecimento progressivo da terra portuguesa e uma beleza cada vez maior da nossa paìsagem.

FRANCISCO CALDEIRA CABRAL

Professor de Arquitectura Paisagista no Instituto Superior de Agronomia

Figure 3.1 Cabral's signature, as Landscape Architecture professor.

Source: BLX-Hemeroteca Municipal de Lisboa.

globe, and also including a national perspective of aspects that could be improved (Figure 3.2).

Designating erosion to be "the greatest factor of impoverishment of the land" (Cabral, 1943a), Cabral offers some possible solutions, namely in terms of soil covering (specially forestry) and land modelling, evoking traditional techniques.

Figure 3.2 Examples of different stages of landscape conservation.

Source: BLX-Hemeroteca Municipal de Lisboa.

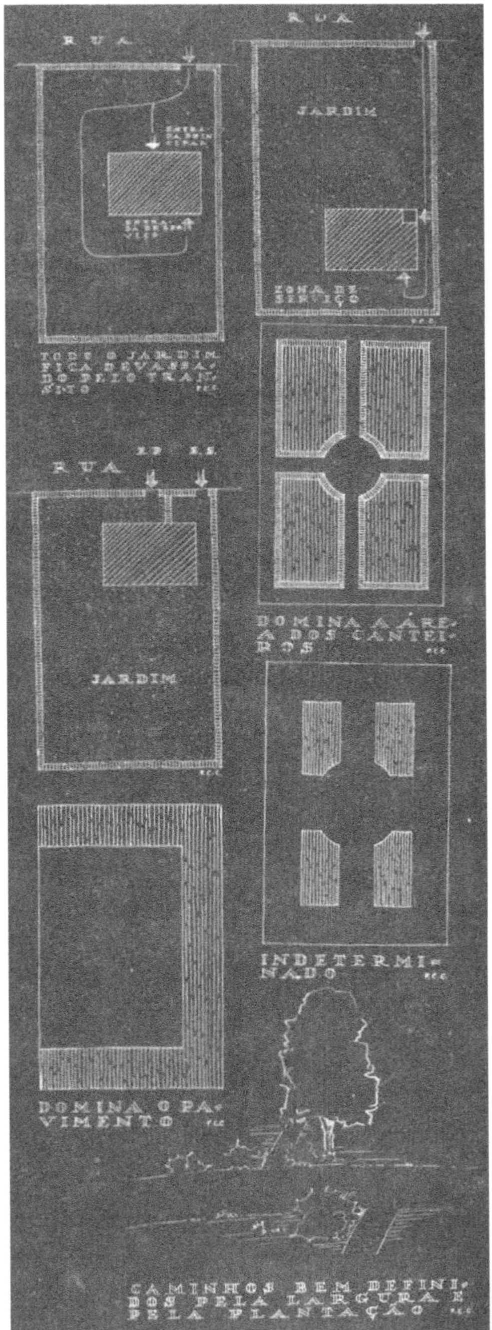

Figure 3.3 Illustration of garden layouts.

Source: BLX-Hemeroteca Municipal de Lisboa.

Cabral's signature in this article is the only one to display the title of Landscape Architecture Professor in the ISA.

The second article was titled "Jardins Portugueses" (Portuguese gardens, issue 15–16, July 1943), in which Cabral reflects upon the Portuguese garden and the existence of a differentiated Portuguese gardening school, where the inputs generated by the 15th century Portuguese sea expansion stand out:

> it is its special atmosphere that gives it character, an atmosphere of light and shade, of contrast between the freshness of the garden and the scorching sun of the landscape, an atmosphere of quietness and peace in the midst of exuberant and varied vegetation, where, alongside our wild plants . . . there are representatives from all continents. . . . Portugal, starting point of the Descobrimentos, . . . was thus able to gather in its gardens plants from all over the world.
>
> (Cabral, 1943b, p. 68)

The third, and final, article repeated the title of the second one, again "Jardins Portugueses" (Portuguese gardens, issue 18, December 1943). But this time, the focus of Cabral's reflections was not on the historic aspects of the garden, but on "general considerations and . . . essential aspects of garden design" (Cabral, 1943c), such as layout, functional zoning and compositional relationships between elements (Figure 3.3).

Cabral's collaboration with "Panorama" magazine served Ferro's idea of the "creation of a 'tourist conscience'" which "had in *Panorama* a precious support, in its component of reporting . . . a cultural and environmental heritage often unknown" (Victorino, 2018, p. 344).

On the other hand, it was an important opportunity to promote Landscape Architecture. These simple articles allowed illustrating, to a broader public, the potential contribution of the emerging profession, in several domains, and showcasing its groundbreaking core values.

Planting the seed of Portuguese Landscape Architecture

Cabral is rightfully considered the "founding father" and pioneer of Portuguese Landscape Architecture.

Already a professor at the Higher Institute of Agronomy (ISA), he was there authorized to test, during the academic year of 1940–1941, a Landscape Architecture course, which functioned as a specialization, available to students of the agricultural and forestry engineering courses from that school.

The students benefited from Cabral's "great experience and culture . . . that mobilized everyone, infecting us with his enthusiasm for the new profession" in classes "where scientific and artistic criteria of landscape architecture mixed in the relationships with other sciences and arts, but also with occasional circumstances of life" (Viana Barreto (one of the first ten students) in Jorge, 2009, pp. 10–11).

A German education

Cabral obtained his degree in landscape architecture in Germany, attending the Institüt für Gartengestaltung, of the Friedrich-Wilhelms-Universität, in Berlin, from 1936 to 1939, after being "invited by Lisbon Municipality to apply for the position of head of the parks and cemeteries department", being obvious that he "would have to specialise abroad" (Andresen, 2001, p. 25). Cabral conducted a research, concluding that "what they want is not a gardener, they want a landscape architect" (Cabral in Jorge, 2009, p. 73), thus convincing the municipality.

Choosing Germany, not only but mostly because of his acquaintance with the German culture, language and academic world, there he met Heinrich Friedrich Wiepking-Jürgensmann, landscape architect, teacher and head of the course at the Institüt, one of the most influential persons of his path, and with whom he would "maintain contact and seek advice" (Antunes, 2019, p. 105).

The rich and intense training, both in theoretical and practical terms, would determine Cabral's profile and action, upon his return to Portugal, to disseminate Landscape Architecture.

German landscape architecture under the III Reich

Under the III Reich, environmental stewardship was an important part of the development policies set in motion for a (re)imagined Germany.

In fact, part of the Reich's infamous eugenics theories was founded on the belief that Aryan supremacy had telluric bases. One such example is the Dauerwald, a forestry model that does not focus on primary productivity, but instead seeks overall improvement of the forest ecosystem. Two of the main tenets of Dauerwald are "continual selective cutting rather than periodic clear-cutting" and "natural regeneration rather than planting" (Imort in Brüggemeier et al., 2005, p. 47), setting a sort of Darwinist approach to forestry.

The Dauerwald allowed the establishment of parallels between trees and people, "emphasizing that . . . it comprised only *bodenständige* (native) species, . . . the Nazis were able to naturalize their ideal of classless, racially pure, and 'eternal' *Volksgemeinschaft* or national community" (Imort in Brüggemeier et al., 2005, p. 52).

Another telluric foundation of the Aryan delusion was Naturbedingtheit, a 19th century based "naturalistic vision of the national community as a cohesive organic entity, effectively a sort of Volk-organism . . . seen as an integral part of the ecology of the natural world" (Bassin in Brüggemeier et al., 2005, p. 206). In light of such an understanding, "a Volk could be genuinely integrated into the matrix of the natural world only to the extent that it was literally anchored in the earth or soil and attuned to their natural rhythms" (Bassin in Brüggemeier et al., 2005, p. 207). Hence the *Blut und Boden* (blood and soil) theory, for which *landschaft* (landscape) was a core concept, operating as the perfect synthesis between people (including culture) and the natural environment.

The passing of the *Reichsnaturschutzgesetz* (RNG, Reich Nature Protection Act), in 1935, "created the possibility for protecting entire landscapes and curbing the destructive effects of economic development on the countryside" (Closmann in Brüggemeier et al., 2005, p. 18), giving legal arguments to partisans of the völkisch movement and also to nature conservation leaders who "hoped to modernize their movement by fusing its aims with the goals of Nazi racial hygiene" (Lekan in Brüggemeier et al., 2005, p. 81). Furthermore, it became an important tool for the "group of environmental officials known as the *Landschaftsanwälte* or 'advocates for the landscape,' whose role was to oversee the ecological impact of public works projects sponsored by Hitler's regime" (Staudenmaier, 2020, p. 272), led by Alwin Seifert, the "most prominent environmentalist in the Third Reich, as expressed in the illustrious yet legally meaningless title, Reichlandschaftsanwalt (Reich landscape advocate)" (Zeller in Brüggemeier et al., 2005, p. 148).

Although there are obvious connections between the Nazi regime and German Landscape Architecture (and other professional areas),

> the emphasis on ideological links between Nazis and conservationists has led environmental historians to depict conservationists of the Nazi era . . . as people with a totally different mindset who, owing to their intellectual disorientation, ended up as accomplices to a genocidal regime. . . . quite the contrary, they were, in a way, "perfectly ordinary conservationists" who, to their own surprise, were handed unprecedented opportunities.
>
> (Uekötter, 2007, pp. 281–282)

Risk of a Portuguese "blood and soil"?

Considering that Cabral had a very influential experience through his Landscape Architecture degree in Germany, and the significance of the Landschaftsanwälte movement, how much could the Portuguese pioneer have been biased by doctrinarian dimensions present in his education?

Cabral's mentor, Wiepking-Jürgensmann, was a somewhat prominent figure in the Landschaftsanwälte movement, although "Seifert's archenemy" (Zeller in Brüggemeier et al., 2005, p. 158). But "the importance of these landscape advocates remains contested. Some historians have downplayed their responsibilities, noting that they were "limited to a consultative function"" (Staudenmaier, 2020, p. 272). In fact, Wiepking's participation in the Landschaftsanwälte left him unscathed, as "in the Federal Republic, the political burden on the landscape designers involved in National Socialist crimes hardly played a limiting role. . . . Wiepking and Seifert were able to successfully defend themselves against all incriminating charges" (Zutz, 2009, p. 133).

An analysis of the principles promoted, taught and practiced by Cabral reveals none of the traits that could place his thought as a form of national adaptation of the "blood and soil" vision.

Even when framing an explanation of the profession of Landscape Architecture within the "A bem da Nação" (for the good of the Nation, staple signature of any official document issued within the Estado Novo), Cabral underlines humanist principles, for instance in a lecture given in 1943, in ISA: "a profession of our time, branded by the concern of common good, defends the primacy of . . . permanent solutions over the narrow vision of the present, . . . to work for the good of the Nation" (Cabral, 1993, p. 35).

Furthermore, Cabral's thoughts and lessons on gardens and landscape (Cabral, 1993, pp. 73–198) divert entirely from the purist proselytism advocated by the völkisch and Landschaftsanwälte movements. In fact, Cabral had a troubled professional relation with the Estado Novo, from the very beginning of his career, with his commitment to Landscape Architecture principles prevailing over political agendas, with significant disadvantage to him.

One example is the project for the National Stadium, in Lisbon. Cabral's involvement began in 1937, following an invitation by José Belard da Fonseca (director of SETH, a large contractor involved in the construction process), before he even graduated in Landscape Architecture, beginning "the practice of landscape architecture in Portugal" (Andresen, 2003, p. 60).

Cabral and co-author Konrad Wiesner, wrote a critical study of an original masterplan, based "on 3 factors: soil, wind and relief and also on the project of the Stadium, from which he demarcates" (Novais, 2020, p. 79), upon which they developed a new and more adaptive masterplan, blending with the landscape attributes of the Jamor valley—specially the river and adjacent slopes—within which a completely new stadium project, although abdicating some monumentality, worked with the biophysical context, applying state-of-the-art concepts "in a natural and organic design that makes expressive use of the relief of the land" (da Cruz, 2005, p. 38).

However, political pressure mounted throughout the years, certainly due to the evolution of the course of World War II, Cabral's German education and lobbying by other architects, closer to Duarte Pacheco, the Minister of Public Works and Communications. In 1941, Cabral and Wiesner are set aside, and Jacobetty Rosa, an architect, signs their project, becoming the designated author. Cabral's memoirs reveal that he was even "told to forget that he had ever worked on the stadium" (Andresen, 2001, p. 93). In fact, "he did not work again for the Portuguese state until the 1974 revolution" (Andresen, 2001, p. 132).

The second example comes from a professional collaboration with the Lisbon municipality, working together with Ribeiro Telles. From 1955 through to 1960, Cabral and Telles developed a remodelling project for the Avenida da Liberdade, one of the main arteries of the city. Here, they introduced "innovative concepts for their time which were not understood" (Câmara, 2021, p. 119) by the public opinion, municipality officials and other professionals. Facing political backlash, they upheld their technical views, refusing to give in to external pressure. As a result, the project would be preterred, leading inclusively to Ribeiro Telles resignation from the Municipality. Cabral would later describe this experience as "one of the most painful periods" (Câmara, 2021, p. 121) of his professional life.

Politically, in spite of an obvious inspiration of the Estado Novo propaganda in some of the Nazi lore, the political-ideological alignment of the Portuguese regime with the Nazi regime "was by no means consensual" (Ninhos, 2017, p. 268), not even to Salazar, who thought that the German chancellor "would go too far, in the economic and social fields. It is of no interest to fight men, but instead ideas, the doctrinaire systems" (Salazar in Ferro, 2007, p. 140).

"Panorama magazine", while drawing general inspiration from foreign propaganda instruments, had its own identity and set of values.

Conclusion

The question of the Nazi environmental legacy "poses a dilemma for historians of Nazi Germany and environmental historians alike" (Staudenmaier, 2020, p. 283). Cabral's work, in light of the German influence of his academic upbringing, can be framed within this historiographical debate.

Although trained in Landscape Architecture in Hitler's Germany, and as a founder of a school of Landscape Architecture in Salazar's Portugal, Cabral managed to isolate the guiding principles of his vision of the profession from any ideological contamination, always following his "solid landscape architecture training and . . . strongly-held convictions" (Andresen, 2001, p. 76).

The exact terms under which his collaboration with "Panorama" magazine occurred remain undisclosed. Although relevant, they are not indispensable for a clear understanding that, in a time when Cabral struggled with difficulties for the affirmation of the Landscape Architecture course in Portugal, every space for its promotion and the demonstration of the groundbreaking nature of the theoretical body, was valuable. Furthermore, in the aftermath of the National Stadium project, all opportunities to gain sympathy from well-placed figures of the Estado Novo were relevant, in a time when engineers and architects, established around Duarte Pacheco, formed a closed circle, in the "meeting point of an ill termed brotherhood" (Costa, 2016, p. 232), perfectly aligned with the Estado Novo.

Elevating above partisan agendas, Cabral and his brief appearance in "Panorama" magazine constitute a lively proof of life for the embryonic Portuguese Landscape Architecture in the 1940s.

Reference list

Andresen, T. (2001). *Francisco Caldeira Cabral*. Surrey: LDT Monographs.

Andresen, T. (ed.). (2003). *From the National Stadium to the Gulbenkian Garden. Francisco Caldeira Cabral and the First Generation of Portuguese Landscape Architects (1940–1970)*. Lisboa: Fundação Calouste Gulbenkian.

Antunes, A.C.D.S. (2019). *A influência alemã na génese da Arquitetura Paisagista em Portugal*. PhD. Universidade do Porto.

Brüggemeier, F.-J., Cioc, M., & Zeller, T. (eds). (2005). *How Green Were the Nazis? Nature, Environment and Nation in the Third Reich*. Athens, Ohio: Ohio University Press.

Cabral, F.C. (1943a). Em defesa da paisagem continental—A erosão e a paisagem. *Revista Panorama*, Issue 14, 5–11.

Cabral, F.C. (1943b). Em defesa da paisagem continental—Jardins portugueses. *Revista Panorama*, Issue 15–16, 66–68.

Cabral, F.C. (1943c). Em defesa da paisagem continental—Jardins portugueses. *Revista Panorama*, Issue 18, 50–51.
Cabral, F.C. (1993). *Fundamentos da Arquitectura Paisagista*. Lisboa: Instituto da Conservação da Natureza.
Câmara, T.B. (2021). *Espaço público de Lisboa. Plano, projeto e obra da primeira geração de arquitetos paisagistas (1950–1970)*. Lisboa: Câmara Municipal de Lisboa.
Costa, S.V. (2016). *O País a Régua e Esquadro—Urbanismo, Arquitetura e Memória na Obra Pública de Duarte Pacheco*. Lisboa: ISt Press.
da Cruz, L.A.S.F. (2005). *O Estádio Nacional e os novos paradigmas do culto—Miguel Jacobetty Rosa e a sua época*. Master. Universidade Lusíada.
Ferro, A. (1941). Apresentação. *Panorama*, Issue 1, 11.
Ferro, A. (2007). *Entrevistas a Salazar*. Lisboa: Parceria A. M. Pereira.
Gomes, G.D. (2022). Turista doméstico, ou o Algarve, um estranho na sua própria paisagem. In *Turismo—Algarve, Segredos por Revelar*. Forte da Casa: Escolar Editora, pp. 167–188.
Jorge, F. (ed.). (2009). *Francisco Caldeira Cabral. Memórias do mestre no centenário do seu nascimento*. Lisboa: Associação Portuguesa dos Arquitectos Paisagistas.
Ninhos, C. (2017). *Portugal e os Nazis. Histórias e segredos de uma aliança*. Lisboa: A Esfera dos Livros.
Novais, M.M.L.M. (2020). *Das hesitações no Estado Novo à construção do lugar: o Estádio Nacional do Jamor*. Mestrado: ISCTE.
Paulo, H. (1994). *Estado Novo e Propaganda. Em Portugal e no Brasil. O SPN/SNI e o DIP*. Coimbra: Livraria Minerva.
Raimundo, O. (2015). *António Ferro: o inventor do Salazarismo—Mitos e falsificações do homem da propaganda da ditadura*. Alfragide: Dom Quixote.
Staudenmaier, P. (2020). Advocates for the landscape: Alwin Seifert and Nazi environmentalism. *German Studies Review*, 43(2), 271–290. Available at: www.researchgate.net/publication/341782511_Advocates_for_the_Landscape_Alwin_Seifert_and_Nazi_Environmentalism [Accessed 25th October 2022].
Uekötter, F. (2007). Green Nazis? Reassessing the environmental history of Nazi Germany. *German Studies Review*, 30(2), 267–287. Available at: https://zeitgeschichte-digital.de/doks/frontdoor/index/index/year/2014/docId/214 [Accessed 25th October 2022].
Victorino, J.G. (2018). *Propaganda e Turismo no Estado Novo. António Ferro e a revista Panorama (1941–1949)*. Lisboa: Alêtheia Editores.
Victorino, J.G. (2019). Do jornal Bandarra à revista Panorama (1935–1941): a estratégia dos projectos editoriais do Secretariado da Propaganda Nacional no plano mediático. In *Projectos Editoriais e Propaganda. Imagens e Contra-Imagens no Estado Novo*. Lisboa: Imprensa de Ciências Sociais, pp. 147–166.
Zutz, A. (2009). Wege grüner Moderne: Praxis und Erfahrung der Landschaftsanwälte des NS-Staates zwischen 1930 und 1960. In *Vom Dritten Reich zur Bundesrepublik: Beiträge einer Tagung zur Geschichte von Raumforschung und Raumplanung*. Hannover: Akademie für Raumforschung und Landesplanung, pp. 107–148. Available at: www.ssoar.info/ssoar/handle/document/35968 [Accessed 6th November 2022].

4 Landscape Architecture education in Portugal and Italy. The pioneering approaches of Francisco Caldeira Cabral and Pietro Porcinai

Ludovica Nardella

Introduction

Cabral and Porcinai were among the leading 20th-century landscape architects in Europe. In 1948, Porcinai was one of the seventeen founders of the International Federation of Landscape Architects (IFLA) in Cambridge (*IFLA Green Book by International Federation of Landscape Architects–Issue*, no date), while Cabral became a member of IFLA in 1952 and was its president from 1962 to 1966 (*Caracterização da Arquitectura Paisagista em Portugal*, 2010, p. 17). Therefore, they met frequently at IFLA conferences and surely had many opportunities to exchange ideas about academic training courses for landscape architects and related teaching programmes and methods. Cabral and Porcinai both believed in the need to combine professional experience with constant research and teaching activities, as well as in the need to organize interdisciplinary and comprehensive academic study plans that included both humanistic and scientific disciplines. They were both influenced by the reformist cultural debate on Landscape Architecture and nature protection ongoing in Germany at the time of their education and training, and this circumstance significantly contributed to their common ecological approach to the Landscape Architecture project and their future proposals for the landscape architect's academic education.

This research was carried out through an exploratory survey of relevant literature to ascertain the current knowledge regarding the proposed topics. Furthermore, results are described, and data, also derived from selected documents coming from historical records and archives, such as letters, memories, and graphic and photographic material, is interpreted. Specifically, the chapter aims to highlight how Cabral and Porcinai, both inextricably linked to their countries' unique natural and cultural contexts and very active within IFLA, have affected the development of innovative approaches to landscape architecture educational programmes in Italy and Portugal, albeit following different pathways and destinies. Moreover, it intends to deepen a little-known chapter of the history of southern European Landscape Architecture by retracing analogies and differences between the educational and professional paths of its main protagonists, Porcinai and Cabral, and proceeding with an interpretative analysis of the events and dynamics that have led to the birth of Landscape Architecture education in Italy and Portugal.

DOI: 10.4324/9781003454007-7

The unique case of Portugal in Southern Europe

Francisco Caldeira Cabral (1908–1992) is unanimously considered to be the founder of the Landscape Architecture school in Portugal. His outstanding contributions, spanning over four decades, have been critical to the birth and evolution of this new profession in his country and around the world. Professor Cabral's teaching principles and methods were strongly influenced by his training as a landscape architect in Berlin at the Agricultural College, which was preceded by Postdam's Royal School of Gardeners (1823) and Berlin's Botanical Garden (1910), both deeply rooted in traditional 18th-century German landscape culture (Antunes, 2019, p. 5). Cabral returned to Portugal intending to establish a high-education programme on Landscape Architecture at the Instituto Superior de Agronomia (ISA) in Lisbon, where he had previously achieved a degree. Hence, in 1940, Cabral launched his first experimental *Curso Livre de Arquitectura Paisagista*, lasting 4 years, which was free and optional and met considerable appreciation from students (Andresen, 2001, pp. 99–100). In 1942, the Minister of Education ratified it as an autonomous course, and the first graduates were granted diplomas in 1946 (Nunes, 2011, p. 109). It was the beginning of a process leading to the birth of the first illustrious generation of landscape architects in Portugal. Some of Cabral's former students found employment in local and central public and administrative bodies (Andresen, 2001, p. 108). Cristina Castel-Branco pointed out that, after graduating, five of his best pupils were entrusted with five missions: "design, applied ecology, landscape heritage, planning, and nature conservation" (Castel-Branco, 2003). However, Cabral's experimental course in Landscape Architecture at ISA received official acknowledgement after a very long process that lasted 40 years (Andresen, 2001, p. 100).

Cabral's course aimed to be comprehensive, and it integrated numerous scientific disciplines that were already taught at ISA. Nevertheless, the specific Landscape Architecture formation required further theoretical studies on art, science, and technical disciplines, with a particular concern for ecology, as well as practical training and fieldwork (Andresen, 2001, pp. 102–108). Cabral argued that the landscape architect should be trained to deeply understand how plants can adapt to environmental factors related to "macro and microclimate, soil", to change them if needed, and to be able to design space effectively (Cabral, 1978). He believed that although the architect and the landscape architect both design man's living environments, the latter mainly uses living materials and operates in contexts that are strongly dynamic, where time is an essential factor (Andresen, 2001, p. 106). Moreover, Cabral often recalled the artistic dimension of the discipline; therefore, in the first 2 years, students achieved drawing and painting abilities (pencil and ink hand-drawing, water colouring, descriptive geometry, and perspective) in addition to the necessary technical and botanical knowledge (garden plants, garden construction, civil construction, and architecture). The last 2 years were dedicated to the in-depth study of Landscape Architecture, the history of art, urban planning, and colonization. Each class included the submission of projects followed by a discussion, considered to be a very important part of the student's formation process

(Andresen, 2001, pp. 102–108). Cabral's educational programme also included foreign student exchanges and study trips abroad (Andresen, 2001, pp. 125–129).

In 1953, Cabral established the first Portuguese institution entirely dedicated to landscape, the *Centro de Estudos de Arquitetura Paisagista* (CEAP), which was formally recognized by IFLA as the first Portuguese professional organization for landscape architects (Andresen & Antunes, 2021). The CEAP had a catalytic function for the transmission of landscape culture among the members of the first generation of Portuguese landscape architects, who were constantly updated with the ideas coming from Cabral's engagement within IFLA. Besides, Cabral was often invited abroad to attend meetings and conferences and to give lectures to foreign academic institutions in Europe and the United States (Andresen et al. 2003, p. 57). Indeed, Cabral's contacts with the contemporary major landscape architects such as Geoffrey Jellicoe, René Pechère, Sylvia Crowe (Andresen, 2003, pp. 50–51), Lawrence Halprin, Lewis Mumford, and Pietro Porcinai (Andresen, 2001, p. 19) were critical in keeping him in touch with the main European landscape architecture trends and novelties.

In his first speech as president of IFLA, during the closing session of the Congress of Jerusalem (1962), Cabral stressed several urgent issues, specifically the need to provide proper education opportunities for landscape architects. He pointed out how the origins of the profession of landscape architect lie in the traditional gardener, who was a repository of basic knowledge about "climate, soil, and plants". However, he argued that the contemporary complex tasks related to "town planning" and "rural planning" should be addressed through an interdisciplinary approach, among various experts such as "biologists, agronomists, engineers, forestry science experts, and historians", and it should also aim to provide for "spiritual, economic, national, and historical" interests ("Personal Archive of Francisco Caldeira Cabral", no date).

According to Cabral, the "landscape is the figuration of the biosphere and results from the complex action of man and all living beings, plants, and animals in balance with the physical factors of the environment" (Cabral, 1973). The emphasis on the importance of anthropic and natural interaction within landscape dynamics, which allow the life of all living species, including humans, foreshadowed the current concept of landscape "sustainability". Besides, he believed that a *continuum naturale* should be preserved through proper territorial planning, in the scope of rural and urban environments, to address the numerous crucial issues affecting the landscape, such as green spaces in metropolitan areas, conservation of nature, and agricultural development (Francisco Caldeira Cabral Centenary. A New Concept of Nature Protection, no date).

Cristina Castel-Branco highlighted that the beginning of the modern movement in Portugal corresponds to the birth of the landscape architect profession. Cabral's first Landscape Architecture degree programme at ISA had a great innovative scope and included, from the beginning, the integrated study of humanistic and scientific disciplines (Castel-Branco, 2003, pp. 118–119). The Teaching Committee of IFLA stated that Cabral established the first university-level landscape architecture programme in Portugal and within Mediterranean countries (Andresen, 2001, p. 112).

Between 1965 and 1977, Cabral was a member of the National Council of Education, and especially in the 50s and 60s, he was greatly involved in writing texts concerning issues related to Landscape Architecture teaching and professional education as well as nature conservation and landscape protection (Pinto, 2014).

The difficult path of Landscape Architecture education in Italy

Pietro Porcinai (1910–1986) has been the Italian landscape architect who mostly contributed to the birth and development of Landscape Architecture at the national and international scales. His biography was included in "The Oxford Companion to Gardens" (Jellicoe et al., 1986) among the leading figures of Landscape Architecture in Europe. His practice as a landscape architect includes 1318 projects, mainly in Italy, but he also received numerous prestigious commissions from abroad, such as the consultancy for the UNESCO project for the relocation of the temples at Abu Simbel, in Egypt (1963–1971) (Medoro, 1991, p. 307). He participated in the historical ICOMOS-IFLA meeting when the International Committee for Historic Gardens drafted the Florence Charter (1981) as an extension of the Venice Charter (1964) (Zangheri, 2006, pp. 41–49).

Porcinai's father, the head gardener at the famous *Villa Gamberaia*, introduced him to famous English-speaking garden designers, such as Cecil Pinsent, who was very active in Tuscany (Latini, 2012, pp. 19–25). Furthermore, the reformist cultural debate in northern European countries, namely Germany, Belgium, and England, played a significant role in his formation. After graduating from the renowned Royal Agrarian School in Florence as an agricultural expert, he carried out his apprenticeship, first at the Draps nursery in Brussels (1928–1929) and then at the Fürstenstein castle park near Breslavia (Zangheri, 1998, pp. 131–138). In the following years, Porcinai was hired by the Royal Agrarian School in Florence as a professor of agricultural techniques, and later, he was employed at the esteemed nursery Bianchi in Pistoia before focusing on his work as a landscape architect.

Porcinai's frequent participation in IFLA's international meetings and conferences allowed him to have fruitful exchanges with many prestigious foreign colleagues, such as Sylvia Crowe, Geoffrey Jellicoe, René Pechère, and Carl Theodor Sørensen (Latini, 2012, p. 24), as well as Thomas Church, Gerda Gollwitzer, and Francisco Caldeira Cabral (Andresen, 2001, p. 19) ("Personal Archive of Francisco Caldeira Cabral", no date) ("Archivio Pietro Porcinai", no date). The first Italian professional association of landscape architects, the *Associazione Italiana degli Architetti del Giardino e del Paesaggio* (1950), was founded by Pietro Porcinai and a few other colleagues, pursuing two relevant IFLA objectives: the recognition of the landscape architect profession and the foundation of Landscape Architecture degree courses in Italy (Guccione, 2019). Porcinai claimed that, compared to other European countries, Italy had a considerable delay concerning Landscape Architecture education and consequently a lack of specialists capable of carrying out effective spatial planning in a historical moment marked by great social, economic, and environmental challenges. He proposed to introduce ten disciplines:

> biological and natural landscape planning; protection of natural heritage; public urban green; implementation of the green areas of the landscape; landscape ecology; land culture in developing areas; history of gardens, urban and rural landscapes, and agriculture; legislation; design of green areas; and landscape economy.
>
> (Porcinai, 1968, pp. 50–56)

Despite his numerous appeals to the representatives of the academic community in Florence, his great commitment to finding autonomous funds, his efforts in preparing a comparison of the subjects taught in other major European universities, and his great personal availability and experience, his initiatives did not yield the expected results. A postgraduate course in Landscape Architecture was successfully introduced later at the Faculty of Architecture of Genoa (1979–1980) by Annalisa Maniglio Calcagno.

Porcinai still aimed to establish an International School of Garden Art in his studio at *Villa Rondinelli* in Fiesole (1957) where the scientific studies of nature and environment, as well as the history of art, were considered equally necessary to support experimental projects. The school was going to host international high-level conferences with the participation of distinguished lecturers and famous landscape architects, but he had to give up his plans because of bureaucratic constraints. However, his studio was still a workshop where young Italian and visiting foreign students learned by doing, collaborating with the *Maestro* and other experts and major landscape architects coming from abroad. In his studio laboratory, the first landscape architecture projects in Italy were realized, aiming to improve people's lives by designing public spaces and architecture, pursuing environmental protection, and achieving harmonious integration in the landscape (Matteini, 2011, pp. 39–45).

Porcinai's "Nordic" attitude and sensitivity towards nature and landscape ecology are probably the lesser-known aspects of his work. Like Burle Marx, he had a deep knowledge of the dynamics and balances among the ecosystems that characterized the garden. Besides, projects built in very different environments always sought an equilibrium between design, the client's demands, and biological rules (Grossoni, 1998, pp. 157–167). Taking up the theories of van Leeuwen on ecosystems, Porcinai warned about the possible future deterioration of the landscape and the consequent risk of depletion of vital resources for human life (Giannini, 1991, pp. 245–251).

Thanks to his deep interest in plant ecology and applied ecology, Porcinai was able to approach several crucial themes, such as

> the conservation of nature and biodiversity protection; landscape ecology and landscape quality conservation; the sanitation role of the vegetation to control air

and water pollution and to improve landscape quality; the ecology of environmental recovery and the use and abuse of soils; the urbanization of the sustainable city, energy savings, uses and abuses of green; the inclusion of landscape architecture and ecology within the disciplines taught in university degrees.

Besides, Porcinai studied the issue of land stabilization through natural engineering techniques, often referring to the work of Hugo Meinhard Schiecht, the Austrian founder of bioengineering. He was very interested in the applications of phytoremediation to polluted soil, water, and air. In this regard, he was able to considerably improve the water quality of pools, ponds, and streams in his gardens by making use of rhizophyltration (Grossoni & Giuntoli, 2017, pp. 10–15).

Porcinai published numerous articles in the international design magazine "Domus" and in other prestigious specialized journals to disseminate his ideas on Landscape Architecture in Italy and abroad, drawing attention to his outstanding professional practice. From the years following his death until today, there have been numerous publications, debates, and conferences regarding Porcinai's work, mainly in Italy. Although Porcinai didn't establish official Landscape Architecture academic education in Italy, his ideas and *modus operandi* undoubtedly kept inspiring the following generations of Landscape Architecture students and professionals in Italy. Therefore, thanks to his avant-garde mindset, most of the core concepts underlying his work went much beyond the spatial and temporal limits within which they were conceived and applied. For instance, his ecological and environmental approaches to the landscape project, along with the consideration of the social aspects and values achievable through the public and private design of green spaces, are undoubtedly still highly topical.

Landscape Architecture in Portugal and Italy. The international context

Ever since its foundation in 1948, IFLA had the merit of establishing connections among the professional associations of its member countries (Caracterização da Arquitectura Paisagista em Portugal, 2010), contributing to the creation of a network of international relationships among landscape architects and thus facilitating their mutual exchange of ideas over the most urgent tasks, including education. Cabral and Porcinai were the first members to represent southern European countries within IFLA (Corbari et al., 2018). Both were internationally renowned landscape architects who were often invited to lecture at prestigious foreign academic institutions and international congresses. In 1959, the Grand Council of IFLA reported on its intention to promote Landscape Architecture education. In 1960, Hubert Owens was appointed to chair the Education Committee, while Cabral coordinated the work to collect data regarding Landscape Architecture education in Italy, Spain, Portugal, and Germany, according to the standards of the American Society of Landscape Architects. The IFLA delegates, among whom Porcinai was Italy's representative, were asked to provide information about the opportunities available in their countries ('Personal Archive of Francisco Caldeira Cabral', no date; 'Archivio Pietro Porcinai', no date).

As highlighted, these initiatives, aiming at the creation of Landscape Architecture academic courses in Portugal and Italy, produced different outcomes over time. Nevertheless, it should be noted that many of the principles outlined by recent international documents, such as the Charter for Landscape Architecture Education, already guided the work of Cabral and Porcinai, thanks to their pioneering attitude and their deep interest in novelties coming from abroad. Specifically, the Charter's emphasis on sustainable social, cultural, and ecological approaches to landscape planning and design practice, the integration of "biological, land, water, and atmospheric systems", the enhancement of biodiversity, and the need for "interdisciplinary" work, both within the scope of educational and professional settings (IFLA/UNESCO Charter for Landscape Architectural Education, 2012, 2017), have been an integral part of Cabral's and Porcinai's approaches to the discipline and profession since a very early stage.

Reference list

Andresen, T. (2001). *Francisco Caldeira Cabral*. Reigate: LDT Monographs, pp. 19, 99, 100, 102–108, 112, 125–129.

Andresen, T. *et al.* (2003) *From the National Stadium to the Gulbenkian Garden*. Lisbon: Calouste Gulbenkian Foundation Fine Arts Department.

Andresen, T., & Antunes, A.C.D.S. (2021). Transmissions through landscape architecture education from Germany to Portugal. In *Garden Transmissions*. Knowledge transfer through gardens. (Abstract). Lisbon: Calouste Gulbenkian Foundation. p. 3, pp. 50–51, pp. 57.

Antunes, A.C. (2019). *A influência alemãna génese da Arquitetura Paisagista em Portugal*. PhD thesis, Faculty of Sciences of the University of Porto, p. 5. Available at: https://repositorio-aberto.up.pt/handle/10216/121946.

Archivio Pietro Porcinai (no date). Fiesole.

Cabral, F.C. (1973). Paisagem. In *Enciclopédia Luso-Brasileira de Cultura*. Lisbon: Editorial Verbo.

Cabral, F.C. (1978). The landscape architect—what is it? The landscape architect's mission the training of landscape architect: The choice of Berlin. Trip to USA traditional Portuguese garden elements. In *Defense of the Continental Landscape. Barcelona*. Available at: http://proffranciscocaldeiracabral.portaldojardim.com/ensino/o-arquitecto-paisagista-—que e/ [Accessed 25th September 2022].

Caracterização da Arquitectura Paisagista em Portugal. (2010). Lisbon. Available at: https://apap.pt/wp-content/uploads/2017/08/Caracterizacao-da-Arquitectura-Paisagista-em Portugal.pdf., p. 17 [Accessed 27th September 2022].

Castel-Branco, C. (2003). The origin of the landscape architecture profession in Portugal during the modern movement. *Academia*, 2(14–21), 118–119. Available at: www.academia.edu/32805662/The_Origin_of_the_Landscape_Architecture_Profession_in_ Portugal_during_the_Modern_Movement [Accessed 27th September 2022].

Corbari, V. *et al.* (2018). *La formazione in architettura del paesaggio in Europa: Riflessioni per un terreno comune, tra sintonie e divergenze Allegato A Casi studio*. Borse di studio 2018 in Teorie e politiche del paesaggio, Fondazione Benetton Studi Ricerche. Available at: https://www.fbsr.it/wp-content/uploads/2020/05/ViolaCorbari_CASI-STUDIO.pdf [Accessed 22nd October 2022].

Francisco Caldeira Cabral Centenary. (no date). *A New Concept of Nature Protection*. Available at: http://proffranciscocaldeiracabral.portaldojardim.com/ambiente/uma-nova-concepcao-de proteccao-da-natureza/ [Accessed 16th September 2022].
Giannini, A. (1991). Gli scritti di Porcinai. In Matteini, M. (ed.), *Pietro Porcinai. Architetto del giardino e del paesaggio*. Porcinai. Milano: Electa, pp. 245–251.
Grossoni, P. (1998). Flora e vegetazione nei giardini di Pietro Porcinai. In Pozzana, M. (ed.), *I giardini del XX secolo: l'opera di Pietro Porcinai*. Firenze: Alinea, pp. 157–167.
Grossoni, P., & Giuntoli, A. (2017). Una rilettura della sensibilità ecologica di Pietro Porcinai (Firenze, 1910–1986) nel progetto di giardino. *Bullettino-Periodico della Società Toscana di Orticoltura*, (1), 10–15. Available at: www.societatoscanaorticultura.it/bullettino/ [Accessed 16th October 2022].
Guccione, B. (2019). L'AIAPP e la storia dell'Architettura del Paesaggio in Italia. Architettura del Paesaggio. Available at: www.academia.edu/45793290/L_AIAPP_e_la_storia_dellArchitettura_del_Paesaggio_in_It a lia [Accessed 22nd October 2022].
IFLA Green Book by International Federation of Landscape Architects—Issue (no date). Available at: https://issuu.com/ifla_publications/docs/ifla_green_book [Accessed 20th September 2022].
IFLA/UNESCO Charter for Landscape Architectural Education. (2012). Available at: https://media.voog.com/0000/0044/0363/files/2_IFLA-Charter-for-Landscape-Architectural-Education-Revised-2012.pdf.
IFLA/UNESCO Charter for Landscape Architectural Education. (2017). Addenda to IFLA UNESCO Charter for Landscape Architectural Education. Available at: https://www.eclas.org/wp-content/uploads/2020/09/E1_IFLA-Charter-for-Landscape-Architectural-Education_2017_with_IFLA-Europe-addenda_2017.pdf [Accessed 15th October 2022].
Jellicoe, G.A. *et al*. (1986). *The Oxford Companion to Gardens*. Edited by G. Jellicoe *et al*. Oxford: Oxford University Press, pp. 450–451.
Latini. (2012). Uno sguardo moderno sui paesaggi italiani. Note introduttive a Porcinai. In Cunico, M., & Latini, L. (eds.), Pietro Porcinai. *Il progetto del paesaggio nel XX secolo*. Venezia: Marsilio, pp. 19–25.
Matteini, M. (2011). La "scuola" a Villa Rondinelli. In Romitti, I. (ed.), Piero Porcinai. *L'identità dei giardini fiesolani. Il paesaggio come immenso giardino*. Firenze: Polistampa, pp. 39–45.
Medoro, G. (1991). Regesto delle opere. In Matteini, M. (ed.), *Pietro Porcinai. Architetto del giardino e del paesaggio*. Milano: Electa, p. 307.
Nunes, O.P. (2011). O Arquitecto Paisagista em Portugal: a construção do grupo profissional e o seu regime justificativo de acção perante a legislação que o "regula". Dissertação de Mestrado Em Sociologia do Território, da Cidade e do Ambiente, Faculdade de Ciências Sociais e Humanas da Universidade Nova de Lisboa, p. 109. Available at: https://www.proquest.com/openview/c309b790818608f9d31201f4f84a8071/1?pq-origsite=gscholar&cbl=2026366&diss=y [Accessed 22nd October 2022].
Personal Archive of Francisco Caldeira Cabral. (no date). Paços da Serra.
Pinto, M.A. (2014). *O legado escrito de Francisco Caldeira Cabral Construção do pensamento teórico em arquitetura paisagista*. Mestrado de Arquitetura Orientador: Maria Teresa Andresen. Faculdade de Ciências da Universidade do Porto.
Porcinai, P. (1968). Per l'insegnamento del "verde", del paesaggio e del giardino in Italia. *Pietro Porcinai. Architetto del giardino e del paesaggio*. Monografia. Edited by Associazione Italiana degli Architetti del Paesaggio. Notiziario AIAPP, (N° 10), 50–56.
Zangheri, L. (1998). Pietro Porcinai e la Gamberaia. In *I giardini del XX secolo: l'opera di Pietro Porcinai*. Firenze: Alinea, pp. 131–138.
Zangheri, L. (2006). La formazione di Pietro Porcinai. In Grifoni, T. (ed.), *L'eclettismo nell'opera di Pietro Porcinai*. Firenze: Polistampa, pp. 41–49.

5 Landscape Architecture in Portugal

The quest for a needed memory

Maria João Fonseca

Introduction

Portugal has one of Europe's longest legacies of Landscape Architecture education and practice, dating back to 1941, following in the footsteps of the United Kingdom, Belgium, and Germany. However, the rich history of Portuguese Landscape Architecture remains largely unexplored and undocumented, presenting a risk of disconnection between theory and practice. As cautioned by De Certeau (1992 [1975], p. 57), this fact will lead the discipline "drift into the dogmatism of 'eternal values' or into an apology for a 'timelessness'". While legal, epistemological, and cultural factors have contributed to the current circumstances (Fonseca, 2022), it is also evident that the discipline, in general, lacks "serious or sustained interest in the history and theory" (Hunt, 2000). Constructing this history relies heavily on research, and its sustained progress depends on ensuring accessibility to primary sources.

The politics of archival memory play a pivotal role in determining who has the right to speak and whose voices are silenced, "both in public life and in archival records" (Cook, 2011, p. 621). Considering the crucial role of archival materials in shaping the future memory of a discipline, this chapter aims to provide a contextualisation of the current situation, present effective approaches to address this issue, and underscore the importance of integrating these practices and primary sources into Landscape Architecture education programmes.

To achieve this, the chapter begins by listing the existing records available in Portuguese institutions. Subsequently, it discusses the influence of these records on the construction of history. From this vantage point, the chapter explores current archival approaches and recommendations for improving archiving practices as well as an analysis of the methods that have already been employed in the academic realm, outlining their strengths and weaknesses. The chapter concludes with a synthesis of the key findings.

By addressing the scarcity of documented history in Portuguese Landscape Architecture and emphasising the significance of primary sources, this research contributes to filling the existing gap in knowledge and understanding. A comprehensive examination of archival practices aims to foster a more informed and integrated approach to Landscape Architecture education and practice.

DOI: 10.4324/9781003454007-8

Landscape Architecture documents in the Portuguese institutions of memory

To accurately survey the existence and status of documents[1] related to the discipline in Portuguese archives, I first used the survey conducted by Zélia Pereira (2018) for her PhD on Portuguese personal archives. According to the "Individuals and Associated Information" list, all archives related to figures in Landscape Architecture[2] are under the custody of the Forte de Sacavém/DGPC archive. Each of these archives contains a brief description of its contents:

1. Araújo, Ilídio de. 1925–2015 (1959–1972): Landscape Architecture drawings and photographs.
2. Barreto, Viana 1924–2012 (1950–2001): Papers from "Projetos de Espaços Verdes, Lda", drawings, and photographs.
3. Cabral, Caldeira. 1908–1992 (1939–1974): Drawings and papers related to his landscape architecture activity.
4. Câmara, Sousa da. 1929–1992: [no information].
5. Castelo Branco, Albano. 1927: [no information].
6. Telles, Ribeiro. 1922– (1950–2004): Drawings related to landscape architecture.

Correlating this data with the information given by the current person responsible for the institution, Ana Paula Figueiredo (2021, p. 131; Archivoz, 2021) adds the Portuguese Association of Landscape Architects' archive handed over in 2006 with documentation from 1967[3] to 1980. Additionally, the Edgar Sampaio Fontes' (1922–2000) archive was incorporated in 2021[4] and still lacks information on its content. It is worth noting that the list presented by Ana Paula Figueiredo (2021, p. 131) does not reference the archive of Sousa da Câmara, contradicting the information provided in Zélia Pereira's (2018) thesis. The incorporation of these funds took place between 2004 and 2021, concentrated on the years up to 2007 (Figueiredo, 2021, p. 131). By cross-referencing this data with information available in the SIPA Inventory (Direção-Geral do Património Cultural, no date), only four of these archives—Viana Barreto, Cabral, Ribeiro Telles, and Ilídio de Araújo—are mentioned.

The Júlio Moreira's archive is also held at the Art Library of the Calouste Gulbenkian Foundation (Biblioteca de Arte e dos Arquivos Gulbenkian, no date). With 14.705 records, this archive consists of the Landscape Architecture professional archive (1966–2017), part of his literary production and publications, and a small specialised library to support the projects.

To summarise the Landscape Architectural documents in Portuguese public institutions, there are eight (personal or professional) archives of landscape architects and the institutional archive of the Portuguese Association of Landscape Architects. Of these, just one had an archival treatment and was made available recently, in 2022.

Archives and the rise of the history

Now that we have an understanding of the current documentation panorama, it is important to consider its role in the process of history-making, particularly when compared to architectural history, which shares several similar particularities. As Mirko Zardini (2021, p. 11) stated, there was a surge of interest in architecture and architectural archives, museums, and institutions in Europe and North America during the 1970s and 1980s, leading to a renewed interest in history. In the Portuguese context, institutions holding archives related to architecture and urbanism emerged in the 1980s (Pereira, 2018, p. 267), with the Calouste Gulbenkian Foundation's incorporation of architects' archives (2018, p. 266). In 1992, the Architectural Heritage Information System was created to accommodate the funds from the public bodies that preceded the Directorate-General for National Buildings and Monuments, currently under the Directorate-General of Cultural Heritage (DGPC) custody and installed in the Forte de Sacavém. In 2001, it hosted its first non-institutional archive from the architect Frederico George. Since then, it has incorporated several archives related to architecture, urbanism, landscape architecture, and design. During this century, the growing interest in archival institutions resulted in the creation of new ones and a thriving interest in the existing ones.

Additionally, an active incorporation policy has been in place.[5] Alfred Willis (1996, p. 195) emphasises the importance of archives in shaping historical narratives and their relationship with users. Willis supports this argument by referring to a study by Eugene E. Matysek, Jr. in 1991, which revealed that architectural historians are the main users of these archives, followed by architects. However, the accessibility of archives is a critical factor in shaping history. The availability of relevant archives is vital for research in architectural history, and the lack of accessibility presents a significant obstacle to the (1996, p. 196).

How to approach the Landscape Architecture records?

An immersion in the archives with Landscape Architecture records calls for a brief discussion on expectations, challenges, and issues and on the existing recommendations for better archiving practices. First, recall the importance of these primary sources, as they may be the only evidence in the case of Landscape Architecture praxis, including gardens or parks design, landscape planning, rules, conceptual thoughts, and related information. These primary sources undergo an essential transition from document to record in a process that delves into two domains of action and inner discourses—archival theory and the built environment—and it is essential to consider them as cross-fertilisation sources. I begin with this challenge as the specific content of these records may demand the knowledge of the language used in Landscape Architecture and the archival treatment based on archival theory. I purposefully bring this issue in first because of the impact of records typology on both the institutional politics of acquisition—frequently very oriented to only choose or privilege the drawings—and because of the specific nature of drawings and processes, which may require specific knowledge beyond archivists. Concerning archival acquisition policies, the pernicious results may be the overvaluation of isolated documents by their format or information nature instead of their contexts of production and accumulation. Another potential issue is the prevalence of only certain aspects of an individual or an area of activity in shaping collective memory while neglecting others of no less importance for understanding society.

Regarding this risk of overestimating certain aspects of an individual's personality and life to others' detriment, placing individual memory in the framework of specific collective discourse has been enhanced in the archive's incorporation in dedicated institutions. As Fernanda Ribeiro (1998) and Malheiro da Silva (1999) expose, it has been a common practice since the 1920s, reminiscent of the methodical classifications of the eighteenth-century intellectual and thematic-based classification. When we look at these issues and their influence on the politics of incorporation, we realise that there may be a selection of materials according to specific events, themes, or research areas, eventually influencing the historiographical discourse built for future collective memory. Discussing the transition from traditional archivists to appraisal theorists, Cook (1996, p. 139) also reflects on the limitations of this approach, which frequently mirrors limited research interests instead of encompassing a diverse range of human experiences. He notes that archivists, by focusing primarily end-product—the actual record—and its potential research value, followed Schellenberg's model of relying on researchers' articulated interests. As a result, they became overly aligned with the academic marketplace. Cook (1996, p. 140) suggests a macro-appraisal strategy to provide adequate archival treatment. Acknowledging that it may not be straightforward for archivists, this approach ensures greater integrity due to the nature of record creation in society and leads to the production of a superior archival record. While I will not delve into the details of the strategy here, it is founded on a comprehensive view of the information universe rather than focusing on individual elements or collections and emphasises appraising the creator's functions instead of solely focusing on records. This strategy also contributes to the long-standing debate on personal archives. However, as Catherine Hobbs (2001) eloquently states, the transactional context, although relevant, does not allow for a comprehensive understanding of personal archives as a whole. People are not entirely planned or rule-driven in their private lives, and personal records can provide valuable insights into how individuals position themselves in society. Examples of such records include personal notes, diaries, lists, and other documentary typologies, which are often dismissed as un-archival (2001, p. 130).

When we transpose Hobbs' example from personal literary archives to personal archives with Landscape Architecture records, we notice that archival institutions also mostly incorporate collections or professional archives, dismissing as un-archival some relevant records. And specialised institutions may encourage this fragmentation of personal archives according to their appraisal and selection policies or just by technical, material, and human constraints. As Zélia Pereira (2018, p. 175) says, "Personal archives can include both textual, iconographic or sound information, as well as various three-dimensional objects, which cannot but be considered as intrinsically linked to the archive". When dealing with personal archives, as opposed to institutional or professional ones, archivists must adopt an approach that reflects the individual's personality rather than solely seeking evidence of historical actions or facts. This approach addresses Cook's challenge to capture the broad spectrum of human experience. In archives containing Landscape Architecture records, it is crucial to recognise that documents such as personal notes, letters, newspaper clippings, and library materials, among other types, hold potential for revealing critical personal and social contexts. Embracing a personal archive approach will not lessen the importance of Landscape Architecture but rather enhance it, contributing to the understanding of the discipline.

Archivists must comprehend the larger historical and social landscape surrounding the records. They should be able to select records that capture not only the public activities but also the personal character of their creators, with an eye toward their representativeness for the future. This approach extends to the appraisal criteria. In practice, archivists should have a documented vita on file, supplemented by biographical and autobiographical details that inform the values used in the appraisal process (Cook, 2006, p. 178). This provides researchers with information about the scope, content, and creator context and the reasons behind the archival choices made. In a broader sense, the archival universe can be defined by employing mapping techniques derived from documentation strategy (Hobbs, 2001), which considers the life cycle of documents and the relationships between creators and other agents involved.

Landscape Architectural records in the academic realm

According to André Tavares (2021, p. 224), "archives need scholars as much as scholars need archives". This mutual dependence arises from the dynamic nature of archives, which provide an interactive learning experience. However, fully harnessing the potential of archives requires what Horsman and Ketelaar (2015, p. 53) term "archival consciousness". This concept encompasses a comprehensive understanding of the purpose behind records and record-keeping systems, their historical uses, and the complex ways in which power dynamics, politics, gender, and social and cultural influences shape them.

Several architectural and Landscape Architectural universities recognise the tremendous value of primary sources in shaping historical understanding and have incorporated archival practices into their curriculum. Alvar Aalto University, for example, provides its students with access to the architectural archive, offering short- and long-term courses on archival collections. This integration of archives into the academic programmes enhances students' abilities to utilise primary sources for creative and theoretical purposes while fostering a deeper appreciation for the archival collections (Kokkinen, 2021, p. 173). Similar approaches have been adopted in Landscape Architecture courses, such as those offered at the Hungarian University of Agriculture and Life Sciences, Budapest (Institute of LA and Urban Planning and Garden Art) or the École Nationale Supérieure de Paysage in Versailles (Lička et al., 2022, p. 55). Although detailed information on the uses of these archives is lacking, they generally appear to primarily serve as repositories for research information, with only a few incorporating the crucial context of production. In Landscape Architecture, the Network of European Landscape Architecture Archives, established in 2019, holds significant potential for increasing awareness about the value of Landscape Architecture records in disciplinary history and education. An example of an educational initiative in the recent workshop "Landscape Architecture Archives and Their Collections" (*Scales of Change. ECLAS 2022*), to explore acquisition strategies and the responsibilities of archives as guardians of professional memory—a critical aspect of ongoing discussions. Knowledge of archives and records management requires critical thinking and specialised expertise.

Conclusions

Integrating archival skills into Landscape Architecture education holds immense potential for shaping the discipline's future. As Lička et al. (2022, p. 61) suggest, early exposure to archives and the development of archival proficiency among students can significantly impact their future careers, fostering a greater utilisation of archives as a working tool. Furthermore, as highlighted by Ann Stoler (2002, p. 90), archives should be regarded not merely as passive spaces for knowledge retrieval but as active sites of knowledge production. Awareness raising in archives' political dimension and harnessing their critical assumptions and uses is only possible through education.

By incorporating the subject of archives and associated knowledge into Landscape Architecture courses, we can address the gaps in the discipline's history and ensure the construction of future memory that relies on critical selection from the present. Landscape architects, as the formative link between client needs and construction technology (Shoskes, 1989 cited in Lowell, 2015, p. 20), are part of a complex system involving multiple actors, producers, and stakeholders (Cook, 1996, p. 138). To analyse such intricate systems, Armstrong (2006) and other scholars advocate for mutual collaboration and an interdisciplinary approach to working with records.

This collaboration between landscape architects, with their proficiency in interpreting internal processes and spatial literacy, and archivists, with their expertise in documentation strategy, is essential to avoid isolating records and provide access to associated information and relationships. By contextualising the production of records, we establish the foundation for constructing multiple narratives and ensuring the efficient preservation of future memory. Education plays a crucial role in considering all these dimensions.

In conclusion, integrating archival skills and knowledge into Landscape Architecture education is crucial for the development of the discipline but also for nurturing a comprehensive understanding of history. By fostering students' proficiency in archives, we empower future professionals to utilise archives as working tools and actively contribute to producing knowledge. Landscape architects and archivists can collaborate through education, leveraging their respective proficiencies to analyse complex systems, preserve context, and construct multiple narratives. By embracing these dimensions, we pave the way for a more informed and connected approach to Landscape Architecture rooted in historical awareness and critical engagement with archives.

Notes

1 I use the term because most still await an archival description.
2 Despite the general term of architecture/Landscape Architecture archives, when considering records produced by an individual creator, I rely much on the term and inner conception of personal archives containing documents of professionals who, in their lives, bequeathed documents related to their practice and the disciplines.
3 Although it mentions the year 1967, it was created in 1976, and it can be a typo.
4 The incorporation took place after the interview.
5 For instance, the Marques da Silva Foundation, Casa da Arquitectura, or archives that are part of the Municipal Archives and Torre do Tombo.

Reference list

Archivoz. (2021). *"Uma imensidão documental que conta a história do Património em Portugal durante o séc. 20.": Entrevista a Ana Paula Figueiredo, Archivoz*. Available at: www.archivozmagazine.org/ [Accessed 15th November 2021].

Armstrong, A. (2006). Architectural archives/archiving architecture: The digital ERA. *Art Documentation: Journal of the Art Libraries Society of North America*, 25(2), 12–17.

Biblioteca de Arte e dos Arquivos Gulbenkian. (no date). *Coleção Júlio Moreira*. Available at: www.biblartepac.gulbenkian.pt/ipac20/ipac.jsp?session=16KJ516531361.1940343&profile=ba&menu=tab13&submenu=subtab86&ts=1663516553097# [Accessed 20th September 2022].

Cook, T. (1996). Building an archives: Appraisal theory for architectural records. *American Archivist*, 59(2),136–143. https://doi.org/10.17723/aarc.59.2.9016827w6t4271wl.

Cook, T. (2006). Remembering the future: Appraisal of records and the role of archives in constructing social memory. In Blouin, F.X., & Rosenberg, W.G. (eds.), *Archives, Documentation, and Institutions of Social Memory: Essays from the Sawyer Seminar*. University of Michigan Press, pp. 169–181.

Cook, T. (2011). The archive(s) is a foreign country: historians, archivists, and the changing archival landscape. *The American Archivist*, 74, 600–632.

da Silva, A.M. (1999). Arquivística e Cultura Popular. Algumas notas sobre o binómio Informação e Cultura. In *Separata das Atas Congresso Cultura popular*. Maia: Câmara Municipal da Maia, pp. 73–92. Available at: https://repositorio-aberto.up.pt/bitstream/10216/27057/2/amalheiroarquivistica000110204.pdf%5Cnhttp://hdl.handle.net/10216/27057.

De Certeau, M. (1992 [1975]). *The Writing of History*. New York: Columbia University Press.

Direção-Geral do Património Cultural. (no date). *Inventário do Património Arquitetónico, Sistema de Informação pra o Patrimínio Arquitectónico*. Available at: www.monumentos.gov.pt/site/app_pagesuser/SitePageContents.aspx?id=3528f11c-be07-4088-91be-4f8ed78738c9 [Accessed 4th January 2022].

Figueiredo, P. (2021). Arquivo de Património Arquitectónico da DGPC. Partilha de conhecimento. In *International Congress on Architectural Archives "Professional Experiences in a Cultural Diversity"*. Braga: Arquivo Distrital de Braga, pp. 129–132.

Fonseca, M.J. (2022). Arquitetura Paisagista: do lugar do documento de arquivo à construção de uma história. [Manuscript submitted for publication].

Hobbs, C. (2001). The character of personal archives: reflections on the value of records of individuals. *Archivaria*, 52(February), 126–135. Available at: https://archivaria.ca/index.php/archivaria/article/view/12817.

Horsman, P., & Ketelaar, E. (2015). Archival history. In Duranti, L., & Franks, P.C. (eds.), *Encyclopedia of Archival Science*. Lanham, Maryland: Rowman & Littlefield.

Hunt, J.D. (2000). *Greater Perfections: The Practice of Garden Theory*. Philadelphia: University of Pennsylvania Press.

Kokkinen, S. (2021). Digitalizing Finnish architectural heritage. In *International Congress on Architectural Archives "Professional Experiences in a Cultural Diversity"*. Braga: Arquivo Distrital de Braga, pp. 164–171.

Lička, L. *et al.* (2022). Building up historical continuity. Landscape architecture archives in education. In Bruns, D., & Hennecke, S. (eds.), *The Routledge Handbook of Landscape Architecture Education*. New York, London: Routledge, pp. 54–61. https://doi.org/10.4324/9781003212645.

Lowell, W. (2015). Architectural records. In Duranti, L., & Franks, P.C. (eds.), *Encyclopedia of Archival Science*. Lanham, Boulder, Nova Iorque, Londres: Rowman & Littlefield.

Pereira, Z.M.C. (2018). *O universo dos arquivos pessoais em Portugal: identificação e valorização—Parte I*. Tese apresentada à Universidade de Évora para obtenção do Grau de Doutor em Ciências da Informação e Documentação.

Ribeiro, F. (1998). *O acesso à informação nos arquivos*. Dissertação de doutoramento em Arquivística, apresentada à Faculdade de Letras da Universidade do Porto.

Scales of Change. ECLAS 2022. University of Ljubljana, Slovenia. Programme. (2022). *ECLAS European Council of Landscape Architecture*. Available at: https://conference.eclas.org/wp-content/uploads/2022/09/Programme_ECLAS22_finalRD-1.pdf [Accessed 30th September 2022].

Stoler, A.L. (2002). Colonial archives and the arts of governance. *Archival Science*, 2(1–2), 87–109. https://doi.org/10.1007/BF02435632.

Tavares, A. (2021). Euphoria and pragmatism: Using Archival Archives. In *International Congress on Architectural Archives "Professional Experiences in a Cultural Diversity"*. Braga: Arquivo Distrital de Braga, pp. 221–225.

Willis, A. (1996). The place of archives in the universe of architectural documentation. *American Archivist*, 59(2),192–198. https://doi.org/10.17723/aarc.59.2.l54510w443504578.

Zardini, M. (2021). Old institutions, new. In *International Congress on Architectural Archives "Professional Experiences in a Cultural Diversity"*. Braga: Arquivo Distrital de Braga, pp. 11–16.

6 Lisbon city gardeners, from horticulture to Landscape Architecture (1840–1960)

Cristina Castel-Branco, Ana Raquel Cunha* and Ana Luísa Soares*

Introduction

During the 20th century, the parks' and gardens' multi-disciplinarity and multi-interest theme was a rare educational purpose at the university level, and the problem of garden success in the city and its intricate multi-dimension issues was seldom tackled in scientific literature. Until the late 20th century garden problems in Portugal were studied and ultimately published through separate themes such as botany (e.g., Palhinha, 1935; Meyer, 1959; Meyer et al., 1960; Tavares, 1967) or the history of garden art (e.g., de Araújo, 1962; Carita et al., 1990).

A more complete answer for the causes of their lasting life seems to require a multi-disciplinary analysis. Few works on the subject have tried to portrait the many aspects that simultaneously address the complexity of garden maintenance and restoration associated with botany, horticulture, the evolution of garden design (e.g., Castel-Branco, 1999), the education for gardeners, the artistic issues that compose the gardens' design, the evolution of political will investing in garden-making, and even the stimulus of scientific knowledge for exchanging and acclimatizing plant species in gardens and parks in Lisbon (e.g., Castel-Branco, 2001).

This research aims at understanding the reasons that have made city gardeners' works in Lisbon rise as a priority in political investment and become a social success, then decline to an almost inexistent urban effort, followed by a new period of rising forces allowing for city gardeners' effective production of qualified parks and gardens in Lisbon. We hypothesize that the introductions in Lisbon, both in the 1840s and in the 1940s, of the know-how, instruction, and experience in horticulture, garden maintenance, botany, and garden art are the backbone for the success, stimulus of a new interest in garden-making development, and the creation of the garden heritage we still have in Lisbon. We confirmed that in both ascending periods, the interest grew, along with instruction, vegetation experiments, new businesses revolving around plant production, innovative garden uses, and risk-taking for new design solutions for urban gardens and parks.

* Equal contribution as first author.

DOI: 10.4324/9781003454007-9

The emerge of ways of teaching a team of followers played a crucial role in the long-lasting success of gardens as living ecosystems, which take decades to achieve maturity. The transmission of this knowledge is essential for garden's success. Additionally, the dissemination of user guides and good practices contributing to the achievements of lasting gardens complete the work.

Twice in the 120 years of our research, gardens became an important issue in the city design. In both periods, a triple endeavour became evident, contributing to the gardens' lasting lives as a gift to the city through centuries: experience/instruction followed by transmission and dissemination. Through the curriculum of the 22 "gardenists" in Lisbon, studied and selected as agents of the garden movement, these three criteria were organized and became evidence of a trend that confirmed our hypothesis.

Methodology

The experience of managing public gardens[1] in an urban setting was the basis of the research method adopted. The Jardim Botânico da Ajuda (3,8 ha) and the Parque Botânico da Tapada da Ajuda (100 ha) are large areas of semi-public access with main scientific and education roles. They both also offer recreational and conservation activities, and for three decades the authors were involved in various experiments within garden/parks management such as heritage restoration, water management, soil improvement, meccanization, plantation plans, gardeners team building, botanical inventory and data base, seed banks, propagation of native species, and new uses for an historical garden. Data was collected that aimed at improving the upkeep of such landscape units in the core of Lisbon. For this purpose all areas of Landscape Architecture teaching, such as soil science, botany, hydraulics, history of garden art, project, sociology, and drawing, were useful and even became indispensable to understand the needs for a sustainable management of park and gardens. Three factors emerged as seminal for the success of parks and garden managers: i) **instruction**, ii) ways of **transmission**, and iii) means of **dissemination**. Subsequently, with this holistic background we approached the information on professionals who had worked from 1840 to 1960 in Lisbon gardens, and we analyzed their contribution to each of three factors identified: experimentation/instruction (botany, horticulture, landscape architecture), ways of transmission (teaching and research), and means for dissemination (publishing magazines, journals, books, and inventories).

Several published sources were consulted, including Cabral (Cabral, 1940, 1956, 1993), Castel-Branco (Castel-Branco, 1999), Andresen (Andresen, 2003), Soares (Soares, 2021), and Cunha (Cunha et al., 2021), as well as non-published research found in Landscape Architecture degree thesis carried at Instituto Superior de Agronomia, namely Sousa da Câmara (Câmara, 1957). The documents rendered 22 key people and around 30 gardens during the study interval (1840 to 1960). After an extensive bibliographic review, these 22 men (Table 6.1) who were politicians, botanists/horticulturists, head gardeners, agronomists, foresters, painters, and landscape architects, were identified and were so varied we coined the word "gardenist"[2] for their outstanding achievements within the world of garden making. With their academical background and/or botanical education, these men have shown a deep understanding of nature and its phenomena, as well as the ability to transmit and disseminate their knowledge and experience for garden-making in Lisbon. As a result, the present gardens under study are a lasting heritage of these gardenists' legacy.

A compared analysis of these gardenists curricula was facilitated by creating *Tables* where the list of the 22 professionals (including relevant information and the respective bibliographic references consulted) in rows crosses the three criteria of their contribution: experience/instruction, transmission, and dissemination. The main trends issued from this analysis led to the results of the research, compiled in Tables 6.2 to 6.5. Their curricula were also examined to identify the ways they acquired and transmitted their knowledge, namely: their background trained in schools and/or empirical knowledge; if they had seen new gardens being built in Europe; establishing schools of gardening; collaborating with local gardeners; writing practical manuals to contribute to the continuity of garden knowledge; and opening offices to design gardens. The dissemination criteria included: publishing books and articles, presenting and preparing exhibitions, or even starting nurseries, etc.

The gardens mentioned in Table 6.1—private gardens and parks, botanical gardens, and public gardens—are directly related to these people and were long-lasting.

Table 6.1 The selected gardenists and gardens during the interval under study, 1840 to 1960 in the city of Lisbon.

Gardenists	***Gardens***
Politicians: King D. Fernando Saxe-Coburgo-Gotha (1816–1885); Francisco de Melo Breyner—4th Count of Ficalho (1837–1903); Francisco Simões Margiochi (1848–1904). **Botanists and Horticulturists:** Jean-Baptiste Desiré Bonnard (1797–1861); Friedrich Welwitsch (1806–1872); Jacob Weiss (1815–1898); Edmond Goeze (1838–1929); Jules Daveau (1852–1929); Henri Cayeux (1869–1963); Henri Navel (1878–1963). **Head gardeners:** Bento António Alves (1796–1878); João Francisco da Silva (s.d.). **Agronomists, Foresters, Painters and Landscape Architects:** Jorge Gomes de Amorim (1900–1943); José Pulido Garcia (1904–1983); Francisco Caldeira Cabral (1908–1992); Francisco Keil do Amaral (1910–1975); Manuel Azevedo Coutinho (1921–1992); Edgar Sampaio Fontes (1922–2000); Gonçalo Ribeiro Telles (1922–2020); António Vianna Barreto (1924–2012); Álvaro Ponce Dentinho (1924–2014); Manuel de Sousa da Câmara (1929–1992).	**Private gardens and parks:** *Tapada das Necessidades*, 1742; *Parque Monteiro-Mor*, séc. XVIII; *Parque das Laranjeiras*, 1779; *Parque de Santa Gertrudes*, 1857 (present *Jardim da Fundação Calouste Gulbenkian*). **Botanical gardens:** *Jardim Botânico da Ajuda*, 1768; *Jardim da Escola Politécnica*, 1842 (present *Jardim Botânico de Lisboa*); *Jardim Colonial*, 1914; *Tapada da Ajuda*, 1910. **Public gardens:** *Passeio Público*, 1766; *São Pedro de Alcântara*, 1840; *Jardim da Estrela*, 1852; *Campo de Santa Clara*, 1862; *Jardim de Santos*, 1873; *Jardim 9 de Abril*, 1879; *Alto de Santa Catarina*, 1883; *Praça D. Luís I*, 1884; *Jardim do Príncipe Real*, 1886; *Avenida da Liberdade*, 1886; *Campo Mártires da Pátria*, 1886; *Passeio do Campo Grande*, 1887; *Jardim Constantino*, 1889; *Parque Eduardo VII*, 1945; *Mata de Alvalade*, 1951; *Jardim da Capela de S. Jerónimo*, 1958; *Jardim da Torre de Belém*, 1958; *Jardim do Castelo de São Jorge*, 1959; *Parque do Vale do Silêncio*, 1968.

Table 6.2 The three politicians' curricula and their contribution according to the three criteria.

6.2. POLITICIANS	
6.2.1. King D. Fernando Saxe-Coburg-Gotha (1816-1885, German) (Ramalho and Costa Ramalho, 1935; Castel-Branco, 2001; Soares, 2021)	
Instruction/ Experience	■ Botanical education in Coburg and Gotha, a strong preparation in art and visited many referenced gardens of Europe. ■ In Portugal, he stood out for his work with his gardener Bonnard in constructing the first Landscape Garden in Portugal (1841), *Quinta Real das Necessidades*.
Transmission	■ *Quinta das Necessidades* became a laboratory for some 450 species. ■ Inspired the design of the *Parque da Pena* in Sintra and served the public gardens of the city of Lisbon. ■ The King commissioned his gardener, Bonnard, at the service of the city, who became both master gardener to the royal house and inspector of the municipality's gardens and public walks. ■ Contributed to the renovation of the *Passeio Público* and the project of the *Passeio da Estrela* and the *Jardim S. Pedro de Alcântara*. ■ Responsible for the first gardening school created in Portugal in 1841, directed by Bonnard. ■ The King's extensive library catalogue included books on botany (*e.g.* Lineu's fundamental works), natural history, chemistry, pharmacy, practical agriculture, forestry, and journals disseminating horticultural knowledge (*e.g.* "*The Garden*").
Dissemination	■ Visits and exchange of plants with other private gardens. ■ Walking with the Royal Family in the city's gardens, introducing this habit in Lisbon. ■ Creation of the *Sociedade Flora e Pomona*, in 1854, inspired by the Paris, where he was responsible for one of the first Portuguese periodicals of horticulture and gardening - "*Flora e Pomona: Jornal de agricultura, horticultura e jardinagem em Portugal*".
6.2.2. Francisco de Mello Breyner – 4th Count of Ficalho (1837-1903, Portuguese) (Ramalho and Costa Ramalho, 1935; Tavares, 1967; Castel-Branco, 1999; Cunha et al., 2021; Soares, 2021)	
Instruction/ Experience	■ He was part of the entourage of King Pedro V and King Luís on their trips to different European courts in 1854 and 1855, where he visited important gardens at the age of 17. ■ He graduated in Botany at the *Escola Politécnica de Lisboa*, where he was a full professor. ■ He encouraged the increase of studies of the Portuguese flora. ■ As Director of the *Instituto Agrícola* (1864 to 1877), he created the *Jardim Botânico da Escola Politécnica* (JBL). ■ With Edmund Goeze, he designed and decided the plantations of the upper deck of the JBL (class), and with Jules Daveau, he made the lower part of the JBL (*Arboretum*).
Transmission	■ Member of the *Academia Real das Ciências*, where was First Class President and Vice-President of the Academy. ■ Full member of the *Sociedade de Geografia de Lisboa*.
Dissemination	■ Published scientific and historical works, about the Portuguese Flora and the history of the introduction of new species in Portugal, such as: "*Apontamentos para o Estudos da Flora Portuguesa*", (1875); "*Useful Plants of Portuguese Africa*", (1884); "*Colóquio dos Simples e Drogas da Índia by Garcia de Orta*", (1891 and 1895).

(*Continued*)

Table 6.2 (Continued)

6.2. POLITICIANS	
6.2.3. Francisco Simões Margiochi (1848-1904, Portuguese) (Oliveira Júnior, 1873; Rodrigues, 2020)	
Instruction/ Experience	▪ Agronomist and forester graduated from the *Instituto Geral de Agricultura.* ▪ Councilor of the *Pelouro dos Passeios e Arvoredos of Câmara Municipal de Lisboa (CML)* between 1872-75. Highlights: tree planting in the city (recognized by Oliveira Júnior in one of his articles in the "*Jornal de Horticultura Pratica*" in 1876); and played an active role in defending green spaces (such as the transformation of *Passeio Público* into a boulevard). ▪ Chairman of the CML administrative committee.
Transmission	▪ He formed a specialized library, acquiring around 57 books and engravings. ▪ He was part of the commission appointed in 1882 to choose the species to plant in Avenida da Liberdade. ▪ As provider of Casa Pia created, in 1895, the *Escola Prática de Agricultura da Real Casa Pia* in Lisbon and a course for horticultural gardeners. ▪ In 1898 he founded the *Real Sociedade Nacional de Horticultura* of Portugal and as president, organized exhibitions and a gardening course.
Dissemination	▪ Launched a magazine, "*Boletim da Real Sociedade Nacional de Horticultura*", the most important publication of the 19th century aimed at the general public. ▪ Published works such as: "*Duas palavras ácerca da Avenida da Liberdade*" (1886); "*A luzerna: sua cultura e vantagens*" (1870); "*Assumptos agrícolas*" (1901).

Table 6.3 The seven botanists' and horticulturists' curricula and their contribution according to the three criteria.

6.3. BOTANISTS AND HORTICULTURISTS	
6.3.1. Jean-Baptiste Desiré Bonnard (1797–1861, French) (Viterbo, 1906; Castel-Branco, 2001; Rosa, 2013; Soares, 2021)	
Instruction/ Experience	▪ He was hired for *Quinta das Necessidades*, by King D. Fernando, due to his knowledge of plant acclimatization. ▪ Together with the King, he was responsible, in 1841, for the construction of the first Landscape garden in Portugal. ▪ Creation of the field of experimentation and plant acclimatisation. ▪ Distinguished himself as a master gardener in the public gardens of the capital, sharing his knowledge of the art of gardens. ▪ Worked in the gardens at the *Palácio das Laranjeiras* (owned by the Count of Farrobo) and *Palácio Daupiás* in Alcântara (owned by the Count of Alcochete).
Transmission	▪ Taught at the *Quinta das Necessidades* gardening school. ▪ Was co-owner of the *Companhia Horticola Lisbonense* nursery, in partnership with Bento Alves.

(*Continued*)

Table 6.3 (Continued)

6.3. BOTANISTS AND HORTICULTURISTS	
Dissemination	■ Collaborated with the *Sociedade Flora e Pomona.* ■ In 1854, he coordinated the 1st Exhibition of Flowers and Plants of the *Sociedade Flora e Pomona* at the Passeio Público. ■ Published works such as: "*Horticulture et Arboriculture. Végétaux introduits dans les cultures du Jardin Royal des Necessidades depuis sa restauration*", (1852); the catalogue "*La Compagnie Horticole*", (1854–55). ■ Promoted Portuguese plant production in France, to sell the products acclimatised here, together with Bento Alves.
6.3.2. Friedrich Welwitsch (1806–1872, Austrian) (CML, 1853; Oliveira Júnior, 1873; Viterbo, 1906, p. 185; Ramalho and Costa Ramalho, 1935; Tavares, 1967; Dolezal, 1974; Castel-Branco, 1999; Cal, 2021)	
Instruction/ Experience	■ Graduated in medicine at the University of Vienna. ■ Settled in Portugal, where he was appointed botany trainer at the *Escola Politécnica.* ■ Conservator of the *Jardim Botânico da Ajuda* (JBA), when integrated in the *Escola Politécnica.* ■ In 1844, he took over the management of *Quinta do Lumiar* (*Parque Monteiro-Mor*) from the Duke of Palmela. ■ He was part of the expedition to Angola, which he led between 1853–61. In this mission, his most famous discovery was *Welwitschia mirabilis* (1859), a desert plant.
Transmission	■ Devoted himself to floristic and phytogeographic research in Portugal. According to Gonçalo Sampaio (1865–1937), from the *Faculdade de Ciências da Universidade do Porto*, he was the most knowledgeable foreign botanist of Portuguese plants.
Dissemination	■ He had relationships with horticulturists and gardeners in Europe, and he maintained these relationships with gardeners and directors in Lisbon. ■ He dedicated some of his discoveries to personalities, such as: *Machadoa* (to Joaquim Machado); *Faroa* (to Lapa Faro); *Alvesia* (to Bento Alves). ■ Published works such as: "*Cultura do algodão em Angola*", (1861); "*Sertum angolense sive stirpium quarundam novarum vel minus cognitarum in itinere per angolam et benguelam observatarum: descriptio iconibus illustrata*", (1869).
6.3.3. Jacob Weiss (1815–1898, Swiss) (CML, 1852, 1853; Viterbo, 1906; Rodrigues, 2020; Cal, 2021; Soares, 2021)	
Instruction/ Experience	■ Trained at the *Jardin des Plantes* in Paris, as a horticulturist and gardener specialising in ornamental herbaceous plants. ■ Arrived in Portugal in 1847 to work for the Duke of Palmela at *Paço do Lumiar*. In 1853 he was promoted to director of the gardens, a position he held until his death. ■ Between 1866–70 directed the construction of the *Parque de Santa Gertrudes*, owned by Eugénio de Almeida (currently the *Jardim da Gulbenkian*). He used trees from French nurseries, introducing new species to Lisbon. ■ Was hired by the CML to manage the municipal nurseries. ■ Was invited to join several experts committees: in 1851 the committee to improve the *Passeio Público*; in 1852 the committee to inspect the state of the trees on the *Passeio Público*; and in 1882 the committee in charge of the tree planting on the *Avenida da Liberdade*.

(*Continued*)

Table 6.3 (Continued)

6.3. BOTANISTS AND HORTICULTURISTS	
Transmission	■ Collaborated with the *Sociedade Flora e Pomona*. ■ Donated 14 books to the *Pelouro de Passeios e Arvoredos* library in 1898.
Dissemination	■ Participated in horticultural exhibitions as an award-winning participant and as a jury. ■ As a horticulturist, he stood out in obtaining cultivars, such as the *Caladium*, advertised in the *Jornal de Horticultura Practica*.
6.3.4. Edmond Goeze (1838–1929, German) (Viterbo, 1906; Ramalho and Costa Ramalho, 1935; Tavares, 1967; Cal, 2021; Cunha et al., 2021; Soares, 2021)	
Instruction/ Experience	■ Trained and was a gardener at the *Jardin des Plantes*, in Paris, and at the Royal Botanic Gardens Kew, in London, and maintained correspondence with their directors, Joseph Descaine (1807–1882) and Dalton Hooker (1817–1911). ■ In 1866, by recommendation of José do Canto (1820–1898), he was hired by the Universidade de Coimbra, where he remained until 1873. ■ He took part in a botanical expedition to the island of S. Miguel in the Azores, where he discovered several private gardens, rich in plant diversity, including the property of José do Canto. ■ He was the first head gardener at *Jardim Botânico da Escola Politécnica* (JBL) and played an important role in this enriching of the botanical collection, having gathered a large number of species, mainly from the Azores, and from the botanical gardens with which he maintained close collaboration (Paris, London, Berlin and Hamburg).
Transmission	■ Contributed to the exchange and dissemination of known species in the Azores. ■ Sends 200 rare species to the *Jardin des Plantes* and around 1500 species of plants and seeds to the Royal Botanic Gardens Kew.
Dissemination	■ Collaborated in several publications, such as: "*Jornal de Horticultura Prática*"; "*The Gardens*"; "*Linnaea*"; "*Sonntags*". As a result of the botanical expedition to the island of S. Miguel, in the Azores, in 1867, he published "*A Ilha de São Miguel e o Jardim Botânico de Coimbra*".
6.3.5. Jules Daveau (1852–1929, French) (Viterbo, 1906; Ramalho and Costa Ramalho, 1935; Tavares, 1967; Rodrigues, 2020; Cal, 2021; Cunha et al., 2021; Soares, 2021)	
Instruction/ Experience	■ Trained at the *Jardin des Plantes* and worked at the Museum of Natural History in Paris. ■ In 1876, in Lisbon, he succeeded Goeze, assuming the position of head gardener of the JBL, until 1892. ■ Installed the irrigation system and created waterfalls and streams, in JBL. ■ The Count of Ficalho commissioned him to conduct botanical explorations in Portugal, and between 1876–92 Daveau dedicated himself to the study of the Portuguese Flora. ■ In 1882, he was part of the commission created to choose the species to plant in *Avenida da Liberdade*. ■ Record of 1886, Daveau offers plants to the municipal gardens (*e.g.* offer of 162 plants of various species *for Praça da Alegria*).

(*Continued*)

Table 6.3 (Continued)

6.3. BOTANISTS AND HORTICULTURISTS	
Transmission	■ Around 1877, he began a seed and plant exchange programme in the garden and published the first *Index seminum* with 1559 species (greatly benefited by Goeze's activity). ■ Was a corresponding member of the *Real Sociedade de Horticultura* of Portugal from 1899.
Dissemination	■ In addition to his gardening work, he contributed over 50 papers and studies on the Portuguese Flora in scientific publications. He published works in several magazines, such as: "*Revue Horticole*" (*e.g.*"*Les Jardins de Lisbonne*", 1879) and in the "*Buletin de la Societé Botanique de France*".
6.3.6. Henri Cayeux (1869–1963, French) (Viterbo, 1906; Tavares, 1967; Cal, 2021; Cunha et al., 2021; Soares, 2021)	
Instruction/ Experience	■ Son of a nurseryman, he trained at the *École Nationale d'Horticulture in Versailles* and at the *Jardin des Plantes* in Paris. ■ He stood out as a specialist in the field of plant propagation. ■ In 1892, hired as JBL's head gardener until 1909, Cayeux dedicated himself to introducing and cultivating ornamental plants in the garden (*e.g. Rosa, Hydrangea, Dahlia, Iris, Chrysanthemum*). ■ Creator of new hybrids such as *Dombeya x cayeuxii* and Rosa Bela Portuguesa. ■ Worked in the gardens of Count of Burnay (1838–1909), Luís Sommer and Henrique Mendonça.
Transmission	■ Member of: *Société Nationale d'Horticulture de France*; *Association des anciens élèves de l'Ecole Nationale d'Horticulture de Versailles; Société d'Horticulture d'Alger*. It should be noted that he was one of the leading French plant breeders in the first half of the 20th century. ■ Founder of the *Société des Chrysanthemistes Français* and the *Sociedade de Horticultura of Portugal*. ■ Recognized as one of the leading French plant breeders of the first half of the 20th century.
Dissemination	■ From 1894 to 1898, he organized annual *Chrysanthemum* exhibitions in the greenhouse of the JBL. ■ Published articles in several periodicals: *Le Jardin, Revue Horticole, Buletin de l'association des anciens élèves de l'Ecole Nationale d'Horticulture de Versailles, Revue Horticole de l'Algérie;* and *Boletim da Real Sociedade de Horticultura de Portugal*.
6.3.7. Henri Navel (1878–1963, French) (Morembert, 1965; Tavares, 1967; Cal, 2021; Cunha et al., 2021; Soares, 2021)	
Instruction/ Experience	■ Trained at the *École Nationale d'Horticulture de Versailles*. ■ Held various positions: Head Gardener at the Whiteley Institute of Horticulture (1899–1901); Gardener at the Royal Botanic Gardens Kew (until 1902); he worked in the greenhouses of the *Jardin Colonial de Nogent Sur Marne* (1902–1906) and was head of cultures in the colonial section of the *École Supérieure de Commerce de Nantes* (1906). ■ In 1909 he was hired to replace Cayeux, becoming the 4th JBL head gardener until 1920. ■ Supervised the transfer of the *Jardim Colonial of Quinta das Laranjeiras* to the nearby *Palácio de Belém* in 1910, where he was responsible for the colonial flora, from the Jardim Colonial in Lisbon (currently *Jardim Botânico Tropical*), until 1917.

(*Continued*)

Table 6.3 (Continued)

6.3. BOTANISTS AND HORTICULTURISTS	
	■ Managed gardens on private *Quintas*, such as Monserrate (owned by Francis Cook) and *Junqueira* (owned by Count of Burnay). ■ In 1919 he headed the agricultural mission of the *Sociedade de Fitopathologia* for emigration to the islands of S. Tomé and Príncipe and Portuguese Guinea, where until 1922 he was responsible for plantations (*e.g.* cocoa, coffee, sugar cane, banana trees), leading thousands of workers.
Transmission	■ Was a scholar who supported the scientific community in advancement of horticulture. ■ President of the *Académie Nacionale of Metz,* between 1953–54. ■ Was distinguished by the French and Portuguese governments.
Dissemination	■ Published in Paris in 1921: "*Les principaux ennemis du cacaoyer aux îles de San-Thomé et de Principe*".

Table 6.4 The two head gardeners' curricula and their contributions according to the three criteria.

6.4. HEAD GARDENERS	
6.4.1. Bento António Alves (1796–1878, Portuguese) (CML, 1853; Viterbo, 1906; Rodrigues, 2020)	
Instruction/ Experience	■ A gardener who learned from Bonnard and worked as a master gardener for the CML's *Pelouro de Passeios e Arvoredos* (1856–59). ■ He worked in the gardens of *Paço do Lumiar*, where he met the naturalist Welwitsch, with whom he learned. ■ In 1871 Bento Alves offered three specimens of *Araucaria* (*Araucaria excelsa, A. bidwilli* and *A. cookii*) from the nursery of his *Quinta* in Portela to the Lisbon gardens. ■ Was part of the commission responsible for inspecting the condition of the trees in *Passeio Público* (1852).
Transmission	■ Was co-owner of the *Companhia Horticola Lisbonense* nursery, in partnership with Bonnard. ■ In 1852 Bento Alves, driven by King Fernando II, they created together with Welwitsch and Councilor Aires de Sá Nogueira the *Sociedade Hortícola, Pomícola e Florícola* known as *Flora e Pomona.*
Dissemination	■ Published articles about the *Quinta do Lumiar* (*e.g.* "*Plantas Florestaes e de ornamento naturalizadas na quinta do Lumiar*", 1858). ■ Launched the catalogue "*La Compagnie Horticole*", which promoted Portuguese plant production in France, to sell national products.

(*Continued*)

Table 6.4 (Continued)

6.4. HEAD GARDENERS	
6.4.2. João Francisco (s.d., Portuguese) (CML, 1852; Castel-Branco, 2001; Rosa, 2013; Rodrigues, 2020; Cal, 2021; Soares, 2021)	
Instruction/ Experience	■ A gardener who started as an apprentice at *Quinta Real das Necessidades*, in 1851 he was already a municipal gardener. ■ Takes up the role of Master Gardener in the *Pelouro de Passeios e Arvoredos* (c.1869 to 1886). ■ Participates in constructing two 19th century gardens in the city, the *Passeio da Estrela* and *Jardim do Príncipe Real*. Contributed to the design and planting of the small Romantic public gardens, created in the capital: *Campo de Santa Clara* (1862); *Santos* (1873); *9 de Abril* (1880); *Praça da Alegria* (1881); *Alto de Santa Catarina* (1883); *Praça D.Luís* (1884); *Praça das Flores* (s.d.); *5 de Outubro* (s.d.); *Constantino* (1889); and *Campo Santana* (s.d.). ■ During the vereignancy of the 1st Viscount of Carriche, he went to France to learn to handle the tree transplanting machine acquired by the CML, with the head gardener of the *Potager Du Roi*, in Versailles. ■ Although there is not much information about him, João Francisco was part of the team of five technicians – Councilor 1st Viscount of Carriche, Daveau, Weiss, João Francisco and Margiochi – created in 1882 to choose the species to plant in the *Avenida da Liberdade*.
Transmission	
Dissemination	

Table 6.5 The ten agronomists', foresters', painters', and landscape architects' curricula and their contributions according to the three criteria.

6.5. AGRONOMISTS, FORESTERS, PAINTERS, AND LANDSCAPE ARCHITECTS	
6.5.1. Jorge Gomes de Amorim (1900–1943, Portuguese) (Antunes, 2019; Câmara, 2021; Soares, 2021)	
Instruction/ Experience	■ A Forestry engineer who joined the CML from 1938–43 as head of the 6th Division – *Repartição de Arborização e Jardinagem* (RAJ) and dedicated himself to the management and maintenance of green spaces. ■ Financed by CML, visited Germany in 1938–39 to acquire machines to transplant trees; as well as France, Belgium, Holland and Luxembourg to study various services related to the afforestation of parks and gardens, roads and streets, fertilisation, plants and seeds, irrigation of trees; he also visited the Hofstanddbloem Exhibition in The Hague. ■ Was responsible for the gardening and tree-planting work at *Praça do Império*, in 1940, for the Portuguese World Exhibition.
Transmission	
Dissemination	■ In 1941, published "*Jardins municipais: conferência*" as part of the 2nd National Exhibition of Floriculture.

(*Continued*)

Table 6.5 (Continued)

6.5. AGRONOMISTS, FORESTERS, PAINTERS, AND LANDSCAPE ARCHITECTS	
6.5.2. José Pulido Garcia (1904–1983, Portuguese) (Antunes, 2019; Câmara, 2021; Soares, 2021)	
Instruction/ Experience	■ Agronomist, forestry engineer, and employee of CML, he stood out as head of the RAJ. ■ Collaborated with the first landscape architects to join CML staff: Manuel de Azevedo Coutinho (1948) and Gonçalo Ribeiro Telles (1950). ■ Was responsible for installing green spaces in the city, such as the *Parque Florestal de Monsanto* and the *Jardim do Castelo de São Jorge*. ■ Worked on constructing the municipal nursery at *Quinta do Conde de Arcos* and remodelled several nurseries by introducing new techniques. ■ Responsible for the: expansion, improvement and new plantations in the *Estufa Fria* and the creation of *Estufa Quente;* creation of the *Mata de Alvalade* and *Mata Madre de Deus*; restoration of *Avenida da Liberdade* after Underground construction, *Avenidas Fontes Pereira de Melo* and *Avenida da República;* and reconstruction of *Campo Grande* and *Parque Eduardo VII.*
Transmission	■ He has collaborated in creating the practical course for gardeners, which took place in *Vila Correia*.
Dissemination	■ In 1940 published "*A reconstituição da Tapada de Mafra*" and in 1961 the lecture "*A Vegetação de Lisboa e os seus parques e jardins*".
6.5.3. Francisco Caldeira Cabral (1908–1992, Portuguese) (Castel-Branco, 1999; Andresen, 2001; Andresen and Tostões, 2003; Antunes, 2019; Câmara, 2021; Soares, 2021)	
Instruction/ Experience	■ An Agronomist graduated from *the Instituto Superior de Agronomia* (ISA) in 1936, and Landscape Architect graduated from the *Technische School* of Berlin, in 1939. ■ In 1935, he received an invitation to occupy the position of head of the *Repartição dos Jardins e Cemitérios* of CML (which he did not accept). ■ Responsible for directing the JBA, where he carried out his training report (1936). ■ The Institute for High Culture awarded him a scholarship to study Landscape Architecture in Berlin. ■ His landscape architecture projects were based on ecology, maintenance, and aesthetics. Its projects covered different areas of intervention: public gardens; historic gardens; private farms and gardens; agricultural planning and development projects; Spatial and rural landscape planning and landscape integration of infrastructures; and research and development (*e.g.* mission to promote floriculture on the island of Madeira, 1973–78). The *Estádio Nacional* (1938–40) is referred to as an emblematic project.

(*Continued*)

Table 6.5 (Continued)

6.5. AGRONOMISTS, FORESTERS, PAINTERS, AND LANDSCAPE ARCHITECTS	
Transmission	■ Founder of the teaching of Landscape Architecture in Portugal. In 1942, the course of Landscape Architecture for Agronomists and Foresters was launched. ■ In 1953, created the *Centro de Estudos de Arquitetura Paisagista* (CEAP) at ISA. ■ According to Teresa Andresen, Caldeira Cabral intended to create a gardening school at the JBA in collaboration with the CML and even prepared a typewritten document. ■ The University of Évora awarded him the degree of Doctor Honoris Causa in 1980, as did the Technical University of Hannover in 1971. ■ In 1958, he was elected Vice-President of the International Federation of Landscape Architects (IFLA) and, four years later, unanimously elected President. ■ His merit was publicly recognized through the attribution of the degree of *Grande Oficial da Ordem da Instrução Pública* (1982) and the *Grã Cruz da Ordem do Infante D. Henrique* (1989).
Dissemination	■ The creation of the teaching of Landscape Architecture in 1942 allowed the systematic transmission of gardening knowledge, which Caldeira Cabral's disciples continued for 80 years with the systematic dissemination of Landscape Architecture in five Portuguese Universities (*Lisboa, Évora, Vila Real, Faro, Porto*). ■ Published several of his studies, such as: "*Construção de Jardins*", (1963); "*História de Arte dos Jardins*", (1962); "*A Árvore*", (1960); e "*Os Fundamentos da Arquitectura Paisagista*", (1993).
6.5.4. Francisco Keil do Amaral (1910–1975, Portuguese) (Tostões, 1992; Andresen, 2001; Andresen and Tostões, 2003; Tostões and Guimarães, 2013; Antunes, 2019; Câmara, 2021; Soares, 2021)	
Instruction/ Experience	■ Painter graduated, in 1936, from the *Escola de Belas Artes de Lisboa*. ■ In 1937, he won the competition for the Portuguese Pavilion at the Paris International Fair. ■ From 1939–49 worked as an urban planner at CML. ■ In 1939, he went to visit the: parks of Paris, London, Amsterdam and The Hague; the Stuttgart exhibition; and the afforestation of some sections of German highways. ■ He was the author of projects such as: *Parque Eduardo VII, Parque Florestal de Monsanto* and *Jardim Campo Grande*. ■ In the atelier, he designed: the Portuguese International Fair (1953) and the Lisbon Underground (1949–59). ■ Was consultant on several projects, namely: Museum of the Calouste Gulbenkian Foundation (1958); Baghdad Football Stadium in Iraq (1961); urban plan for the Tróia Peninsula.
Transmission	■ He taught the Art and Architecture course at the *Universidade Popular* (1943) and at the *Instituto de Arte e Decoração* (IADE). ■ Received the Municipal Architecture Award in 1951 and the Valmor Prize in 1962, both distinctions for individual housing projects in *Restelo*.

(*Continued*)

Table 6.5 (Continued)

6.5. AGRONOMISTS, FORESTERS, PAINTERS, AND LANDSCAPE ARCHITECTS	
Dissemination	■ Presented conferences and published several articles in specialist magazines, as well as books, such as: "*A arquitectura e a vida*", (1942); "*A moderna arquitectura holandesa*", (1943); and "*O problema da habitação*", (1945).
6.5.5. Manuel Azevedo Coutinho (1921–1992, Portuguese) (Andresen, 2003; Câmara, 2015)	
Instruction/ Experience	■ Agronomist and landscape architect, graduated from ISA, in 1948, with the final work "*O Jardim Botânico da Ajuda*"and was the first landscape architect trained in Portugal by the school of Caldeira Cabral. ■ Started working as a Landscape Architect at CML in 1950. ■ Author of projects of gardens and avenues in *Alvalade*, *Avenida Guerra Junqueiro*, *Avenida da Igreja*, designed the *Jardim da Praça de Londres*. ■ Responsible for the renovation of the municipal nurseries.
Transmission	■ Teaches at ISA between 1950 and 1953.
Dissemination	
6.5.6. Edgar Sampaio Fontes (1922–2000, Portuguese) (Andresen, 2003; Câmara, 2015)	
Instruction/ Experience	■ Agronomist and landscape architect, graduated from ISA, in 1951, with the final work "*A vinha na paisagem do Minho*". ■ Joined CML landscape architects' team in 1953, holding directive positions up to: Chief of RAJ (1970–76); to Director of DSTE (1982–89). ■ His design projects include Bairro da *Encarnação, Parada dos Prazeres, Bairro do Grilo, Estufa Fria*, and *Parque Eduardo VII's* Upper Area.
Transmission	■ He began teaching at ISA in 1957 and became assistant professor until 1974, and as guest professor from 1977/78. ■ Co-founded the "*Associação Portuguesa dos Arquitectos Paisagistas*" in 1976 and served in leadership positions. ■ Represented Portugal at the IFLA from 1965 to 1977, later coordinating IFLA activities in the Mediterranean zone. ■ Received an award from the "*Associação Industrial Portuguesa*" for the project of recovery and landscaping of the Secil quarries, in 1987.
Dissemination	■ Was recognized for expertise in floriculture, especially roses, and participated, as a jury and/or president, in national and international rose competitions. ■ Authored books and articles on landscape recovery, planning, and horticulture (*e.g.* "*Recuperação paisagística das pedreiras. Caso da Secil*", 1986).
6.5.7. Gonçalo Ribeiro Telles (1922–2020, Portuguese) (Andresen, 2003; Saraiva, 2004; Cunha, 2014; Câmara, 2015)	
Instruction/ Experience	■ Graduated as agronomist and landscape architecture at ISA, in 1950, with the final work "*Um caso concreto de Ordenamento Paisagístico. Estudo de três herdades no concelho de Coruche*".

(*Continued*)

Table 6.5 (Continued)

6.5. AGRONOMISTS, FORESTERS, PAINTERS, AND LANDSCAPE ARCHITECTS	
	■ Worked at CML from 1953 to 1960, designing over 60 public spaces, notably contributing to urban expansion during the 1950s. ■ As a private practice designer he undertook around 350 projects, including the Fundação Calouste Gulbenkian gardens in collaboration with Vianna Barreto. ■ Played a key role in landscape planning and design, contributing to the city plans of Lisbon and Coimbra.
Transmission	■ Established the Landscape Architecture program at the University of Évora, in 1976, and served as a professor and degree coordinator ■ Active in politics, became as Minister of State and Quality of Life in the 8th Constitutional Government of Portugal and launched the National Ecological Reserve Act protecting fragile landscapes and seaside areas. ■ Honoured with significant degrees and recognition for his contributions to environmental and landscape preservation, including: "*Grã-Cruz da Ordem Militar de Nosso Senhor Jesus*" (1988); "*Grã-Cruz da Ordem da Liberdade*" (1990); and "*Grã-Cruz da Ordem do Infante D. Henrique*" (2017). ■ Received the Sir Geoffrey Jellicoe Award for Landscape Architecture in 2013.
Dissemination	■ Authored 4 books and more than 100 articles on: Landscape Planning, Environment, Urbanism, Visual Analysis, Landscape, Design, Agriculture, Resource Management, Education and Politics.
6.5.8. António Vianna Barreto (1924–2012, Portuguese) (Andresen, 2003; Barreto, 2011; Câmara, 2015)	
Instruction/ Experience	■ Graduated as a Forest and Agronomy engineer, and in landscape architecture, from ISA, in 1952, with the final work "*O Parque de Monsanto e a cidade de Lisboa*". ■ Joined the technical staff of the CML, in 1953, becoming the first landscape architect to work for the government, holding various leadership positions in the CML. ■ His projects in Lisbon include the landscape design of: Fundação Caloute Gulbenkian, *Jardim da Torre de Belém*, *Cidade Universitária*, and the terraces of Hotel Ritz. ■ Coordinated the report "*Defesa da Paisagem*"in 1962, which laid the foundation for territorial planning figures like the National Ecological/ Agricultural Reserves (REN and RAN).
Transmission	■ Taught: territorial planning at *Instituto Superior Técnico* (1979 and 1980), *Universidade de Évora* (1981–84); *ISA* (1984–88). ■ Member of APAP and served in leadership positions. ■ Received the prestigious "*Prémio Valmor*" in 1975 for the garden project at Fundação Calouste Gulbenkian. Recognized with commendations from Ministerial Dispatches in 1957 and 1975; and awarded the *Prémio Quercus* in 2009.
Dissemination	■ Authored books (*e.g.* "*O Enquadramento da Torre de Belém*", 1956), articles in specialized articles (*e.g.* "*Sede e Museu da Fundação Calouste Gulbenkian*", 1969, *in Arquitectura*), and technical reports (*e.g.* "*Ordenamento Paisagístico do Algarve*", 1969, DGSU).

(*Continued*)

Table 6.5 (Continued)

6.5. AGRONOMISTS, FORESTERS, PAINTERS, AND LANDSCAPE ARCHITECTS	
6.5.9. Álvaro Ponce Dentinho (1924–2014, Portuguese) (Andresen, 2003; Câmara, 2015)	
Instruction/ Experience	▪ Graduated in Forestry and Landscape Architecture, in 1959, with the final work "*Modulação na composição das estradas*". ▪ Joined CML in 1959, collaborating on urban planning projects ▪ While at CML's GEU, contributed to urban development design with squares in the garden city of *Olivais, Chelas, Marvila*, and the hillside of *Restelo*. ▪ Undertook urban and territorial planning projects in Portugal: Algarve Spatial Plan (1965) and the Municipal Master Plan for Lourenço Marques (1969) presently Maputo, in Mozambique.
Transmission	▪ Served as assistant professor in the Landscape Architecture course at ISA from 1964 to 1967. ▪ Actively participated in IFLA International Congresses and was named honorary secretary during the 1970s. ▪ Member of the "*Centro Nacional de Cultura*".
Dissemination	▪ Authored articles in specialized magazines (*e.g.* "*Arranjo dos espaços exteriores*", 1964, in *Arquitetura*) and other periodicals (*e.g.* "*Bairros construídos ao abrigo do Decreto-lei 42 454 no período de 1961 a 1969*", 1971, *in Boletim GTH*).
6.5.10. Manuel Sousa da Câmara (1929–1992, Portuguese) (Andresen, 2003; Câmara, 2015; Cunha, 2015; Sousa da Câmara, s.d.)	
Instruction/ Experience	▪ Agronomist and landscape architect, graduated from ISA, in 1957, producing the project *"A Tapada Real de Vila Viçosa – Anteprojeto do seu ordenamento paisagístico"*. ▪ Worked in CML (1961–71), where he designed parks, gardens and squares such as: *Mata de Alvalade; Parque do Vale do Silêncio*; *Praça das Flores;* and *Jardim da Estrela* renovation. ▪ Established a private practice *Gabinete de Projetos Sousa da Câmara, Lda*., in 1982, and led numerous groundbreaking investigations and projects in Lisbon, including urban planning in several neighbourhoods. Pioneered biological engineering methods, computer-assisted design, and geographical information systems applied to landscape architecture. Developed original models and computer programs for use in your own studio (*e.g.* Model to calculate the man/power necessary for the maintenance of the public garden system).
Transmission	▪ Taught at ISA (1976–88) contributing significantly to landscape architecture education; and at the *Escola de Belas Artes de Lisboa*, in the graduate program in Urban Studies. ▪ Active member of professional associations and received recognition from CML for his work on the new "*Acessos da nova Ponte sobre o Tejo*".
Dissemination	▪ In 1965, an exhibicion: "*Seminary of Landscape Architecture*" in Fundação Calouste Gulbenkian. ▪ Authored several articles (*e.g.* "O turismo e a Paisaem", 1951, *in Agros*) and technical reports (*e.g.* "*Informação sobre o custo de construção de espaços verdes urbanos*", 1969, *GTH*).

Results

The local expertise pertaining to horticulture and water management in Portugal was significantly enhanced during two periods of garden expansion in Lisbon by two trends. The first was a growing political will to follow the international park movement, while the second corresponds to the introduction of a systematic knowledge of horticulture and gardening from various European countries in the 1840s and later from Germany beginning in the 1940s, during which the empirical knowledge of exotic plants, derived from Portuguese voyages across the oceans, became strengthened and professionalized.

Throughout the research period (1840–1890), Portuguese "gardenist" leaders hired experts from France, Austria, Switzerland, and Germany to work in Lisbon gardens—conditional on each having a thorough understanding of plants, horticulture, and gardening. Notably, all seven foreign gardenists discussed herein held diplomas from renowned European schools, reinforcing the essential preparation required for managing the complex activities of urban parks and gardens in Lisbon (Table 6.3). Later, King D. Fernando's strategy to promote gardens included the renovation of the central promenade, Passeio Público, coupled with his penchant for promenading with his family in this new venue, thereby creating there a new tradition for social gatherings in public gardens (Table 6.2).

The case of Count of Ficalho is unique, as he belonged to the Portuguese aristocracy but actively sought training as a botanist. Credited in large part with creating the second botanical garden in Lisbon, Ficalho approached the project with a strong emphasis on scientific botanic experimentation, coupled with innovative garden design; indeed, his contributions led to a rapid increase in the number of plant species in Lisbon, particularly palms and other exotic varieties (Table 6.2).

By comparing the documents of this period, we have identified a rise in interest for gardens and parks in the period from 1840 to 1890—both in the city and in private realm. After this vibrant period in the history of garden-making in Lisbon, however, further innovations and even maintenance of existing landscapes suffered a strong setback. This drift was likely promoted by the Portuguese bankruptcy of 1892 and the instability that led to the republic replacement of the monarchic system in 1910 and the First World War (1914–1918), in which Portugal was involved. Gardenists seem to have disappeared from the records of the early 20th century. The loss of the articulated efforts—be they municipality, academy, or private—in this period of economic and political instability led to the extinction of the "Pelouro dos Passeios e Arvoredos" (Department of trees and promenades) and the suspension of two main park projects: Parque Eduardo VII and Campo Grande, which were only to be completed in 1938 when Keil do Amaral was tasked with the redevelopment of Lisbon's green spaces.

The beginning of the 20th century was characterized by both economic and political instability, which took its toll on Lisbon's green spaces. It wasn't until the 1940s that an interest in public gardens once again took root. In fact, this decade also witnessed the arrival of the skilled Portuguese landscape architect Cabral and the painter Keil do Amaral, which gave impulse to the political intention led by José Pulido Garcia (Table 6.5).

Discounting those periods of instability and inertia, the gardenists of Lisbon can be viewed as wonderfully influential in shaping the city's landscape heritage, leaving behind an impressive botanical and landscape legacy. Initially hired to work in botanical gardens and scientifically-oriented private gardens, where they laid the foundations of professionalism in this realm, they contributed to the dissemination of knowledge and practice in the municipality in various ways: by introducing and acclimatizing new plant species, advancing horticulture techniques, publishing botanical works, establishing herbaria and *Index seminum*, collaborating in specialized journals, organizing and participating in plant exhibitions, and creating a new professional degree, which responded to the needs of greening Lisbon in the second growing period of the 1940s and 1950s.

The success of plant experimentation and knowledge transmission in Lisbon played a pivotal role in the image of the city and in garden-making as a practice. For example, the collaboration between Bonnard (Table 6.3) and João Francisco (Table 6.4) resulted in the creation of the Passeio da Estrela. Consider also the robust exchange of plants and seeds between botanical gardens, private properties, and the municipality; as a result, public areas such as Passeio Público, Estrela, and Campo Grande received new species of blooming trees (Jacaranda mimosifolia and Tipuana tipu) that changed the image of the city with the creation of vibrant coloured public gardens that became favoured recreation destinations for the local population.

The second boom in the 1940s was led by Francisco Caldeira Cabral who established the first degree of Landscape Architecture at the School of Agriculture (present Instituto Superior de Agronomia) in 1942 from where he himself held a 5-year degree as an agronomist. From 1936–1939 he studied at the Institute for Garden Design (Institut fur Gartengestaltung) at the College of Agriculture (Landwirtschaftlich-Gärtnerische Fakultät), whose Landscape Architecture curriculum he later adopted at the Lisbon School of Agriculture, while also adding location-specific aesthetic, ecological, and project preparation for his students. The 1950s marked a new era (Andresen, 2003) for gardenists and their work, notably six Portuguese agents of transmission—Manuel Azevedo Coutinho, Gonçalo Ribeiro Telles, António Vianna Barreto, Edgar Sampaio Fontes, Álvaro Dentinho, and Manuel de Sousa da Câmara (Table 6.5)—who can be credited with expanding the diaspora of Landscape Architecture in the growing and increasingly modernist city of Lisbon. We include the impact of their gardens in the comparative table and confirmed that they introduced major changes in professionalizing the field of horticulture and garden-making in Lisbon, in the maintenance and machinery used in parks, and in creating a new era in Portugal's urban design, parks conservation, ecological legislation, and academic training protocols.

During these important periods of garden-making, these *gardenists* also contributed significantly to the written record through both national and international publications. Some of them wrote books and published articles in magazines such as the "Jornal de Horticultura Pratica", an esteemed academic journal issued in Porto (1870–1892), in catalogs, and in technical reports. King D. Fernando also played an important role in establishing the first periodic publication on horticulture and gardening in Portugal, called "Flora e Pomona: Jornal de agricultura, horticultura e jardinagem em Portugal". Additionally, King D. Fernando and Francisco Margiochi, Councilor of the *Pelouro dos Passeios e Arvoredos*, established specialized libraries that served as important repositories of knowledge (Table 6.2).

Most of these men actively participated in scientific discussion forums, either by establishing them (e.g., King D. Fernando with the *Sociedade Flora e Pomona* and Margiochi with the *Real Sociedade Nacional de Horticultura*) or by being involved in their activities (e.g., Count of Ficalho as a member and president of the *Academia Real das Ciências* and a member of *the Sociedade de Geografia de Lisboa*, and Francisco Caldeira Cabral as a member and president of the International Federation of Landscape Architects). The rise of various garden societies as discussion forums and platforms for knowledge exchange became crucial social meeting points to promote garden knowledge. These forums facilitated the sharing of expertise, disseminating advancements in horticulture and gardening, and contributing to the overall development and improvement of garden practices and landscaping.

The growing interest in plants as aesthetic tools accessible to anyone led to the organization of flower shows imitating those taking place in Britain, France, and Germany. Catalogues and other sources of news and information confirm the success of these garden events in Lisbon at the Tapada das Necessidades and the Tapada da Ajuda. To reiterate, these major horticultural events came to represent important venues for both disseminating knowledge about plant species, as well as conveying the value of public and private gardens (e.g., nursery owners were able to improve their profits significantly by taking part in flower shows).

Regarding the transmission of knowledge, King D. Fernando established the first gardening school in 1841 at the *Real Tapada das Necessidades*, directed by Bonnard; in 1895, Margiochi created the "*Escola Prática de Agricultura da Real Casa Pia*" and introduced a course for horticultural gardeners; José Pulido Garcia contributed to the creation of a practical course for gardeners in *Vila Correia*; and in 1942 Cabral founded the Course of Landscape Architecture for Agronomists and Foresters. The beginning of the teaching of Landscape Architecture in Portugal in 1942 also contributed to the rearticulation of the aforementioned three key factors. A new stage of green spaces in Portugal was inaugurated with the integration of the first generation of landscape architects in the CML staff, from the 1950s on.

In terms of individual achievements, Dentinho, Sousa da Câmara, and Sampaio Fontes—three talented landscape architects—designed a new neighbourhood, Olivais, as a garden city that remains to this day a reference in landscape design. Some new parks have received international recognition, such as the Gulbenkian Foundation gardens designed by Ribeiro Telles and Viana Barreto and the Jardim da Torre de Belém created by the latter. Also, Ribeiro Telles designed 60 green areas in Lisbon. Sousa da Câmara and Azevedo Coutinho also optimized production processes at municipal nurseries.

Conclusions

The key impact of this research is that the art and skill of garden-making and the subsequent maintenance of Lisbon's parks and green spaces has required (and continues to demand) the input of qualified professionals at all levels who are prepared scientifically and aesthetically. Increasingly, the city's trees symbolize a critical ecological infrastructure, while parks and gardens in Lisbon serve as open-air meeting rooms for the urban population—showcasing an immerse biodiversity and helping to offset increasing temperatures stemming from climate change. These public and private spaces represent a precious legacy from gifted and highly qualified gardenists—a word coined to signify gardeners, botanists, horticulturists, forestry engineers, and landscape architects who are well trained in the work of garden-making, upkeep, and restoration.

We draw our conclusions from the data collected that signify incremental growth in garden expansion, interest, and growing use from the 1840s on and later from the 1940s on. We attribute this growth to i) the formal preparation of gardenists in both instruction and on-the-job experience. Subsequently, the capacity to maintain their urban "experiments" can be attributed to ii) their transmission's strategy through teaching, instructing gardeners, exchange knowledge internationally, and finally through iii) various means of knowledge dissemination: publications, conferences, horticultural events, and the exchange of plants and seeds in-country and abroad.

The art and skill of combining the three factors, which we gleaned from studying the 22 gardenists described in the research, embody the key for the growth and endurance of Lisbon's parks and gardens. Notably, the second boom of garden-making in the city in the 1940s and 1950s can be linked to the establishment of a Landscape Architecture degree at the Technical University of Lisbon, imported from an earlier Germany model, but expanded in terms of science, technology, and engineering to reflect local aesthetic and ecological needs. Much more research, however, is needed on this subject and the role that these three criteria have played in promoting the growth and importance of gardens in the 19th/early 20th century in Lisbon. In closing, we assert that these intertwined factors represent indispensable elements behind the world of garden-making and upkeep.

Acknowledgments

The authors thank to FCT for the A.R.C. PhD scholarship (Ref. 2020/04824/BD). We are greatly indebted to the two reviewers whose comments substantially helped to improve the paper.

Abbreviations

APAP—*Associação Portuguesa dos Arquitetos Paisagistas* (Portuguese Association of Landscape Architects)
CEAP—*Centro de Estudos de Arquitetura Paisagista* (Center for Landscape Architecture Studies)
CML—*Câmara Municipal de Lisboa* (Lisbon City Hall)
DSTE—*Direção dos Serviços Técnicos Especiais* (Direction of Special Technical Services)
GEU—*Gabinete de Estudos de Urbanização* (Urbanization Studies Office)
GTH—*Gabinete Técnico da Habitação* (Housing Technical Office)

IADE—*Instituto de Arte e Decoração* (Institute of Art and Decoration)
IFLA—International Federation of Landscape Architects
ISA—*Instituto Superior de Agronomia* (School of Agronomy of Lisbon)
JBA—*Jardim Botânico da Ajuda* (Ajuda Botanical Garden)
JBL—*Jardim Botânico da Escola Politécnica* (current Lisbon Botanical Garden)
RAJ—*Repartição de Arborização e Jardinagem* (Division of Arboriculture and Gardening)
RAN—*Reserva Agrícola Nacional* (National Agricultural Reserve)
REN—*Reserva Ecológica Nacional* (National Ecological Reserve)

Notes

1 Cristina Castel-Branco was part of the Management Board of the School of Agronomy (ISA/ULisboa), in charge of the *Tapada da Ajuda* green spaces (1992–1995); and was director of the *Jardim Botânico da Ajuda* (1997–2002). She edited the book "Jardim Botânico da Ajuda" and organized (1993) the submission process to the National Heritage Department (IPAR) for the classification of *Tapada da Ajuda* as cultural landscape, thus defending the land from a high construction and real estate allotments. Ana Luísa Soares was part of the Management Board of ISA/ULisboa, in charge of *Tapada da Ajuda* (2009–2014); and is the coordinator of *Jardim Botânico da Ajuda* (2019-present). She headed the *Project LX GARDENS—Lisbon's Historic Gardens and Parks* and the publication of the books ("O arvoredo, os jardins e parques públicos de Lisboa (1755–1965)". With Ana Raquel Cunha she collaborates monthly, in "Revista Jardins", the leading magazine in the field of gardening in Portugal. Ana Raquel Cunha has co-authored the scientific paper, "Natural and Historical Heritage of the Lisbon Botanical Gardens: An Integrative Approach with Tree Collections".

2 **Gardenist**: the word gardenist was coined to signify people prepared to work for garden making, up-keeping, restoring, and defending their high quality level, including botanists, horticulturists, forestry engineers, agronomists, painters, and landscape architects.

References list

Andresen, T. (2001). *Francisco Caldeira Cabral*. 1ª Ed. Richmond, Surrey, U.K: Landscape Design Trust.

Andresen, T., & Tostões, A. (eds). (2003). *From the National Stadium to the Gulbenkian Garden: Francisco Caldeira Cabral and the First Generation of Portuguese Landscape Architects,1940–1970*. Lisbon: Calouste Gulbenkian Foundation, Fine Arts Department.

Antunes, A.C. (2019). *A influência alemã na génese da Arquitetura Paisagista em Portugal*. Tese de Doutoramento em Arquitetura Paisagista e Ecologia Urbana. Universidade do Porto. Available at: https://repositorio-aberto.up.pt/handle/10216/121946.

Barreto, F.M.M. de A.S. (2011). *Contributo para a história da arquitectura paisagista em Portugal: arquitecto paisagista António Facco Vianna Barreto*. Dissertação de Mestrado em Arquitetura Paisagista, Instituto Superior de Agronomia, Universidade de Lisboa. Available at: www.repository.utl.pt/handle/10400.5/4146.

Cabral, F.C. (1940). *Jardins*. Lisboa: Publicações Culturais da Câmara Municipal de Lisboa.

Cabral, F.C. (1956). *Protecção à natureza e arquitectura paisagista*. Lisboa: Emp. Tip. Casa Portuguesa.

Cabral, F.C. (1993). *Fundamentos da Arquitectura Paisagista*. Lisboa: Instituto de Conservação da Natureza.

Cal, P.S. (2021). *Jardineiros-Botânicos em Lisboa (1764–1919). Contributos para a história dos seus jardins*. Dissertação de Mestrado em Arquitetura Paisagista. Instituto

Superior de Agronomia, Universidade de Lisboa. Available at: www.repository.utl.pt/handle/10400.5/23593 [Accessed 30th November 2022].

Câmara, M.P.N.S. (1957). *A Tapada Real de Vila Viçosa. Ante projecto de seu ordenamento paisagistico*. S.l.: s.n.

Câmara, M.T. (2015). *Contributos da Arquitetura Paisagista para o espaço público de Lisboa (1940–1970)*. Tese de Doutoramento para obtenção do Grau de Doutor em Arquitetura Paisagista e Ecologia Urbana. Faculdade de Ciências, Universidade do Porto. Available at: https://hdl.handle.net/10216/85668.

Câmara, M.T. (2021). *Espaço público de Lisboa: plano, projeto e obra da primeira geração de arquitetos paisagistas, 1950–1970*. Lisboa: Câmara Municipal.

Carita, H., Cardoso, A.H., & Cardoso, M.E. (1990). *Tratado da grandeza dos jardins em Portugal ou da originalidade e desaires desta arte*. Lisboa: Círculo de Leitores.

Castel-Branco, C. (ed.). (1999). *Botanic Gardens of Ajuda*. Lisbon: Livros Horizonte e Associação dos Amigos do Jardim Botânico da Ajuda.

Castel-Branco, C. (ed.). (2001). *Necessidades: Jardins e Cerca*. Lisboa: Edição conjunta de Jardim Botânico da Ajuda e Livros Horizonte, Lda.

CML. (1852). *Synopse dos principaes actos administrativos da Camara Municipal de Lisboa no anno de 1851*. Lisboa: Imprensa Nacional. Available at: https://hemerotecadigital.cm-lisboa.pt/Periodicos/Sinopse/1851/1851_master/Sinopse1851.pdf.

CML. (1853). *Synopse dos principaes actos administrativos da Camara Municipal de Lisboa no anno de 1852*. Lisboa: Imprensa Nacional. Available at: https://hemerotecadigital.cm-lisboa.pt/Periodicos/Sinopse/1852/1852_master/Sinopse1852.pdf.

Cunha, A.F. (2015). *Análise e interpretação de obras de arquitectura paisagista de Manuel de Sousa da Câmara (1929–1992)*. Dissertação de Mestrado em Arquitetura Paisagista. Instituto Superior de Agronomia, Universidade de Lisboa. Available at: www.repository.utl.pt/handle/10400.5/8499.

Cunha, A.R. (2014). *Os planos de urbanização dos bairros da Encarnação e Olivais. O modernismo, a expansão da cidade de Lisboa e a Arquitetura Paisagista*. Dissertação de Mestrado em Arquitetura Paisagista. Instituto Superior de Agronomia, Universidade de Lisboa. Available at: http://hdl.handle.net/10400.5/6776.

Cunha, A.R. *et al.* (2021). Natural and historical heritage of the Lisbon botanical gardens: An integrative approach with tree collections. *Plants*, 10, 1367. https://doi.org/10.3390/plants10071367.

de Araújo, I. (1962). *Arte paisagista e arte dos jardins em Portugal*. Lisboa: Direcção Geral dos Serviços de Urbanização.

Dolezal, H. (1974). *Friedrich welwitsch: vida e obra*. Lisboa: Junta de Invest. Científica do Ultramar.

Meyer, F.G. (1959). *Plant Explorations—Ornamentals in Italy, Southern France, Spain, Portugal, England and Scotland*. Washington, D.C: U.S. Dept. of Agriculture. Available at: https://archive.org/details/plantexploration349meye/page/n1/mode/2up.

Meyer, F.G. *et al.* (1960). *Gardens of Western Europe Plants & Gardens*. S.l.: s.n.

Palhinha, R.T. (1935). Jardim Botânico da Faculdade de Ciências. In M.C. Ramalho (ed.), *Guia de Portugal artístico: Jardins Parques e Tapadas*. Lisboa: Of. Gráficas da Casa Portuguesa, pp. 15–22.

Morembert, M.T. de (1965). M. Henri Navel (1878–1963). In *Mémoires de l'Académie Nationale de Metz*. s.l. Available at: http://documents.irevues.inist.fr/bitstream/handle/2042/34102/ANM_1963-1964_1.pdf [Accessed 7th September 2022].

Oliveira Júnior. (1873). Chronica Horticolo-Agricola. *Jornal de Horticultura Pratica*, IV, 56–60.

Ramalho, R. de S.L., & Costa Ramalho, M. (1935). *Guia de Portugal Artístico: Jardins, Parques e Tapadas*. Lisboa: Of. Gráficas da Casa Portuguesa.

Rodrigues, A.D. (2020). *O triunfo dos jardins: o pelouro dos passeios e arvoredos de Lisboa (1840–1900)*. Lisboa: Biblioteca Nacional de Portugal (Estudos).

Rosa, M.E.F. (2013). *As árvores ornamentais introduzidas nos jardins de Lisboa: uma perspectiva histórica (séc. 18–19)*. Dissertação de Mestrado em Arquitetura Paisagista. Instituto Superior de Agronomia. Available at: www.repository.utl.pt/handle/10400.5/5661 [Accessed 7th September 2022].

Saraiva, J.H. (ed.). (2004). Gonçalo Ribeiro Telles. In *História de Portugal e Dicionário de Personalidades*. Matosinhos: QuidNovi.

Soares, A.L. (ed.). (2021). *O arvoredo, os jardins e parques públicos de Lisboa (1755–1965), três séculos de um património botânico, paisagístico e cultural*. 1ªEd. Lisboa: Câmara Municipal de Lisboa. Available at: www.repository.utl.pt/handle/10400.5/23418.

Sousa da Câmara, M. (s.d.). Curriculum vitae Manuel Pereira da Nóbrega de Sousa da Câmara. Available at: https://80anosap.isa.ulisboa.pt/wp-content/uploads/2022/05/CVManuel_Sousa_da_Camara.pdf.

Tavares, C. (1967). *Jardim Botânico da Faculdade de Ciências de Lisboa: guia*. Porto: Imprensa Portuguesa.

Tostões, A. (1992). *Monsanto, Parque Eduardo VII, Campo Grande: Keil do Amaral, arquitecto dos espaços verdes de Lisboa*. Lisboa: Salamandra.

Tostões, A., & Guimarães, M. (2013). *Francisco Keil do Amaral, 1910–1975*. Vila do Conde: Verso da História (Arquitectos Portugueses, 11).

Viterbo, S. (1906). *A jardinagem em Portugal: apontamentos para a sua historia*. Coimbra: Imp. da Universidade (I).

7 From historical research to urban ecology

A new perspective on the evolution of landscape art

Ana Raquel Cunha, Maria Manuel Romeiras, Miguel Brilhante, Francisco Castro Rego and Ana Luísa Soares

Introduction

By the 19th century, the awareness of social issues encouraged the development of utopian planning models with a strong emphasis on hygiene. Notable examples include Charles Fourier's *Fourier's Phalanstère* designed as an integrated building blending urban and rural elements (Baridon, 2017) and Soria y Mata's *Linear City* proposing to ruralize the city and urbanize the countryside (Batty, 2022). Scientific advancements such as Darwin's theory of evolution, Pasteur's discovery of bacteria, and Lister's asepsis concept significantly influenced the perception of population health, ushering in an era of heightened hygiene awareness. This shift led to the incorporation of ecological elements like ecology, vegetation, sunlight, and space into urban and land use theories, as outlined in the Charter of Athens (Fallmann & Emeis, 2020). The concept of landscape homeostasis, introduced by Cannon in 1929, underscored the importance of maintaining natural equilibrium despite human intervention. Therefore, the urban planning models began to include vegetation, and, in the mid-20th century in Portugal, Francisco Caldeira Cabral (1908–1992) laid the groundwork of the Portuguese environmental movement, proposing to "work with nature and not against it". Cabral, the pioneer in the practice, study, and teaching of Landscape Architecture (hereafter, LA), argued that "culture improves the quality of the project"; therefore, together with the concept of ecological suitability of the landscape, he always made the cultural interpretation (Cunha, 2014; Jørgensen et al., 2022).

Cabral and his students played pivotal roles in transforming Lisbon's urban green spaces, revolutionizing traditional block gardens with their naturalistic approach. Notable projects such as *Mata de Alvalade*, *Parque do Vale do Silêncio*, *Jardim da Torre de Belém*, and *Jardim da Fundação Calouste Gulbenkian* significantly contributed to the city's aesthetic appeal and environmental quality (Soares, 2021).

Over the past 80 years, LA education in Portugal has witnessed the continued evolution of garden designs, adapting to changing trends and priorities. This study aims to analyze tree collections in Lisbon's gardens during two distinct historical periods, comparing tree inventories from 1929 and 2014, extracted respectively from the "*LX GARDENS project*" and the "*Inventário do arvoredo existente nos vários*

DOI: 10.4324/9781003454007-10

logradouros públicos da cidade de Lisboa e nos viveiros municipais: relativamente a setembro de 1929". Taking as a starting point the data of the 64 gardens inventoried in 2014, this work allowed us to analyze and compare the information of both inventories. Incorporating a narrative that links botanical and historical evolution, the research provides valuable insights into the landscape art's development in Lisbon, serving as a foundation for future LA studies and urban planning endeavours.

Methodology

A comprehensive dataset regarding the tree species present in Lisbon's green spaces in the years 1929 and 2014 was compiled through an extensive literature review (e.g., Câmara Municipal de Lisboa, 1929; Soares et al., 2011; Vasconcelos et al., 2017; Soares, 2021) and the dataset that resulted from the project "*LX GARDENS—Gardens and Historical Parks of Lisbon: Study and Inventory of Landscape Heritage*". This research project (2011–2014) made a historical-artistic and botanical approach to the evolution of 64 historic gardens and parks in the city of Lisbon that were created from the 18th century to the 1960s, as summarized in Table 7.1.

As depicted in Figure 7.1, the first step was to determine which of the 64 gardens inventoried in 2014 were also documented in the 1929 inventory, carried out by CML. A total of 28 green spaces can be found in both inventories (group1). Next, and after careful consideration, it was decided that eight of the 36 gardens exclusively presented in the 2014 inventory (Table 7.1 and Figure 7.1) would not be included in the present study—the four botanical gardens, due to their great botanical diversity; and *Parque Monteiro-Mor*; *Palácio Fronteira*; *Quinta das Conchas*; and *Mata de Alvalade*—due to their tree density (part of the inventory of these spaces only includes a list of species and not their quantification or location, which hinders accurate comparisons with other gardens). So the gardens exclusively present in the 2014 inventory were also 28 (group2).

This dataset allowed us to: a) compare the botanic diversity of 28 public gardens between 1929 and 2014; b) understand, from the botanical and historical perspectives, the "new" green spaces that have enriched the city—some newly established, others converted from private to public gardens; and c) construct a narrative that offers a fresh perspective on the evolution of landscape art in Lisbon and reveals 85 years of tree plantations in the city.

The statistical analysis of the data, by examining variables such as the period of garden construction, or its area, enabled us to track their evolution and perceive the contribution of the "new profession" of landscape architect during a period of significant urban expansion. Consequently, this study encompasses two different approaches: the botanical and the historical ones.

For the botanic comparison, it was necessary to computerize the data from 1929, as well as to compile the information regarding the plants available in the municipal nurseries for plantation in city gardens and streets. It was noted that the 1929 list holds 50 scientific names—30 of them at the species level, 18 at the genus level, and two categories (i.e., "Palms" and "Others" that may include several genera) —while the 2014 list contains all the scientific names at the species level. To allow a comparison between the two inventories, the data analysis in this chapter will be done at the genus level. The final dataset includes 38 tree genera and the two aforementioned categories, as well as, for each genus/category, the distribution and relative abundance in Lisbon green spaces (1929 and 2014) and nurseries (1929). The family and genus names were updated according to the Plants of the World Online database. The data were analyzed in three different ways: a) tree inventory of 1929 (Figure 7.3): comparison of the genera and respective abundance of trees in the municipal gardens (group1) and nurseries; b) tree inventory of 1929 versus 2014 (Figure 7.4): analysis of the most relevant genera in the gardens belonging to group1 (i.e., the 12 genera with the greatest number of trees in each of the inventories); and c) tree inventory of 2014 (Figure 7.5): comparing the gardens in group1 with those of group2 through the analysis of the most relevant species (i.e., the 12 species with the highest number of trees of both groups).

For the historical study, the green spaces included in group1 and group2 (Table 7.1 and Figure 7.1) were broadly contextualized within the emerging landscape and their roles in the city's development. This was achieved using historical literature of recognized relevance, as well as the analysis and interpretation of relevant cartographic information and diagrams: a) current cartography (Figure 7.2) illustrates the geographical distribution of the 64 gardens examined in the LX GARDENS project within the urban structure of Lisbon; b) historical cartography (Figure 7.6) refers to the "Master Plan of Lisbon (1938–48)" created by the urban planner Étienne de Gröer (this plan served as basis for the expansion and transformation of Lisbon); and c) bar charts (Figures 7.7 and 7.8) illustrating the evolution of public green spaces in terms of number of gardens and areas, comparing the 28 gardens described in the 1929 inventory with the 64 gardens studied in 2014. For this last comparison, the gardens were grouped into area categories (<1 ha; 1–10 ha; and >10 ha), the same as those considered in the "Master Plan of Lisbon (1938–48)" (Brito & Camarinhas, 2007). With these resources, the historical study supplies a comprehensive understanding of the historical context, spatial distribution, and evolution of the green spaces under examination.

Table 7.1 List of the 64 green spaces inventoried in the LX GARDENS project in 2014, by typology, garden name, garden code, Lisbon parish and area; gardens included in group1 (28 gardens present in 1929 and 2014 inventories) and in group2 (28 gardens present only in the 2014 inventory) are indicated; the remaining eight gardens were not considered for the present study.

Green space typology	***Garden name***	***Garden code***	***Lisbon parish***	***Area (ha)***	***Group1***	***Group2***
	Alameda D. Afonso Henriques	ALA	*Several parishes: Areeiro, Arroios and Penha de França*	3.3		X
	Avenida da Liberdade	LIB	*Santo António*	4.5	X	
	Campo Mártires da Pátria	BRA	*Arroios*	2.0	X	

(*Continued*)

Table 7.1 (Continued)

Green space typology	***Garden name***	***Garden code***	***Lisbon parish***	***Area (ha)***	***Group1***	***Group2***
GARDENS	*Jardim 5 de Outubro*	5OU	*Estrela*	0.2	X	
	Jardim Avelar Brotero	AVE	*Alcântara*	0.6	X	
	Jardim Camilo Castelo Branco	CAM	*Santo António*	0.1		X
	Jardim Cesário Verde	CES	*Arroios*	0.3	X	
	Jardim Constantino	CON	*Arroios*	0.3	X	
	Jardim da Biblioteca Nacional	BNA	*Alvalade*	5.6		X
	Jardim da Capela de São Jerónimo	CSJ	*Belém*	1.0		X
	Jardim da Estrela	EST	*Estrela*	4.5	X	
	Jardim da Fundação Calouste Gulbenkian	GUL	*Avenidas Novas*	8.0		X
	Jardim da Parada de Campo de Ourique	PAR	*Campo de Ourique*	0.5	X	
	Jardim da Torre de Belém	TBE	*Belém*	6.0		X
	Jardim das Amoreiras	AMO	*Santo António*	0.6	X	
	Jardim das Damas do Palácio da Ajuda	DAM	*Ajuda*	0.3		X
	Jardim de Santos	SAN	*Estrela*	0.4	X	
	Jardim do Campo Grande	CGR	*Alvalade*	13.4		X
	Jardim do Campo Pequeno	CPE	*Avenidas Novas*	1.0		X
	Jardim do Largo da Luz	LUZ	*Carnide*	0.9	X	
	Jardim do Palacete de São Bento	PSB	*Estrela*	2.5		X
	Jardim do Palácio Beau Séjour	PBS	*São Domingos de Benfica*	0.8		X
	Jardim do Palácio Burnay	PBU	*Alcântara*	1.0		X
	Jardim do Palácio Galveias	PGA	*Avenidas Novas*	0.2		X
	Jardim do Príncipe Real	PRI	*Misericórdia*	1.2	X	
	Jardim dos Anjos	ANJ	*Arroios*	0.3	X	
	Jardim Elisa Baptista de Sousa Pedroso	EBP	*Estrela*	0.4		X

(*Continued*)

Table 7.1 (Continued)

Green space typology	Garden name	Garden code	Lisbon parish	Area (ha)	Group1	Group2
	Jardim Lisboa Antiga	LIS	*Estrela*	0.6		X
	Jardim Roque Gameiro	CAI	*Misericórdia*	0.1	X	
	Jardim Vasco da Gama	VGA	*Belém*	4.2	X	
	Jardins do Palácio Nacional de Belém	PNB	*Belém*	2.5		X
	Jardins do Palácio Pimenta	MCI	*Alvalade*	1.5		X
BOTANICAL GARDENS	*Jardim Botânico da Ajuda**	JBA	*Ajuda*	3.8		
	*Jardim Botânico de Lisboa**	JBL	*Santo António*	5.6		
	*Jardim Botânico Tropical**	TRO	*Belém*	7.0		
	*Parque Botânico da Tapada da Ajuda**	AJU	*Alcântara*	100.0		
BELVEDERES	*Jardim 9 de Abril*	9AB	*Estrela*	0.4	X	
	Jardim Botto Machado	BOT	*São Vicente*	0.5	X	
	Jardim do Castelo de São Jorge	CAS	*Santa Maria Maior*	4.0		X
	Jardim do Largo da Graça	GRA	*São Vicente*	0.1	X	
	Jardim do Largo das Necessidades	LNE	*Estrela*	0.4	X	
	Jardim do Torel	TOR	*Santo António*	1.0		X
	Miradouro Alto de Santa Catarina	ASC	*Misericórdia*	0.1	X	
	Miradouro da Nossa Senhora do Monte	MSM	*São Vicente*	0.1		X
	Miradouro da Penha França	MPF	*Penha de França*	0.2	X	
	Miradouro de Santa Luzia	MSL	*Santa Maria Maior*	0.2		X
	Miradouro do Monte Agudo	MMA	*Arroios*	1.2		X
	Miradouro São Pedro de Alcântara	SPA	*Misericórdia*	1.2	X	
	*Jardins do Palácio dos Marqueses de Fronteira**	PMF	*São Domingos de Benfica*	3.7		
	*Parque do Monteiro-Mor**	PMM	*Lumiar*	11.0		

(*Continued*)

Table 7.1 (Continued)

Green space typology	***Garden name***	***Garden code***	***Lisbon parish***	***Area (ha)***	***Group1***	***Group2***
PARKS	*Parque do Vale do Silêncio*	VSI	*Olivais*	8.5		X
	Parque Eduardo VII e Estufa Fria	VII	*Avenidas Novas*	25.0	X	
	*Parque José Gomes Ferreira**	ALV	*Alvalade*	11.0		
SQUARES	*Jardim da Praça Afonso de Albuquerque*	PAA	*Belém*	1.4	X	
	Jardim da Praça D. Luís I	DLI	*Misericórdia*	0.5	X	
	Jardim da Praça da Alegria	ALE	*Santo António*	0.5	X	
	Jardim da Praça das Flores	FLO	*Misericórdia*	0.1	X	
	Jardim da Praça de Londres	LON	*Areeiro*	0.2		X
	Jardim da Praça do Império	IMP	*Belém*	2.7		X
	Jardim da Praça José Fontana	FON	*Arroios*	0.5	X	
"QUINTAS"	*Quinta das Conchas**	QCO	*Lumiar*	14.1		
	Quinta de Santa Clara	QSC	*Santa Clara*	1.8		X
	Quinta dos Lilases	QLI	*Lumiar*	4.3		X
	Tapada das Necessidades	NEC	*Estrela*	10.0		X

* not considered in this study

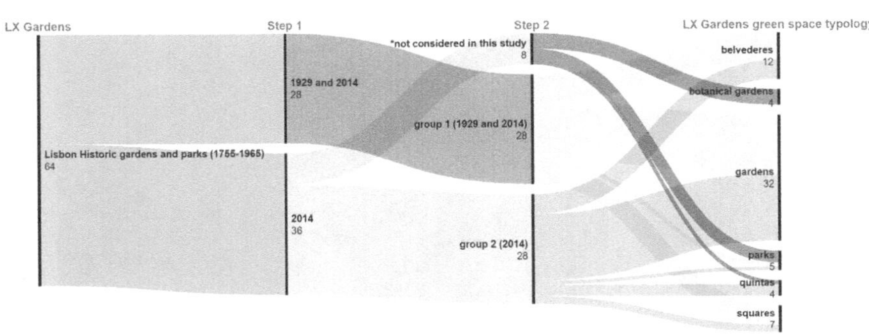

Figure 7.1 Alluvial diagram of the methodology of this work, displaying the flows between categorical variables: how many green spaces, studied in the LX GARDENS project, are in the 1929 inventory or exclusively in the 2014 inventory (step 1); how these green spaces were grouped for this study (step 2); and how they are distributed among the different LX GARDENS typologies. The green spaces represented in purple were not considered in the present study (four botanical gardens, tree parks and one "quinta"). Made withRAWGraphs https://app.rawgraphs.io/.

Figure 7.2 Comparative map showing the location of the 64 gardens studied in the LX GARDENS project in the Lisbon settlement structure: gardens of group1 in green; gardens of group2 in blue; gardens not considered in the present study in purple.

Results and discussion

Botanical study

Tree inventory of 1929

The 1929 inventory indicates 5010 trees in the 28 gardens under study (group1), and 45213 trees in the municipal nurseries. Our results, in Figure 7.3, showed that in gardens the most abundant tree genera/category was *Celtis* (15.37%), followed by *Ulmus* (14.81%), "Others" (13.11%), *Cupressus* (7.25%), and "Palms *s.l.*" (7.07%). It should be noted that despite its abundance, the categories "Others" and "Palms *s.l.*" include several genera that were not identified in this inventory and, therefore, cannot be compared. The least abundant genera, with less than 0.1% of dominance, were *Olea* (0.06%), *Eucalyptus* (0.04%), and *Grevillea* (0.02%). Regarding the municipal nurseries, the most dominant genera were *Cercis* (18.15%), followed by *Celtis* (15.95%), *Broussonetia* (10.48%), *Fraxinus* (9.63%), and *Styphnolobium* (8.80%). Of the 38 genera recorded in the 1929 inventory, 17 were not represented in the public nurseries, that is, despite their abundance, half of the genera were not available in the municipal nurseries. Moreover, it should be noted that the *Crataegus* and *Ceratonia* genera, both native to Portugal, are absent in this inventory.

Tree inventory of 1929 versus 2014

Considering the 28 gardens addressed in group1, in 2014 there were more 475 trees than in 1929. Figure 7.4 shows the 12 genera/categories with the greatest abundances in both inventories, resulting in a list of 17 genera/categories. The genera/categories showing a greater decrease in the number of specimens from 1929 to 2014 were: *Ulmus* (−14.66%), *Cercis* (−4.05%) and *Cupressus* (−3.93%). The large decrease in *Ulmus* and *Cupressus* is related to the pests and diseases often affecting these genera. The decrease in the number of individuals of *Robinia* (−1.89%) and "Palms *s.l.*" (−2.58%) should also be noted, most probably because *Robinia* became considered an invasive species in Portugal and because of the predation of palms by the plague *Rhynchophorus ferrugineus*. The "Others" category presented the greatest increase in tree abundance (+19.05%), probably influenced by LA, which promoted an increased diversity of species, especially of native ones. The other genera with great increases were *Platanus* (+7.03%), *Celtis* (+6.33%), and *Jacaranda* (+3.68%), which may have been used as alternatives, being resistant to the aforementioned pests and diseases.

Tree inventory of 2014

In 2014, the gardens of group1 had 5485 trees of 215 species, and those of group2 had 13810 trees of 257 species. Figure 7.5 shows the most abundant tree species of the two garden groups, globally corresponding to 17 species. In the gardens of group1, the most abundant species were: *Celtis australis* (21.70%), *Platanus*

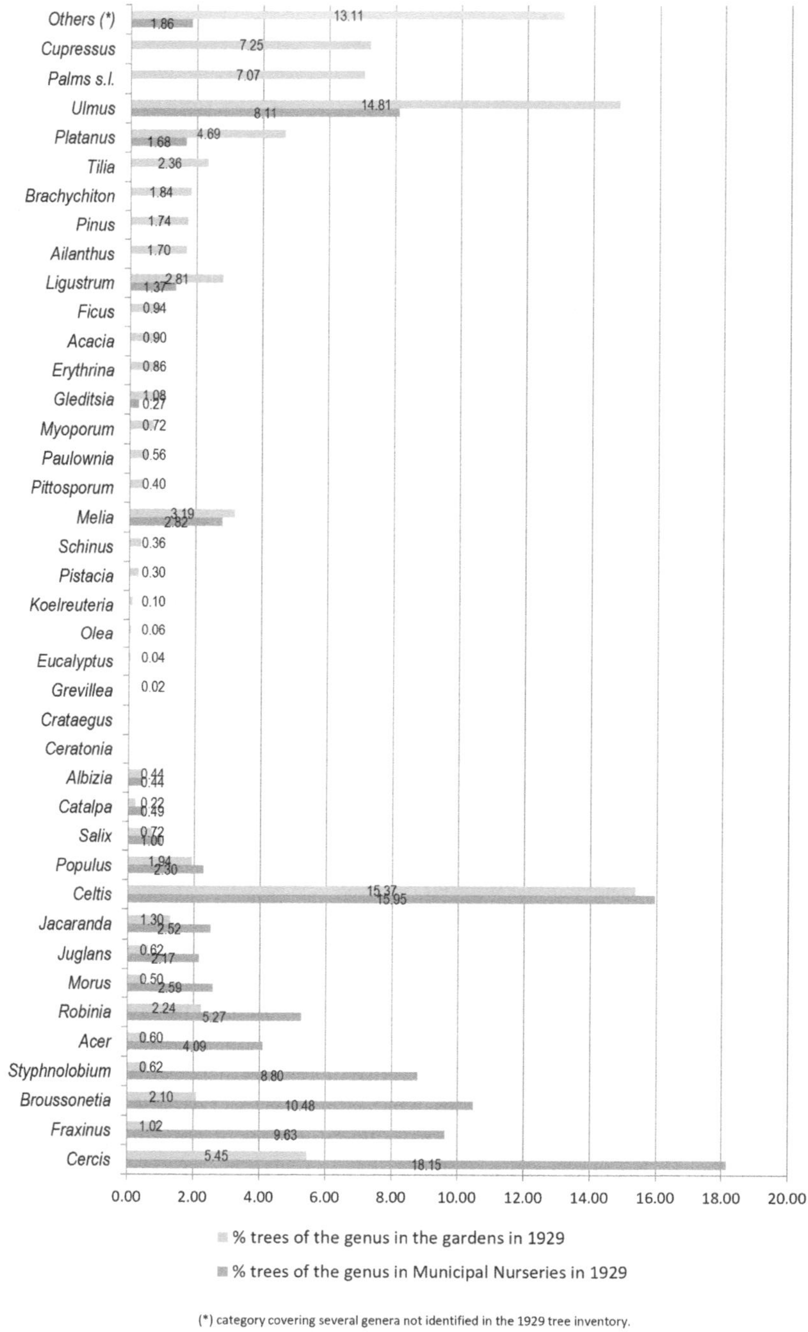

Figure 7.3 Comparison of genera and their relative abundances, in the 1929 inventory, between gardens of group1 and municipal nurseries.

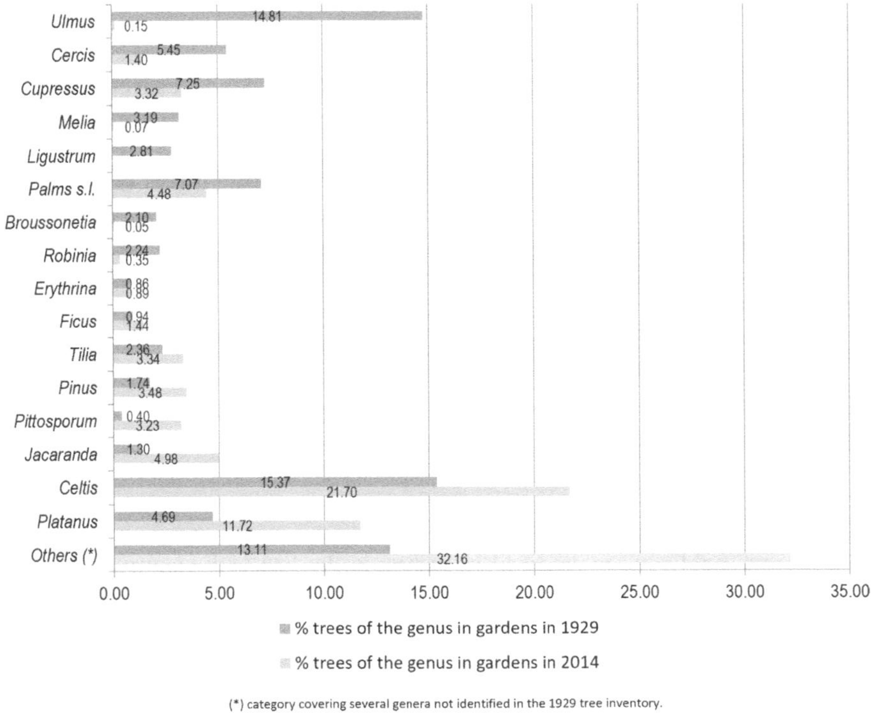

Figure 7.4 Comparison of the most relevant genera present in the gardens of group1 between 1929 and 2014.

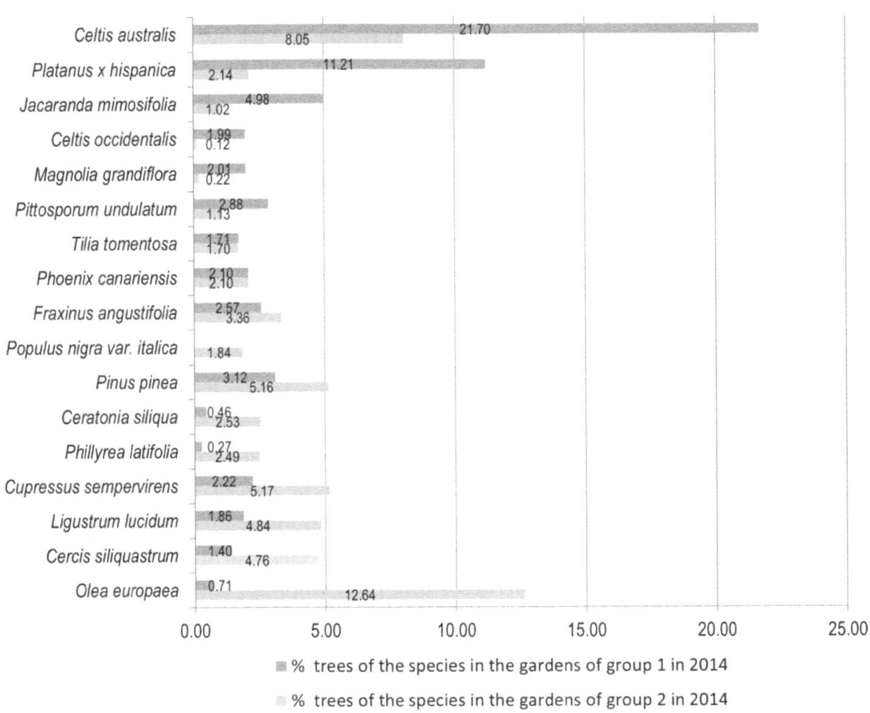

Figure 7.5 Comparison of the most relevant species in the 2014 inventory between gardens of group1 and group2.

×*hispanica* (11.21%), and *Jacaranda mimosifolia* (4.98%). The species with the highest abundance of trees in the gardens of group2 were: *Olea europaea* (12.64%), *Celtis australis* (8.05%), and *Cupressus sempervirens* (5.17%). By comparing the abundance of species between the two groups, the decreased use of the following species is noticeable: *Celtis australis* (−13.64%), *Platanus ×hispanica* (−9.07%) and *Jacaranda mimosifolia* (−3.96%). However, the exotic species have a significant presence in Lisbon, which might be explained not only by the mild climate but also because they are a legacy of the historic gardens. Moreover, group2 had a higher abundance of native species than group1: *Olea europaea* (+11.92%), *Cercis siliquastrum* (+3.35%), *Ligustrum lucidum* (+2.98%), *Cupressus sempervirens* (+2.95%), *Phillyrea latifolia* (+2.22%), *Ceratonia siliqua* (+2.07%), and *Pinus pinea* (+2.04%).

Historical study

The "green" public space emerged in Lisbon in the late 18th century, soon after the big earthquake of 1755. Even a very concise history of the evolution of public green spaces in Lisbon cannot fail to refer the green spaces of private use—e.g., the *Burnay*, *Palmela*, and *Farrobo* houses—which benefited from the services and knowledge of landscape gardeners, horticulturists, and nurserymen and were therefore important places of experimentation and of subsequent dissemination of horticultural, botanical, and artistic gardening knowledge, thus contributing to the botanical and landscape heritage of the city. Also worth mentioning are the public educational spaces, like the four botanical gardens of Lisbon (JBA, JBL, TRO, and AJU), which much contributed and still contribute to that heritage (Cunha et al., 2021; Soares, 2021).

The list of gardens (Table 7.1) and their distribution throughout the urban space (Figure 7.2) show that only 28 of the gardens evaluated in 2014 are found in the 1929 inventory. This does not necessarily mean that these were the only green areas of Lisbon then, but rather that others were private or that CML was not in charge of them. Examples of these are: a) the spaces belonging to the Portuguese Royal House or to high aristocracy, such as the *Real Tapada das Necessidades*, *Real Tapada da Ajuda*, *Real Jardim Botânico da Ajuda*, *Jardins do Palácio de Belém* (actually PNB and TRO), *Jardim do Palácio Burnay*, *Parque Monteiro-Mor*, or *Parque de Santa Gertrudes* (actually GUL); b) lands owned by religious orders that, upon the extinction of the latter, became public property where several green spaces were created, as in the case of the Monastery of *São Bento da Saúde*, where several gardens were created, for example those of *Palacete de São Bento*, *Assembleia da República*, or *Lisboa Antiga*. It could be expected that the 36 green spaces exclusively found in the 2014 inventory would have been constructed after 1929, but this is not so. The oldest ones correspond to private spaces that were later converted into public spaces and were not included in the 1929 for that reason.

Concerning the belvederes, the inventory of 1929 refers five of the eight found in the 2014 inventory (*Alto de Santa Catarina, Graça, Largo das Necessidades, Penha de França, São Pedro de Alcântara)*; for the other three (*Nossa Senhora do Monte, Santa Luzia, and Monte Agudo)* although already existing as public spaces, we only find records of trees for later dates (Soares, 2021). In the first decades of the 20th century, due to the expansion of the modern movement and the new concepts of "city garden", Lisbon begins to incorporate gardens, parks, and squares as a measure to promote hygiene, salubrity, health, and ecology. This development of the "green city" and its distinct green image were only possible due to the political and urbanistic initiative of Duarte Pacheco, the minister of public works, supported by the "Master Plan of Lisbon (1938–48)" (Figure 7.6). The new politics, grounded on severe expropriation and demolition measures, resulted in the expansion of Lisbon and in the transformation of its urbanistic, allowing for the construction of new neighbourhoods, structuring arteries, and public green spaces such as *Alameda D. Afonso Henriques*, *Praça do Império*, *Jardim do Campo Grande*, or *Parque Florestal de Monsanto* (the last is not included in the LX GARDENS inventory but is highly relevant for the history of the town) (Cunha, 2014). The conclusion of the *Parque Eduardo VII* in 1945 should also be mentioned because it explains why this green space is incompletely described (*Parque Eduardo VII—Lagos*) in the 1929 inventory (Soares, 2021).

The appearance of the new "landscape architect" profession in Portugal, in these times of strong urban growth, and their hiring by CML, opened a new stage for the green spaces of Lisbon. Several projects of this period should be highlighted: *Jardim da Praça de Londres*, *Mata de Alvalade*, *Parque do Vale do Silêncio*, *Jardim da Capela de São Jerónimo*, *Torre de Belém*, *Biblioteca Nacional*, and *Castelo de São Jorge*. The project of the *Jardim da Fundação Calouste Gulbenkian*, in the late 1950s, fully assumes the modernity; Vianna Barreto and Ribeiro Telles designed an innovating garden with all the characteristics of the modern movement to host the main office of the Foundation (Andresen, 2003).

Figures 7.7 and 7.8 show the evolution of public green spaces considering their area and number. To compare the 28 studied gardens of 1929 with the 64 gardens studied in 2014, three area classes were considered: <1 ha; 1–10 ha; and >10 ha. In order to obtain more realistic results, the same eight green spaces not included in the botanical study (i.e., the four botanical gardens and four *quintas* and parks, all of considerable size) were used in the analysis of figures Figures 7.7 and 7.8. The public gardens referred in the 1929 inventory were predominantly small (<1 ha: 20 gardens, total area 7.5 ha); the second most frequent class was that of areas between 1 and 10 ha (7 gardens, total area 19.0 ha); and only one green space was larger than 10 ha (25.0 ha). In 2014, the most abundant were the gardens with less than 1 ha (29 gardens, total area 10.4 ha), followed by the gardens with 1 to 10 ha (28 gardens, total area 95.0 ha), and the green areas larger than 10 ha were represented by 7 gardens (total area 184.5 ha). The average area of the public green spaces increased by 60% from 1929 to 2014 (i.e., from 1.8 ha to 4.5 ha).

Figure 7.6 Master Plan of Lisbon (1938–1948).

Source: Plano Director de Urbanização de Lisboa [Material cartográfico]; Autor: Étienne De Gröer; Publicação: CML, 1947; Descrição física: 1 planta, color, 34 x 49 cm; Cota: DP 1272 CMLEO; Arquivo: Gabinete de Estudos Olisiponenses.

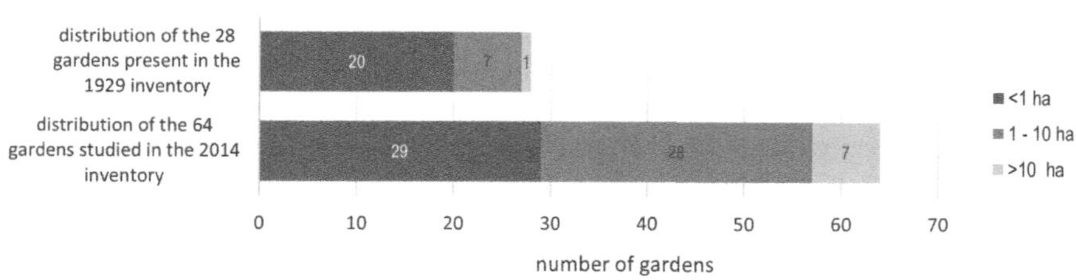

Figure 7.7 Comparison between the number of gardens presented in the 1929 inventory and the 64 gardens studied in 2014, according to three area classes (<1 ha; 1 to 10 ha; >10 ha).

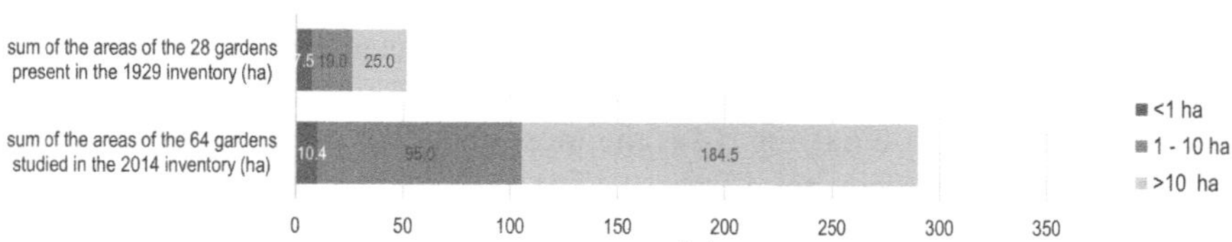

Figure 7.8 Comparison between the sum of the areas of the 28 gardens presented in the 1929 inventory and the 64 gardens studied in 2014, according to three area classes (<1 ha; 1 to 10 ha; >10 ha).

Conclusions

Since the mid-19th century, the interest in "novelties" (i.e., exotic plant species) was reflected on street trees and Lisbon public gardens. These species still have a significant presence in the city, not only favoured by climate, but also because they are a legacy of the historic gardens. Concerning the presence of native plant genera (e.g., *Olea*, *Pinus*, *Fraxinus*), there is an increase from 1929 to 2014 because of LA interventions since the 1950s.

Over the last 80 years of LA in Portugal, professionals have made several important contributions to the city of Lisbon, such as the opening of new streets and the use of native vegetation, as a way to ensure the species' best adaptation and lower maintenance costs, especially in the urbanization plans drawn up for the city's new neighbourhoods (i.e., *Alvalade, Restelo, Encarnação, Olivais,* and *Chelas*).

The distinction of Lisboa Capital Verde 2020 acknowledged the commitment of the municipality to create, preserve, and valorize the city's trees and public gardens. The comparison here presented between 1929 and 2014 confirms that effort in terms of the number and diversity of trees, as well as in the number and area of such spaces.

Finally, the arboreal diversity and the increase of green spaces (number and area), contribute to urban ecosystem services and comfort, preserving and valuing biodiversity, and making Lisbon more sustainable and resilient to climate change.

Acknowledgments

The authors thanks FCT for funding the project "LX GARDENS—Lisbon's Historic Gardens and Parks: Study and Landscape Heritage Inventory" (PTDC/EAT-EAT/110826/2009); and the PhD scholarships to A.R.C. (2020/04824/BD) and to M.B. (UI/BD/151188/2021).

We also thank to Dr. Anabela Valente from GEO/CML for allow us to publish the Master Plan of Lisbon (1938–48).

Reference list

Andresen, T. (2003). Três Décadas de Arquitectura Paisagista em Portugal: 1940–1970. In Andresen, T. (Coord.), *Do Estádio Nacional ao Jardim Gulbenkian: Francisco Caldeira Cabral e a primeira geração de arquitectos paisagistas (1940–1970)*. Lisboa: Fundação Calouste Gulbenkian, pp. 18–97.

Baridon, L. (2017). The Fourierist Phalanstère: Building a new society through architecture? In Mallgrave, H.F. (ed.), *Companion to the History of Architecture*. West Sussex, UK: John Wiley & Sons, Ltd, pp. 1–20. https://doi.org/10.1002/9781118887226.wbcha096.

Batty, M. (2022). The linear city: Illustrating the logic of spatial equilibrium. *Computational Urban Science*, 2(1), 8. https://doi.org/10.1007/s43762-022-00036-z.

Brito, V., & Camarinhas, C. (2007). Elementos para o estudo do Plano de Urbanização da cidade de Lisboa (1938). *Cadernos do Arquivo Municipal—CML*, pp. 163–189.

Câmara Municipal de Lisboa. (1929). *Inventário do arvoredo existente nos vários logradouros públicos da cidade de Lisboa e nos viveiros municipais: relativamente a Setembro de 1929*. Lisboa: Tipografia Municipal.

Cunha, A.R. (2014). *Os planos de urbanização dos bairros da Encarnação e Olivais. O modernismo, a expansão da cidade de Lisboa e a Arquitetura Paisagista*. Dissertação de Mestrado em Arquitetura Paisagista. Instituto Superior de Agronomia, Universidade de Lisboa. Available at: http://hdl.handle.net/10400.5/6776.

Cunha, A.R. *et al*. (2021). Natural and historical heritage of the Lisbon botanical gardens: An integrative approach with tree collections. *Plants*, 10(7), 1367. https://doi.org/10.3390/plants10071367.

Fallmann, J., & Emeis, S. (2020). How to bring urban and global climate studies together with urban planning and architecture? *Developments in the Built Environment*, 4, p. 100023. https://doi.org/10.1016/j.dibe.2020.100023.

Jørgensen, K. *et al*. (2022). Teaching landscape architecture: A discipline comes of age. *Landscape Research*, 47(2), 167–178. https://doi.org/10.1080/01426397.2020.1849588.

Plants of the world online | Kew science (no date). *Plants of the World Online*. Available at: https://powo.science.kew.org/ [Accessed 23th October 2022].

Soares, A.L. *et al*. (2011). Benefits and costs of street trees in Lisbon, Portugal. *Urban Forestry & Urban Greening*, 10(2), 69–78. https://doi.org/10.1016/j.ufug.2010.12.001.

Soares, A.L. (ed.). (2021). *O arvoredo, os jardins e parques públicos de Lisboa (1755–1965), três séculos de um património botânico, paisagístico e cultural*. 1ªEd. Lisboa: Câmara Municipal de Lisboa. Available at: www.repository.utl.pt/handle/10400.5/23418.

Vasconcelos, T. *et al*. (eds). (2017). *Levantamento arbóreo dos Jardins e Parques Históricos de Lisboa*. Lisboa: ISA Press. Available at: www.repository.utl.pt/handle/10400.5/21597?mode=full [Accessed 8th July 2021].

8 Thirty years of research in green infrastructure and landscape planning

Inês Adagói, Natália Cunha, Ana Müller, João Silva, Leonor T. Barata, Manuela R. Magalhães, Selma B. Pena and Luísa Franco

Introduction

The teaching of Landscape Architecture in Portugal began in 1941, with Caldeira Cabral, at the School of Agriculture (ISA), in Lisbon, with the 'Curso Livre' of Landscape Architecture (CLAP) (CEAP-PFCC, 2010). In the 1950s the first projects of Landscape Architecture began to be developed in Portugal, as well as the participation of these professionals in international conferences and scientific meetings (Telles et al., 1992).

The relationship between Landscape Architecture education and Landscape Architecture research is a convergent process that is needed for the success of both. The first landscape architecture research centre, the Centro de Estudos de Arquitectura Paisagista (CEAP), founded in 1953 by Caldeira Cabral and worked until 1975 (CEAP-PFCC, 2010; CEAP, 1968). According to Caldeira Cabral "it is undoubtedly impossible to teach effectively without research and study being carried out alongside pedagogical activity, which is, after all, the life-giving source of teaching itself" (Antunes, 2019). The CEAP provided support for developing international relations, academic and professional work, and the organisation of conferences, field trips and exhibitions, and it was here that the first research projects in Landscape Architecture were developed (Antunes, 2019).

In 1981, the Landscape Architecture degree was created at ISA and the PhD in the fields of Landscape Architecture, Landscape History and Landscape Aesthetics was created in 1989, allowing the development of an academic career in this area (Telles et al., 1992).

The Landscape Architecture Autonomous Section (SAAP) was created in 1993, at ISA, to ensure greater participation and democracy in the management of the teaching of Landscape Architecture. It was an organic unit of teaching, research, development support and provision of services to the community in the fields of Landscape Architecture and Landscape Planning (SAAP, 1993).

Although research in Landscape Architecture was carried out at SAAP, there was a need to institutionalise the research that was being done, hence, in 2003 the Centro de Estudos de Arquitetura Paisagista "Prof. Caldeira Cabral" (CEAP-PFCC) was established (Figures 8.1 and 8.2) under the Foundation for Science and Technology (FCT). The coordination of CEAP-PFCC was by Manuela Raposo Magalhães,

DOI: 10.4324/9781003454007-11

Figure 8.1 The Abegoaria building (ISA) was the workplace of the landscape architecture education and research. The Library building (ISA) has been the workplace of the landscape architecture research centre since 2009.

Source: GBI-TL

Figure 8.2 Photo of the CEAP-PFCC team members (2012). From left to right: Jorge Capelo, Luísa Franco, Manuel Leitão, Ana Müller, Nuno Cortez, Maria Manuela Abreu, Ana Cristina Lourenço, Ribeiro Telles, Sandra Mesquita, Natália Cunha, Selma B. Pena, Teresa Alfaiate, João Ferreira Silva, Andreia Saavedra, Tiago Rodrigues, Gonçalo Abrunhosa, Manuela Raposo Magalhães.

Source: GBI-TL

professor of landscape planning and former student of Cabral and disciple of Ribeiro Telles. The main researchers were Luís Paulo Ribeiro, Teresa Alfaiate and António Muñoz Cardoso, who worked together with young researchers, providing linkages with final-year students and along the landscape planning courses.

In 2013, due to a restructuring order of ISA research units promoted by FCT, the Linking Landscape, Environment, Agriculture and Food (LEAF) was created from the merger of the CEAP-PFCC along with three other R&D units from ISA. In this context, a specific thematic line, the Green and Blue Infrastructure Thematic Line (GBI-TL), was created under the same coordination of Manuela Raposo Magalhães to continue the work developed by CEAP-PFCC. In 2020, this thematic line became coordinated by Selma B. Pena. The main team is composed by landscape architects, working together in a transdisciplinary team of agronomists, phytosociologists, silviculturists, architects, economists, philosophers and other earth or social scientists (Figures 8.2 and 8.3).

The Figure 8.4 frames, in a timeline, the history of the landscape architecture research centre at ISA.

Figure 8.3 Photo of the GBI-TL team members in a participative workshop in Leiria (2022). From left to right: Inês Adagói, Rita Lopes, Selma B. Pena, João Sampaio, Natália Cunha, Manuela Raposo Magalhães, Luísa Franco, Ana Müller, Leonor T. Barata.

Source: GBI-TL

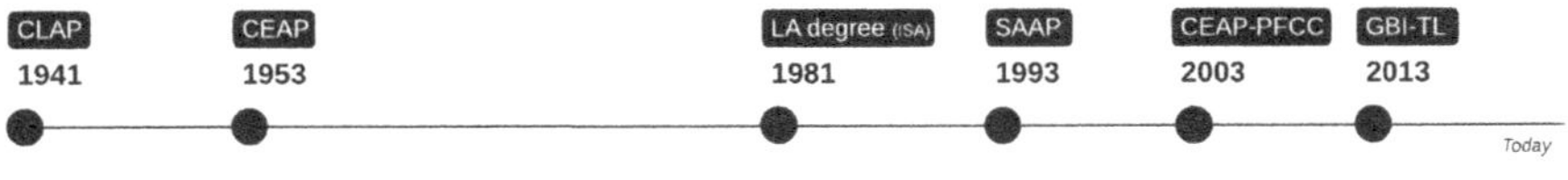

Figure 8.4 Timeline framing the history of the landscape architecture research centre at ISA.

Source: (CLAP: 'Curso Livre' of Landscape Architecture; CEAP: Centro de Estudos de Arquitetura Paisagista, the first centre for landscape architecture studies; LA: Landscape Architecture; SAAP: Landscape Architecture Autonomous Section; CEAP-PFCC: Centro de Estudos de Arquitetura Paisagista "Prof. Caldeira Cabral"; GBI-TL: the Green and Blue Infrastructure Thematic Line).

Landscape planning—the research approach

The landscape planning research brought some concepts that today are fundamental to landscape planning, design and construction. Starting with the concept of *continuum naturale*, brought by Caldeira Cabral but only translated into Portuguese legislation in 1987, in the Environment Framework Law (Law No. 11/87), which recognizes the need for a landscape management policy (CEAP-PFCC, 2010).

The Lisbon Master Plan coordinated by Guimarães Lobato (1958), was the first plan with the collaboration of landscape architects in Portugal (Caldeira Cabral, Ribeiro Telles and Álvaro Ponce Dentinho) and that integrated environmental components, namely green structures in the "Lisbon Green Structure Plan" (Magalhães, 1993; Telles, 1997).

The Lisbon Municipal Master Plan (1993), coordinated by Ribeiro Telles, proposed the Lisbon Ecological Network that constitutes a proposal to adapt the National Ecological Reserve regime to the urban environment (Magalhães, 2001). The concept of ecological network was later included in Portuguese law, in 1999 (Decree law n.° 380/99 of the 22nd September).

Since the early days, the research in landscape architecture was focused in developing studies in landscape planning at several scales, including green and blue infrastructures, which are spatial ecological infrastructures strategically designed and managed to deliver a wide range of ecosystem services.

The research developed in CEAP-PFCC and GBI-TL is based in the "landscape-system" methodology, which defines landscape as a spatial and multifunctional system, constituted by two main subsystems: the structure and complementary areas (Figure 8.5). The structure brings together essential areas, or resources, fundamental to achieve specific landscape planning targets or challenges. The complementary areas are the interstices that result from the network and support more flexible land uses than the structure (Magalhães et al., 2007).

There are different ecological and cultural challenges addressed by the research team, such as floods and droughts, groundwater depth, land degradation, desertification, biodiversity losses, rural fires, forest planning and management, coastal dynamics, agro-food planning, rural-urban linkages, urban heat-island effect, soft mobility, public spaces quality and well-being. As expected, the research developed is in the domain of applied research and aims to contribute to the improvement of public policies and contributes to policy change and knowledge exchange.

The beginning of CEAP-PFCC is characterized by projects funded mainly by local administration, and the scale of intervention was mostly municipal. Some examples are Lisbon's Municipal Green Plan (1991–1993, 2000–2001) (Telles, 1997), Seixal's Municipal Green Plan (1993–1996), Loures' Municipal Green Plan (2001–2003) (Magalhães et al., 2007), Almada's Municipal Landscape Structures (2004–2006) and Sintra's Municipal Green Plan (2004–2008) also where important methodologies were tested, namely the mapping of ecological network and soft mobility plans, using geographical information systems (GIS).

On a regional scale, to mention, Lisbon Metropolitan Area's (LMA) Ecological Network (Magalhães et al., 2007), the Portuguese centre region (SCAPEFIRE

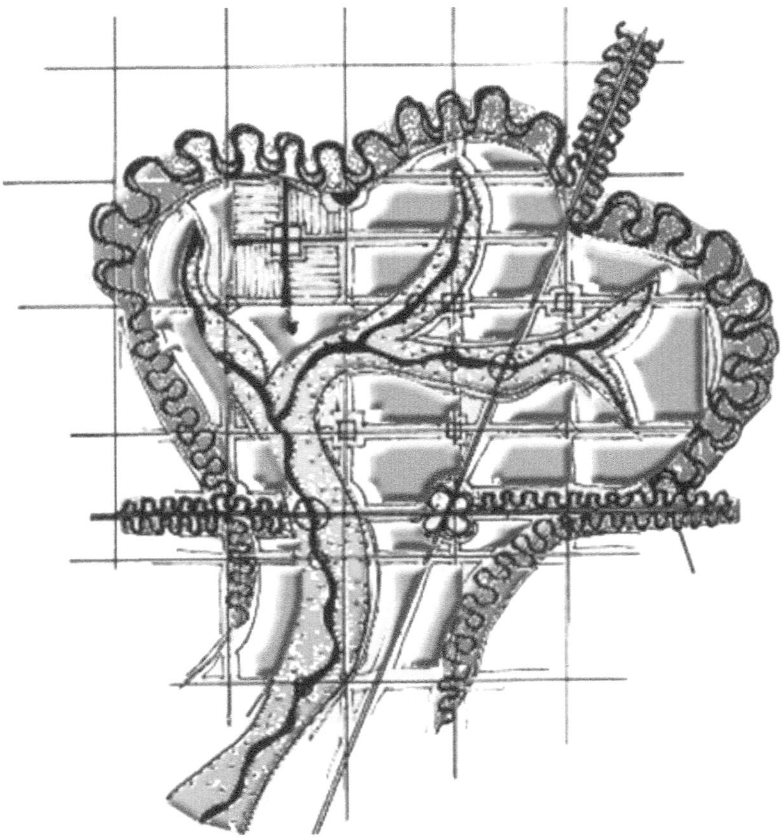

Figure 8.5 Sketch of the concept of "landscape-system".

Source: (Magalhães et al., 2007)

case study) (2019–2023) and, on an intermunicipal scale, Serras da Lousã e Açor (PRGP) (2021–2022) (Magalhães et al., 2022). On a local scale, some examples are the environmental and landscape restoration of the flood plain of Coina River (2010–2011) and Revision of the Biodiversity Action Plan of Monsanto Forest Park 2030 (Lisbon) (2019).

In the transition to GBI-TL two important national projects, financed by FCT, were developed. The first was "National Ecological Network—a proposal of mapping and policies" (EEN) (2010–2013) (Magalhães, 2013), with the aim of delimiting the National Ecological Network. For this purpose, a vast team of scientific experts was assembled for the study and interpretation of the subsystems of the ecological network. The second one was the "Potential Land-Use Ecological Plan. Application to Portugal" (POP), (2012–2015) (Magalhães, 2016) with the aim of proposing an ecological-based potential landscape plan for mainland Portugal and the discussion of development strategies.

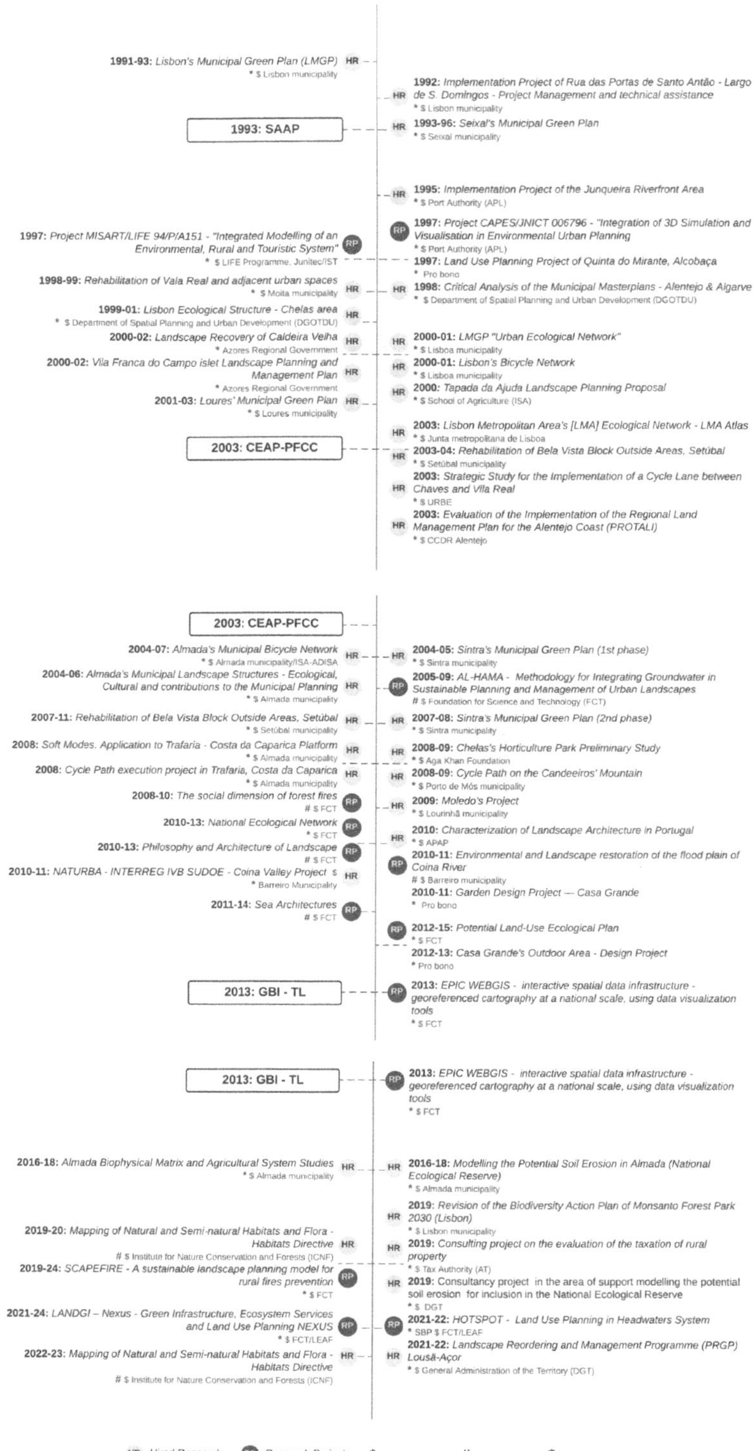

Figure 8.6 Timeline of the projects developed, since the 1990s, in SAAP, CEAP-PFCC and GBI-TL.

These two projects have contributed to enrichen landscape planning courses and the developed cartography was made available through EPIC WebGIS (http://epic-webgis-portugal.isa.ulisboa.pt/). Now it is used by technicians and entities involved in the elaboration of landscape and land use plans, but also in the areas of education, research, management, agricultural and forestry production, nature conservation, cultural promotion and activities that require building.

The SCAPEFIRE project (2019–2023) (https://www.isa.ulisboa.pt/proj/scapefire/) proposes a landscape planning model that contributes to the prevention of rural fires, considering the ecological, economic and social sustainability of the landscape. Since SCAPEFIRE, the GBI-TL has expanded the transdisciplinary approach and research into the following matters: stakeholder participation, ethics of intervention and ecological economics, namely the ecosystem services approach.

Some of these project outputs, are used as pedagogical resources, such as national and international scientific papers, books and chapters and master's and PhD dissertations, and in the scope of the EEN and the POP projects, two major seminars were organised, where the participation of students was encouraged.

The different research projects developed, since the 1990s, are identified in Figure 8.6.

Landscape planning—linking research and education

The CEAP-PFCC/GBI-TL research is closely related to the teaching of landscape planning courses. There are three landscape planning courses, along the Landscape Architecture degree and master's degree at ISA, following a theory which was progressively built based on the developed projects in SAAP, later in CEAP-PFCC and GBI-TL.

The construction of this landscape planning courses theory started by approaching the delimitation of ecological network and cultural network, based on the *continuum naturale* and *continuum culturale* concepts from Caldeira Cabral (Cabral, 1980), further integrated into the "landscape-system" methodology developed in the CEAP-PFCC/GBI-TL (Magalhães et al., 2007). To these was added the concept of soft mobility, green infrastructure, ecosystem services and nature-based solutions.

Through these landscape planning courses, the Landscape Architecture students develop applied knowledge in "landscape-system" planning methodology in a multiscale approach (from local to national) to propose landscape and land use plans and policy changes. The experience-based approach guides students to the need to defining a vision for the future of the landscape dealing with several challenges: ecological suitability of human activities location, rural fire resilience, flood mitigation, soil, water and biodiversity conservation and urban green spaces scarcity, among others.

The landscape planning courses and GBI-TL research projects beneficiated from several partnerships that have been developed through the years, such as with rural sociology courses of ISA, landscape philosophy (School of Arts and Humanities, University of Lisbon), architecture of the city and buildings (Lisbon School of Architecture), ecological economics (Instituto Superior Técnico) and public participation (Nova University de Lisboa). More specifically, in terms of forestry, with Universidade de Trás-os-Montes e Alto Douro (UTAD), the University of Évora, and rural fires with the University of Coimbra, ADAI, municipalities and companies.

Some of GBI-TL's researchers are also professors of the Landscape Architecture degree/master's, and the interconnection between a Landscape Architecture research unit and a Landscape Architecture education programme is crucial, being itself a place of teaching and experimentation par excellence. In the teaching of Landscape Architecture, the research centre is helpful for the GBI-TL's lecturers, keeping the discussion of issues alive and updated, as well for the students who can

Figure 8.7 Workshop in 2006 with between the University of Nürtingen-Geislingen (Germany) and ISA students.

Figure 8.8 Field trip of "Landscape Planning–Municipal Level" course from Landscape Architecture master's at ISA.

Source: GBI-TL

Figure 8.9 GBI-TL team and other landscape architects, at a field trip to Boston Emerald Necklace, at the Fábos Conference on Landscape and Greenway Planning: Pathways to Sustainability, 2013.

do internships, like Erasmus+ or PIIC/LEAF programme or participate in student competitions, PhD programmes, master's thesis, international knowledge exchange, workshops participation (Figure 8.7) or field trips to raise awareness of landscape problems and discuss solutions (Figure 8.8). The landscape architecture research centre (CEAP-PFCC and GBI-TL) is a place of knowledge creation and exchange.

The relationship of research centres with foreign countries has also been through intensive Erasmus and ELEE (European Landscape Education Exchanges) programmes, which began in 1989. The learning experience of researchers from GBI-TL is also done by the different field trips, national and international, and congress participation to exchanges experiences (Figure 8.9).

The PhD students are a living part of the research ongoing in GBI-TL. The different developed theses are identified in Figure 8.10.

Conclusions

The convergence of practice, research and education has improved the landscape architect's capacity to propose intervention methodologies, explore creative design processes at the landscape scale and has enhanced students' and researchers' critical thinking about landscape approaches. The relationship between research and landscape architecture education programs are therefore fundamental to be encouraged.

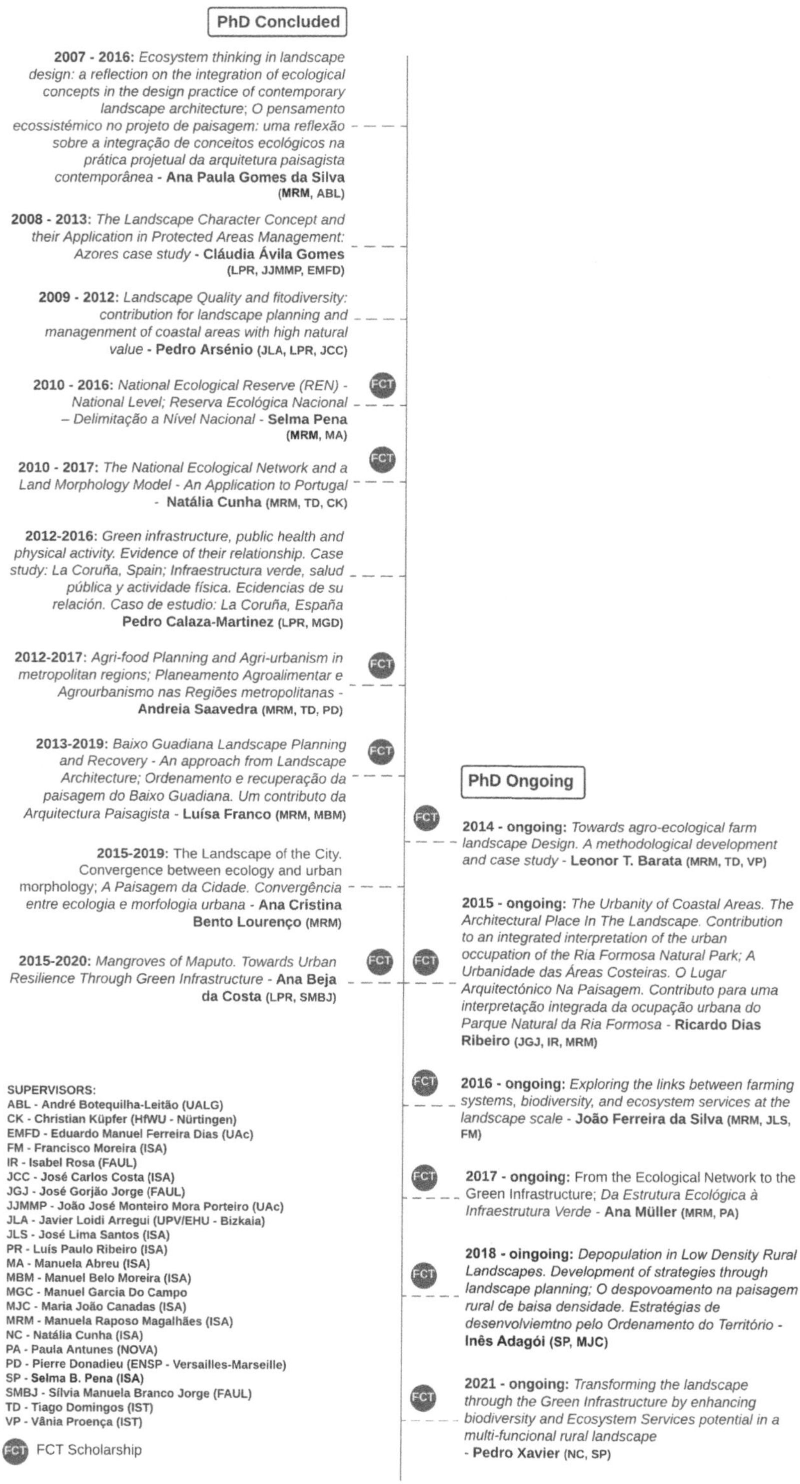

Figure 8.10 Timeline of the PhD thesis developed in CEAP-PFCC/GBI-TL since 2007.

Source: GBI-TL

Over the last 30 years, since the Lisbon Municipal Green Plan (1993), research in green infrastructure and landscape planning has been developed by the landscape architecture research centre (from SAAP, to CEAP-PFCC and GBI-TL). The linkage between the research projects, the landscape architecture courses and students, has contributed significantly to the development of education, research and the landscape architecture profession. And, as Caldeira Cabral said, teaching couldn't be done without research.

Reference list

Antunes, A.C.D.S.A. (2019). *A influência alemã na génese da Arquitetura Paisagista em Portugal*. PhD thesis. Porto, Faculdade de Ciências da Universidade do Porto.

Cabral, F.C. (1980). O "continuum naturale" e a conservação da natureza. *Serviços de Estudos do Ambiente*, 35–54. Available at: http://hdl.handle.net/10400.5/9238

CEAP. (1968). *Memorial. Curso Livre de Arquitectura Paisagista*. Lisboa: Instituto Superior de Agronomia. Available at: https://80anosap.isa.ulisboa.pt/wp-content/uploads/2022/09/Memorial.-Centro-de-Estudos-AP-ISA.-pdf.pdf

CEAP-PFCC. (2010). *Caracterização da Arquitectura Paisagista em Portugal*. Lisboa: Instituto Superior de Agronomia, Universidade Técnica de Lisboa. Available at: https://apap.pt/wp-content/uploads/2017/08/Caracterizacao-da-Arquitectura-Paisagista-em-Portugal.pdf

Magalhães, M.R. (1993). The ecological structure of Lisbon. Ekistics – Athens: The problems and science of human settlements. *The Landscape: Design and Planning I*, 60(360–361), 159–166. ISSN-0013-2942.

Magalhães, M.R. (2001). *A Arquitectura Paisagista. Morfologia e Complexidade*. Lisboa: Editorial Estampa.

Magalhães, M.R. (Coord.). (2013). *Estrutura Ecológica Nacional—uma proposta de delimitação e regulamentação*. Lisboa: ISAPress, pp. 193–200.

Magalhães, M.R. (Coord.). (2016). *Ordem Ecológica e Desenvolvimento. O futuro do território português*. Lisboa: ISAPress, pp. 86–88.

Magalhães, M.R., Abreu, M.M., Lousã, M., & Cortez, N. (2007). *Estrutura Ecológica da Paisagem. Conceitos e Delimitação—escalas regional e municipal*. 1st Edn. Lisboa: ISAPress.

Magalhães, M.R., Pena, S.B., Cunha, N.S., Franco, L., Müller, A. (2022). Transforming rural landscape towards fire resilience and landscape sustainability. In Domingos Xavier Viegas & Luís Mário Ribeiro (eds.), *Advances in Forest Fire Research 2022*. DOI: 10.14195/978-989-26-2298-9_68. Coimbra: Imprensa da Universidade de Coimbra.

SAAP. (1993). *Regulamento da Secção Autónoma de Arquitectura Paisagista* [WWW Document]. Available at: https://arca.isa.ulisboa.pt/saap/1.1_reg_SAAP.htm, [Accessed November 2022].

Telles, G.R. (Coord.) (1997). *Plano Verde de Lisboa. Componente do Plano Director Municipal de Lisboa*. Livro 1. Lisboa: Edições colibri. ISBN 9789728288747.

Telles, G.R., Castel-Branco, C., Fontes, E., Ribeiro, L.P., Magalhães, M., & Saraiva, M. da G. (1992). Cinquenta Anos de Ensino de Arquitectura Paisagista do Instituto Superior de Agronomia 1942–1992. *AGROS*, 1, 42–44.

Part II

A Portuguese context on education and pedagogy in Landscape Architecture

Heading to a comprehensive logical/ rational, aesthetic, and ethic approach

Luís Paulo Ribeiro

The first steps

Landscape Architecture education in Portugal had its foundations on the pioneer work of Francisco Caldeira Cabral, an agronomic engineer, with a broad education on sciences, arts, music, and humanities. He graduates in Landscape Architecture from *Friedrich-Wilhelms-Universität, Berlin*. Cabral immediately understood the significance of such an innovative approach of Landscape Architecture in a country, with the capital of Lisbon, the wanted to become a contemporary European capital. A great number of urban facilities and developments were happening, namely new urban areas, the airport, and the trade port, together with initiatives aiming to prove its significance as the capital of the last European overseas empire such as the imperial Portuguese World Exhibition (*Exposição do Mundo Português*). The National Stadium became its first work, the first Landscape Architecture project built in Portugal, which was truly a reference in ecological design. His alternative location for the new National Stadium, shifting its initial proposed location from the valley flood plain to the valley hillsides, provided new opportunities for natural resource conservation (mainly soil and water), a new park on the valley, and integration of the stadium itself into the hillside, hence contributing to a more balanced and harmonious landscape.

However, to fully achieve the advantage of the profession's implementation, Cabral concentrated efforts in Landscape Architecture education to have colleagues that could establish the new profession fully. Such a task was not easy for the Portuguese in the 40s, as they were dominated by a traditional society and a totalitarian regime. Eventually, the first degree started as an Autonomous Course (*Curso Livre*), approved by the National Ministry of Education, at the Higher Institute of Agronomy (Instituto Superior de Agronomia, ISA), Technical University of Lisbon (Andresen, 2001).

DOI: 10.4324/9781003454007-12

This new profession, aiming to combine art, both in an urban and regional scale, with a wise natural resources management and cultural landscape (understood as the viable evolution of human settlement over the territory) found the appropriate academic context in Lisbon at the Higher Institute of Agronomy.

The first generation of landscape architects (Andresen, 2003) were first educated in this engineering school, as agronomists or forest engineers, under a traditional education, mainly dominated by science, technology, and very much focus on how to increase production. Simultaneously the small number of new students would take classes with Cabral, learning the new landscape design methodologies through a deep understanding of world landscape art history, land cultural evolution, and ecological equilibrium. Although the engineering degrees would focus on production, a good and comprehensive understanding of earth sciences, hydraulics, hydrology, botany, and native vegetation, among others provided the needed knowledge to support landscape design and planning. Education would take place in small classes and in close contact with Cabral, his projects, and works. Gradually, and until the emergence of the first official degrees in the 70s, the first professionals of Landscape Architecture started to participate in the education programs while practicing simultaneously in public entities (mainly Lisbon Municipality and urban development agencies). Therefore, education was performed through a very tight studio environment, where theory was tough in line with design exercises by a faculty that was simultaneously composed of practitioners. This was possible due to the small number of students attending, which enabled a steady formation of professionals that became very competent in demonstrating the value of the profession.

In the first generation of landscape architects, names such as Álvaro Ponce Dentinho, António Viana Barreto, Edgar Sampaio Fontes, Gonçalo Ribeiro Telles, José Marques Moreira, and Manuel de Sousa da Câmara should be mentioned for their efforts and significance in taking ahead Cabral's goal in education. They not only assured the functioning of the Autonomous Course (*Curso Livre*) in Landscape Architecture, but they were also the leading figures, later in the 70s, of the first official degrees at Évora University and the Technical University of Lisbon.

The Autonomous course in Landscape Architecture (Curso Livre de Arquitectura Paisagista, CLAP), during the 60s and 70s, gave rise to what can be named as the second generation of landscape architects. These were the two last decades of the dictatorship, in which internal and international critics of the totalitarian imperial regime were growing with the approach to 1974, the year of the democratic revolution. This second generation, still educated in an agronomy engineering school, the Higher Institute of Agronomy (Instituto Superior de Agronomia, ISA), Technical University of Lisbon, was in line with the critics' movements and in what concerns the development plans and infrastructure of the country. Also, these landscape architects were very much open and willing to understand the world of the 60s and 70s. They fostered more comprehensive development plans, social concerns, and contemporaneity. Their endeavours made them very significant in the Portuguese public entities, and their research and practice were of the upmost influence in landscape design and planning approaches in the public realm, as well as in education with the dawn of democratic regime in Portugal.

This group was further larger and more diversified, with a prominent presence of women that did influence the track into a more participative and comprehensive Landscape Architecture. They took the legacy further, focusing on landscape planning, urban design, and public spaces qualification by working together with architects and other professions, as well as in the public agencies of planning and development. Many of these second-generation landscape architects ended up embracing the professional degrees that opened, later, in Portuguese universities, working together with the pioneers from the first generation.

With the dawn of democracy with 1974's revolution, and the consequent urban development, the second generation of landscape architects, of which women were taking a growing significance, devoted their enthusiasm and believed in the importance that the profession could have in the social and landscape transformations of the country. The focus was very much on municipal planning and resource management at a higher scale and were stimulated by growing social and public concerns. Their participation on development plans, social housing plans, and planning and nature conservation legislative tools widely open the path for landscape architects into public agencies of planning, including the foundation of the National Service for Parks and Reservations, which underwent the strategy for nature and landscape conservation in Portugal, and the General Directorate for Landscape Planning (Direcção Geral do Ordenamento do Território).

For the purpose of this chapter, some of the second-generation landscape architects that took a very significative role in education should be mentioned: Maria Antónia Castro e Almeida (the first Portuguese landscape architect to achieve a doctoral degree), Elias Gonçalves (coordinator of social development plans in the Agency for Residential Areas), Alexandre Cancela d'Abreu (the first doctor in Landscape Architecture), Manuela Raposo Magalhães (which re-opened in the 90s and coordinated the research centre for Landscape Architecture at Lisbon University), Margarida Cancela d'Abreu (director of the Regional Planning Agency for Alentejo/south Portugal), Maria da Graça Saraiva (coordinator of water resources management projects at the Agency for Natural Resources Management), Nuno Lecoq (director of the Natural Park of *Dunas de São Jacinto* in the north coast, first, and later of the Natural Park of *Ria Formosa* in the south province of Algarve), Robert de Moura (director of Natural Park of Alvão in the interior north of the country), Teresa Andresen (president of the National Service of Parks and Reservations, and the first full professor in Landscape Architecture), among others. Most of these landscape architects end up achieving a doctoral degree, mostly in Portugal, becoming full faculty members.

Eventually, the enthusiastic efforts of these two first generations of landscape architects led to the opening of professional degrees in Évora University (late 70s) and the Technical University of Lisbon (early 80s), as well as the creation of the Portuguese Association of Landscape Architects (Associação Portuguese dos Arquitectos Paisagistas–APAP). From now on, the professional title shift from agronomic or forestry engineers with a complementary Autonomous Course in Landscape Architecture (Curso Livre de Arquitectura Paisagista, CLAP), being able to achieve eventually the title of landscape architects.

Evolution and the consolidation of Landscape Architecture education in Portugal

With the creation of the Landscape Architecture degree at two Portuguese universities, landscape education saw its place in Portuguese society conclusively established and officially approved.

The two aforementioned generations, that since the 60s and 70s have opened the way for Landscape Architecture in public agencies, together with the ones that kept their work in private offices, embraced education. They were first invited faculty, and eventually many shifted their professional careers into academia. Therefore, Landscape Architecture education was very much influenced by the professionals practice and skills. This fact had a tremendous impact in education, enabling Landscape Architecture to suddenly become a most influential profession in the country.

In this process, Gonçalo Ribeiro Telles, one of Cabral's first disciples, becomes a well-known public figure. His vision for the significant role of nature protection, ecology, and landscape planning in the country development brought him to integrated early Portuguese governments after the revolution, as the Ministry of Quality of Life. This fact, together with his outstanding park projects, was of most significance in the public acquaintance, diffusion, and recognition of the profession in the country (Carapinha, 2003). Telles also coordinated the university degrees, first in Évora University and after in the Technical University of Lisbon, which was successful in the approval of Landscape Architecture doctorate programs in both universities, hence creating the needed conditions for landscape architects to pursue academic careers. Gonçalo Ribeiro Telles was latter awarded the Jellicoe Prize in 2013 for its tremendous impact in the profession.

The complexity of the Portuguese territory and society after the revolution—in which the return of thousands of people from Africa and Asia and the Portuguese ex-colonies cannot be forgotten—led to the need of more rigorous, efficient, and updated strategies and technologies to support landscape and urban planning.

Manuel de Sousa da Câmara, a first generation landscape architect that had conducted professional work in his private office, was invited to join the degree at the Higher Institute of Agronomy *(Instituto Superior de Agronomia, ISA)*, Technical University of Lisbon, in the early 80s and became its coordinator. His visionary work, thoughts, and reflections were of the upmost importance to the consolidation and evolution of education. He expanded the contacts abroad to address the issues of research, rigor, and the support of computers in landscape design and planning since the early 80s. Rational methodologies and quantitative data collection and processing were some of his powerful contributions to Landscape Architecture education as a means to better reconcile creativity with quantitative/

objective understanding of natural processes—as eloquently defended by Farinha-Marques (Chapter 10) in the article "Nature-based solutions in the teaching of landscape design by Manuel Sousa da Câmara". His contacts, in the early 80s, with the University of Massachusetts in Amherst, namely with Julius Gy. Fábos, strongly contributed to, improved, and evolved these themes in Landscape Architecture education in Portugal, enabling landscape architects in Portugal to advance their educations in master's and doctorate programs overseas.

Câmara's vision in the 80s was also crucial to embrace computer technologies and research strategies to strengthen the professional practice and recognition by the country, as well as to contribute to the outreach of the profession and its education internationally.

This was not a simple and straightforward process. Reactions to ecologic determinism in design, planning, and quantitative approaches, influenced by an academic awakening to postmodern theories, fostered the search for new approaches that intended to flexible design and strongly addressed diversity and creativity in landscape cultural identity—see Carvalho, Marques and Castel-Branco (Chapter 14) in the article "From the concept of Cultural Landscape to its application in conservation policies and higher education in Portugal".

Education witnessed a decrease from the focus on scientific/rational approaches together with an increase of qualitative and heritage approaches in university curricula (Telles et al., 1992)—as clearly demonstrated by Castel-Branco, Azambuja (Chapter 9) in the article "Teaching Restoration of Historical Gardens: Research Through Design Experience".

This process was quite passionate and vibrant, involving different generations and the supporters of different approaches, namely more artistic or more rational, scientific, and technical ones. Significant and stimulating discussions happened inside the profession, taking place in professional meetings and symposiums. In this process, it should be mentioned that the OTAC Symposium (Castelo-Branco, 1986), very much influenced by Câmara's vision, with the coordination of Julius Fábos (from the University of Massachusetts, Amherst, and invited by the landscape planning public agency of Portugal), involved universities, different public agencies, military entities, and a very comprehensive group of leading landscape architects in the country and other professionals dealing with planning, such as environmental engineers, biologists, and geographers, among others.

The entrance of Portugal into the European Union in 1986, fostered the collaboration with university exchange programs and with diverse university programs abroad. The degrees in the universities of Évora and Lisbon engaged in exchange programs of education and research, such as ELEES, Erasmus, and EFLA conferences, among other initiatives. This not only enabled the collaboration of universities abroad, but also the intensification of collaboration between the two Portuguese educational university programs in the country. The year of 1992 saw the creation of the national association of students of Landscape Architecture. This collaboration also influences and contributes to the endurance and international recognition of Portuguese education in Landscape Architecture.

Throughout the 90s, contacts with other European universities (Technische Universität Berlin, Universitat de Barcelona, Nuertingen-Geislingen University, Wageningen University & Research, *Université de Versailles*, University of Sheffield) and American universities (University of Massachusetts, University of California/Berkeley, Harvard University, among others), intensified and strengthened the international collaboration, influencing education through the exchange of educational experiences. This also enabled Portuguese curricula on Landscape Architecture to enlarge their contents and objectives, becoming comprehensive in art, science, technics, and humanities and was eventually approved in Portuguese universities. This was especially significant to a balanced evolution of the Lisbon University programme since it evolved in a school of agriculture and forestry (Higher Institute of Agronomy (*Instituto Superior de Agronomia, ISA*), Technical University of Lisbon), dominated by engineering, technics, and science education.

The creation of the European Foundation of Landscape Architecture led to an agreement between universities with landscape education all over Europe on a common ground in their curricula, enabling the co-validation of the different degrees in Europe. This came in line with the evolution of education in Portugal from a high presence of sciences and technology into a more comprehensive and balanced curricula merging art, humanities, sciences, and technology, that, after all, gave continuation to Cabral's education strategies initiated in the 40s and 50s and reaffirmed by the following generations—as argued by Pinto (Chapter 11) in the article "The Legacy of Three Generations of Architects-Professors: Francisco Caldeira Cabral, Manuel Sousa da Câmara and João Nunes".

The end of 20th century, and into 21st century, Portuguese landscape architects, both faculty and professionals, fully embraced the contacts and collaboration with Europe, Americas, Asia, and Oceania, ultimately stating—despite the extraordinary academic outreach of Cabral in Europe and North America in the 50s and 60s as IFLA President from 1962 to 1966—the presence of Portuguese Landscape Architecture education in the international realm.

The 21st century saw landscape education expanded to three more Portuguese public universities—*Universidade do Algarve*, *Universidade do Porto*, and *Universidade de Trás-os-Montes e Alto Douro*—corroborating the significance and public recognition of the profession and the maturity of its education. All these five universities offer degrees in Landscape Architecture, from undergraduate to master's and doctorate degrees, exchanging their students and faculty worldwide.

Lisbon University, together with Oporto University and Coimbra University, under the coordination of Cristina Castel-Branco, opened the first inter-university doctoral programme in Landscape Architecture and Urban Ecology, with the collaboration of Carl Steinitz/Harvard University. Its impact extends beyond research and doctoral degrees, reaching the overall education and bringing and exposing landscape challenges on education to worldwide issues. Landscape Architecture education does have to deal with Portugal's current expectations on the international scene. Carl Steinitz became the first landscape architect awarded with an *Honoris Causa* by the University of Lisbon (Universidade de Lisboa, ISA, 2023).

Landscape Architecture faculty from Portuguese universities search for active collaborations with universities out of the country, inviting faculty and being invited by an increasing number of countries such as Brazil, China, European Union countries, New Zealand, the United States, and others. Landscape architects educated in Portugal also collaborate abroad, becoming faculty in different universities outside the country, such as in Baltic countries, Czech Republic, Iceland, Italy, New Zealand, and Scotland, among others.

Such a context endows Portuguese landscape education with a significant responsibility in the formation of landscape architects who should be adequately prepared to face the challenges that the current world demands from these professionals, as landscape planners, designers, and managers. They must be able to contribute to the well-being and common good of world community, through their proposals of parks, gardens, ecological infrastructure/greenways, developments, and landscape planning, which have to address globalization trends, the ecological balance of the planet, site specificity resources management, cultural identity respect, and aesthetics with sustainable and innovative development.

Current challenges

It should be recognized that Landscape Architecture education in Portugal always stands on the objective of leaning on curriculum contents that always included arts, sciences, techniques, humanities to support design, planning and management education, avoiding specialization (for instance in design or planning or ecology or historic restauration, just to mention some). Also, it always embraced and included notable professionals as faculty or invited faculty in their education programs. The continuation of this trend and specificity is not easy since public and administrative pressure occurs toward specialization, hence becoming a significant challenge in education.

Currently education challenges in Landscape Architecture education are focused on the preparation of professionals with the adequate skills to warrant and explain the importance of the profession in multidisciplinary teams but also on addressing the increasing competition of other professions in public space, landscape planning, and resources management, among others.

The challenge is now to educate future landscape architects with the needed skills to develop proposals that address the worldwide environmental, social, and economic issues, bringing about solutions that also became, with globalization, local issues—as argued by Marques (Chapter 13) in the article "Breaking siloes and embracing the future: Western and Indigenous lenses in the education of Landscape Architects"; and Costa (Chapter 12) in the article "Landscape Architecture and Urban Ecology research as means for knowledge transfer on urban resilience in Maputo, Mozambique".

Currently, society expects from landscape architects' proposals that significantly promote a more pleasant, beautiful and equative ambience. The goal is to contribute to achieve a territory, a landscape, and a structure of public spaces and parks that consciously aim the public good. This is to be achieved through balanced curricula in logic/reason, aesthetics, and ethics. This must be achieved with education programs that foster the development of self-confidence in its proposals together with an increasing capacity to understand other professions, as well as the world's diversity and the inevitable and desirable capacity to work in multidisciplinary teams.

Reference list

Andresen, T. (2001). *Francisco Caldeira Cabral*. Surrey: Landscape Design Trust.

Andresen, T. (ed.). (2003). *Do Estádio Nacional ao Jardim Gulbenkian—Francisco Caldeira Cabral e a primeira geração de arquitectos paisagistas (1940–1970)*. Lisboa: Fundação Calouste Gulbenkian; Serviço de Belas Artes, Lisboa.

Carapinha, A. (2003). Apresentação. In Caetano, J.O. *et al.* (eds.), *A Utopia e os Pés na Terra—Gonçalo Ribeiro Telles. Exhibition catalogue*. Lisboa: Instituto Português de Museus.

Castelo-Branco, A. (ed.). (1986). *Ordenamento do Território Assistido por Computador—OTAC. Symposium Proceedings*. Lisboa: Direcção Geral do Ordenamento do Território, Ministério do Planeamento e da Administração do Território.

Telles, G., Castel-Branco, C., Fontes, E., Ribeiro, L., Magalhães, M., & Saraiva, M. (1992). Cinquenta anos de ensino de arquitectura paisagista no Instituto Superior de Agronomia. *Agros—Revista da Associação de Estudantes do Instituto Superior de Agronomia*, A.75(1), Janeiro-Junho, pp. 42–44.

Universidade de Lisboa, ISA. (2023). *Honoris Causa—Carl Steinitz*.

9 Teaching restoration of historical gardens

Research through design experience

Cristina Castel-Branco and Sónia Talhé Azambuja

Introduction

In the 1980s and 1990s, the field of Landscape Architecture focused on developing theories and methods for restoring historic gardens. These efforts were supported by universities, heritage protection institutions, and professionals, and they served as a basis for more rigorous and authentic restoration interventions. The International Committee for Historic Gardens, part of the International Council on Monuments and Sites (ICOMOS/IFLA), created the Charter of Florence in 1981, and it was adopted by ICOMOS in 1982, establishing principles for the protection and preservation of historic gardens. In 1992, the cultural landscape category was also established as part of the World Heritage List created by UNESCO, and the Historic Gardens Committee became part of the Scientific Committee for Cultural Landscapes at ICOMOS (Fowler, 2003). In Portugal, theoretical principles based on the Florence Charter were discussed and used to restore historic gardens (Castel-Branco, 1999) and incorporated in the Universities of Lisbon and Évora teaching programmes. In 1989, the Instituto Superior de Agronomia (ISA), School of Agriculture of the University of Lisbon also began teaching the restoration of historic gardens[1] within the course of History of Garden Art II in the Landscape Architecture programme, drawing on the theories and practices of the National Trust of England (Watkins & Wright, 2007; Harney, 2014), a private foundation in the United Kingdom with expertise in restoration, management, and enhancement of properties in England, as well as the concepts of the Florence Charter (ICOMOS, 1981/1982).

Teaching methodologies on historic garden restoration at the University of Lisbon (1989–2022)

At ISA, students in the Landscape Architecture programme learn how to design restoration plans for historic gardens. Homogeneous regions in Portugal were selected where gardens with similar ecological characteristics and the same time periods could be studied. The case studies could vary from historical estates (*quintas*[2]) and their gardens, botanical gardens, urban parks and gardens, sanctuaries, and convents' enclosures, and the methods have been applied and used as 'hands-on' exercises. Each year, students work in a different region and are divided into

DOI: 10.4324/9781003454007-13

teams to focus on the main aspects of the analysis for a historic garden: 1) history; 2) space and composition; 3) botany; 4) hydraulics; 5) surroundings. These lectures aim to give students a comprehensive understanding of the place so that they can interpret it and create effective restoration plans.

Field trips (Figure 9.1) are organised for the case studies to five to seven historic gardens within the selected region (Araújo, 1962; Castel-Branco, 2014, 2017). In these field trips, the students visit the place, take notes, draw, conduct interviews with the owners and head gardeners, and perform several types of surveys (botanic, hydraulic system, uses, functions, etc.). During each academic year, instructors organise visits to some of the most significant historic gardens in the Lisbon region, including Quinta da Bacalhoa (16th century), Fronteira Palace Garden (18th century), Botanical Garden of Ajuda (18th century), Estate of Necessidades (18th–19th century) and Tropical Botanical Garden (18th–20th century) (Castel-Branco, 1999, 2002, 2008, 2017). These field trips are intended to create a visual archive in the student's memory and help learn the characteristics of Portuguese gardens and their adaptations to the natural environment.

Simultaneously, students research in the Portuguese National Archives (Arquivo Nacional da Torre do Tombo), National Library of Portugal, owners' private archives, etc. (Viterbo, 1906; Serrão, 2003; França, 2004). The availability of

Figure 9.1 Field trip to visit a historic 'quintas', Arrábida Natural Park, Setúbal, with students of Landscape Architecture of ISA/ULisboa, and Professors Cristina Castel-Branco and Sónia Talhé Azambuja, Spring 2011.

historical sources, maps, photographs, and reports is essential for the analysis phase and allows a quick synthesis of the historical value of *quintas*, convents, gardens, and parks (Treib, 2019). This traditional research process is needed as most of this information is not on the internet. A thorough survey and technical drawing of one selected built structure, from fountains, pergolas, walls, railings, stairs, and paths in woods, is part of the exercise to present within each group's master plan. Learning how to represent a garden feature by measuring and drawing is complemented by the use of 3D simulation and renderings required for each team of students (from 4 to 6 students) (Figures 9.2 and 9.3). For the entire hydraulic system of the garden (Castel-Branco, 2010), which usually needs restoration, a diagnostic survey is carried out to prepare the design pieces for the works that will lead to the restoration. A vegetation survey is also carried out to identify the species and provide information for labelling the most significant specimens (Azambuja,2009, 2021). This survey serves as the basis for the map of outstanding specimens later highlighted in the brochure that students prepare as part of the final presentation. This will be delivered to the owner and possibly to future visitors (Andresen & Matos Silva, 2022). Students will gain additional learning experience by maintaining frequent contact with the garden owners and gaining valuable insights into the client/designer dynamics, which will benefit their future careers as landscape architects.

Figure 9.2 Master's students of Landscape Architecture making presentations of the analyses of the historic gardens case studies during a studio class, ISA/ULisboa, Lisbon, Spring 2022.

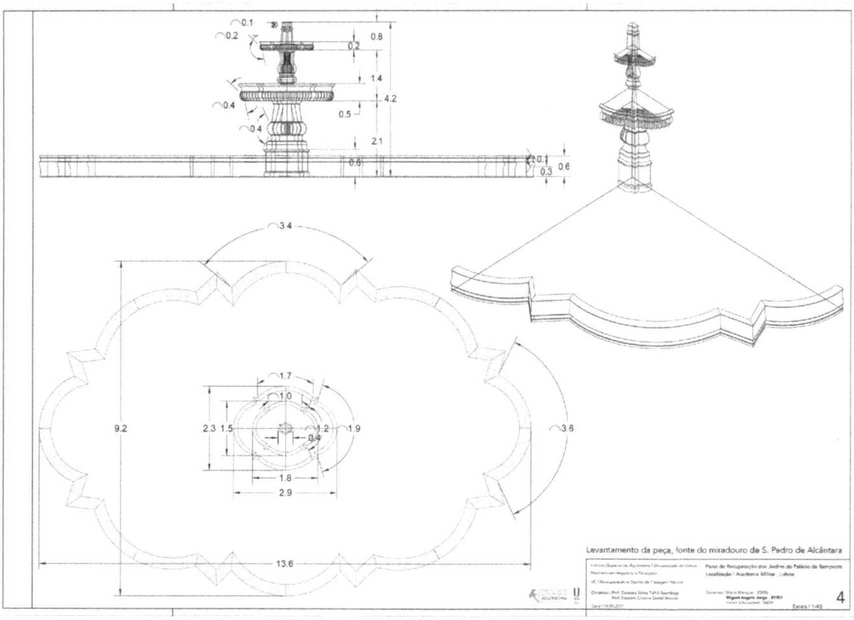

Figure 9.3 Survey of the fountain of S. Pedro de Alcântara Garden, carried out by the students Maria Marques, Miguel Angelo Jorge, and Helen Schwochert under the supervision of Professors Cristina Castel-Branco and Sónia Talhé Azambuja, within the scope of the discipline of recuperation and management of the cultural landscape of the master's in Landscape Architecture at ISA/ULisboa, May 2021.

Following the methodology of the National Trust of England, historical uses of the garden are identified, and surveys of gardens, estates, or heritage sites are recorded using international methods and guidelines (Watkins & Wright, 2007; Fieldhouse & Woudstra, 2000). The Charter of Florence, Charter of Burra, and the Document of Nara define restoration principles for historic gardens, which are presented to the students, who are then encouraged to use them in their proposals (ICOMOS, 1981/1982, 1994; Australia ICOMOS, 2013). The students also follow Carmen Añon Feliú's (1993) four postulates for historic garden restoration 'Respect the Existing Design/Composition', 'Value the Inputs', 'Avoid Dissonances', and 'Find Restoration Solutions within the Garden'. The first postulate emphasises the importance of respecting the existing design of a historic garden and the interventions and maintenance carried out by its various owners over time (Gothein 1928; Jellicoe & Jellicoe, 1975; Girot, 2016). The second postulate stresses the value of all the contributions and layers of time in a historic garden and the need to find a strategy to present them in harmony. The third postulate advocates that we should create an atmosphere in the garden that goes back to the time of its origin to achieve a peaceful and harmonious environment (Castel-Branco, 1999). The fourth postulate advises starting the restoration process with an in-depth study of the garden's history and understanding the motivations behind its creation. By following these

Figure 9.4 Master's class with visiting Professor Timothy Baird from Pennsylvania State University, attended by students and professors of Landscape Architecture at ISA/ULisboa, Lisbon, Spring 2016.

postulates, students can ensure that the plan they propose to restore a historic garden is respectful of its past and sensitive to its cultural and historical significance and meaning (Treib, 2011).

Since 2010, Carl Steinitz's (1990, 2012) six-step method framework for complex and large areas of landscape design, including sequence models, was adapted to garden restoration plans, and the final outputs produced by each group of students follow a multistage structure that includes representation, processes, diagnosis, changes, evaluation, and decision-making. Using this sequence and conducting research while proposing a restoration master plan involves identifying a real-world need for a project in a historical garden, having students conduct research to gather background information on the problem while encouraging them to develop new solutions based on their research findings (Jørgensen et al., 2019, 2022; Castel-Branco & Azambuja, 2020; Azambuja, 2002, 2023). In a studio class, students present their proposals to the class or to a panel of experts and receive feedback and discussion to refine their ideas (Figure 9.2). From 1989 to 2022, specialists such as Ilídio de Araújo, José-Augusto França, Carmen Añón Feliú, John Sales, Thomas Wright, David Jacques, Monica Luengo, Pierre-André Lablaude, Timothy Baird, Marc Treib, Charles Birnbaum, and Carl Steinitz were invited to hold master's classes at ISA/University of Lisbon and comment on proposals for the restoration of historic gardens developed by students (Figure 9.4). Such visiting professors introduce new perspectives during the class's final presentation of the restoration master plan, and students gain important feedback for their work and better understand the

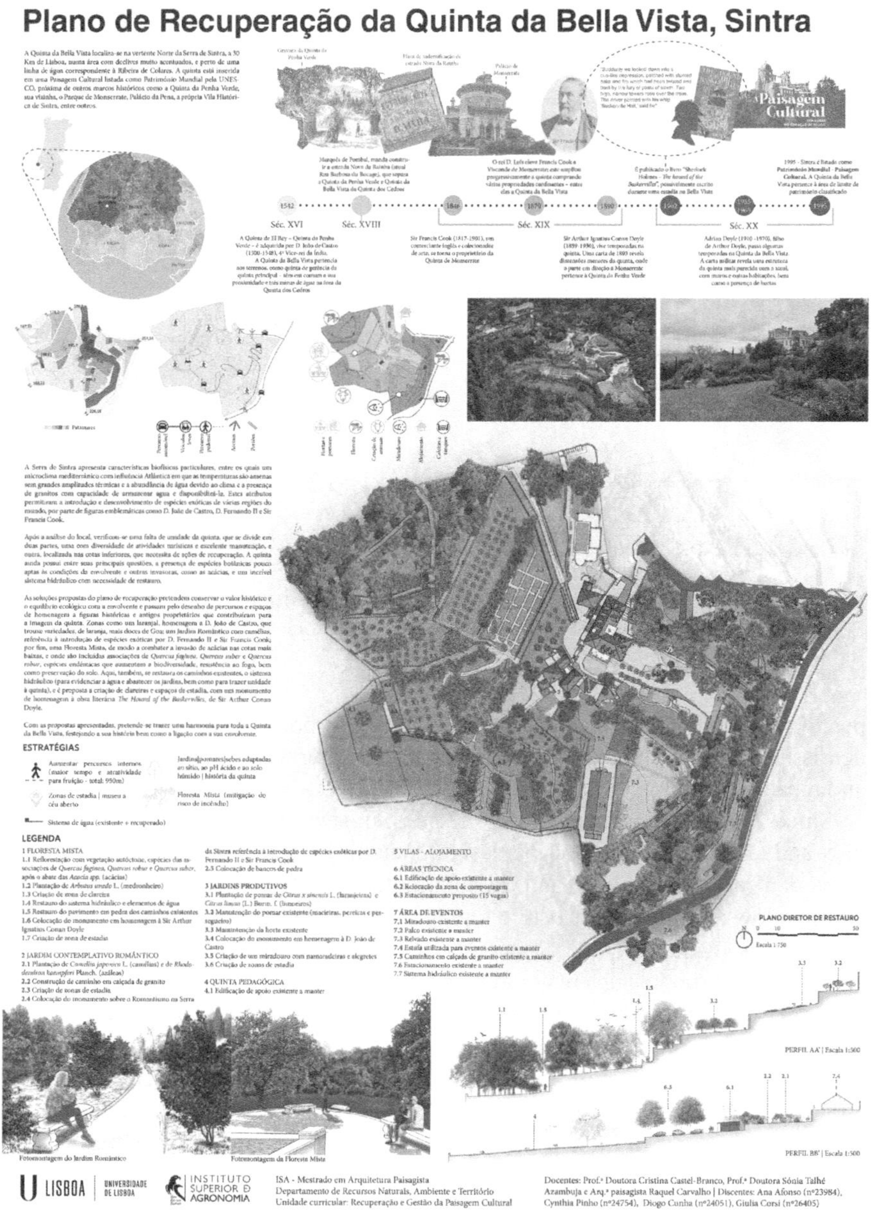

Figure 9.5 Poster with a proposal for the restoration of Quinta da Bella Vista, Sintra, developed by Ana Afonso, Cynthia Pinho, Diogo Cunha, and Giulia Corsi, master's students of Landscape Architecture at ISA/ULisboa, under the supervision of Professor Cristina Castel-Branco, Professor Sónia Talhé Azambuja, and Assistant Raquel Carvalho, 2022.

Figure 9.6 Posters with proposals of restoration for Portuguese historic gardens displayed at the Exhibition 80 Years of Teaching Landscape Architecture in Portugal (1942–2022): Art and Ecology, ISA/ULisboa, 1–9 October 2022, Curator: Invited Professor Sónia Talhé Azambuja, Commissioners: Professor Cristina Castel-Branco and Professor Teresa Andresen.

impact of their efforts. Owners and faculty members are invited, and we believe (Figure 9.5 and Figure 9.6) this approach helps students develop the skills and knowledge needed to restore historical gardens accurately and authentically.

Innovative perspectives explored over the 30 years of teaching practice

Innovative perspectives have been explored over the 33 years of teaching practice, with findings that have improved the result of the exercise. Adding ecology, social issues, and data collected from internet maps, aerial photography, and online research from archives have improved the result of the exercise. Ecology became a requirement, namely in the analysis phase; a thorough landscape analysis of geomorphology, hydrology, ecosystems, climate, natural vegetation, and soils was included. The ecological approach has proven to be very useful for interpreting cultural landscapes. For this reason, we have refined the research question procedure and the result of this teaching exercise over the decades. The diagram from the National Trust of England (Figure 9.7 [Cobham, 1984, p. 5]) has been serving since the 90s as both an academic exercise in teaching and a professional practice of restoration and is recommended to students for preparing the restoration master plan, and it has proved very useful as a 'road map' of good practices in garden restoration.

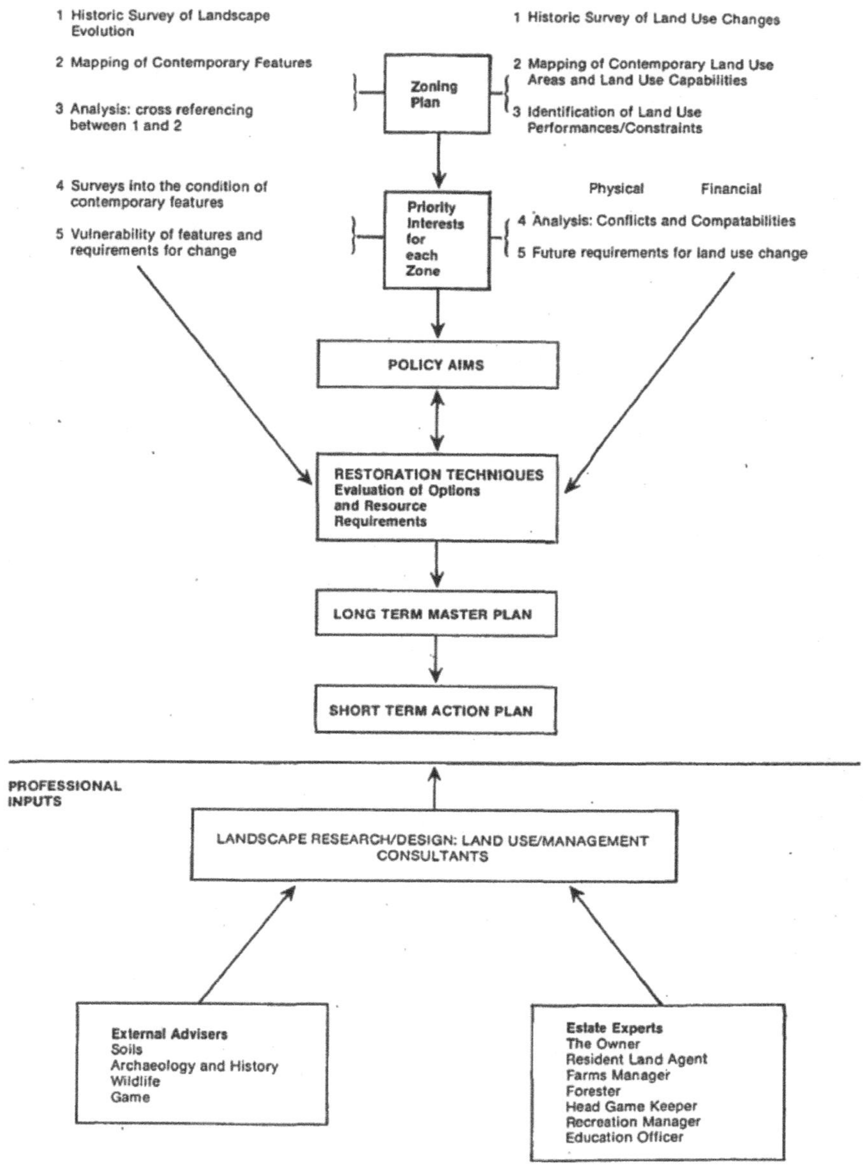

Figure 9.7 The processes involved in producing the landscape restoration plan, National Trust of England methodology.

During the past 33 years, classes have studied approximately 500 historic gardens, and we may have enlarged the scope of the garden restoration research by adding awareness to the ecological processes within the gardens and emphasising the value of interpreting the landscape where the garden was 'sculptured'.

We consider historic gardens 'replacement ecosystems', and though changed by humankind, they still work in balance with nature. They also have proved through two, three, or four centuries that they could withstand the test of time and still display inner sustainable and durable solutions. Analysing and understanding these solutions is an exercise for students that can be used as references for fossil-free outdoor design solutions, inspiring future compositions and processes in garden making. This exercise of finding, understanding, explaining, and finally seeing how it can be used is referred to by students as an important integration of history and heritage into their designer's life.

Two extra learning experiences derived from this 'case study' process emerged, as, during the semester, students maintain frequent contact with the garden owners, gardeners, or people involved with the site and go there by themselves, learning the nature of a first client/designer experience, which is useful for landscape architects. Usually, historical surveys are accompanied by a person responsible for the families' archives, and they are usually asked to come and collaborate in a class where the problems of the garden under study, its history, constraints, and elements that are threatened or in ruin are presented. The availability of elements, maps, photographs, and reports is essential for the analysis phase. Since the internet became an available source, an abundant amount of good data has become available much faster, allowing for a quick synthesis of the case study's historical value without replacing the archives/library research. Following the diagram (Figure 9.7) of the National Trust of England (Cobham, 1984, p. 5), historical uses of the garden are identified, and surveys of the state of gardens, estates, or heritage sites are recorded.

Finally, for the restoration of the garden, 'it is necessary to ally oneself with the mark of whoever designed it, who built it, and whoever maintained it, and respect the mark as an essential factor, as important as the natural processes acting on the site' (Monteiro et al., 1999, p. 143). It becomes more important to show the qualities of the initial design than the design qualities of those who restore the garden or park.

Acknowledgements

The authors would like to thank the students of Landscape Architecture of ISA/ULisboa (1989–2022).

Notes

1 The introduction of this new course was suggested in 1989 by Cristina Castel-Branco and accepted by Gonçalo Ribeiro Telles, who was the course director at ISA. It adapted the findings of her master's thesis, coordinated by John Martin at the University of Massachusetts, Landscape Architecture and Regional Planning master's degree. In 2008 Sónia Talhé Azambuja joined as an invited professor and added her research area as PhD in History of Art.

2 'Quinta' is a property of varied sizes with more than 10 hectares and usually less than 100 hectares, where agriculture is the main use; a manor house creates the nucleus, and usually a garden completes it. Spread within the whole 'quinta', ornamental elements of the garden create an atmosphere of artistic intentions where the views outward are a final complement.

References and further reading

Andresen, T., & Matos Silva, M. (2022). Historic gardens of Portugal: Qualification and structuring of a new touristic offer. In *Proceedings of the Fábos Conference on Landscape and Greenway Planning* (Vol. 7, No. 1, Article 54). Available at: https://scholarworks.umass.edu/fabos/vol7/iss1/54 [Accessed 2023].

Añón Feliú, C. (1993). El jardín Histórico: Notas para una Metodología Previa al Proyecto de Recuperación. In *ICOMOS – Journal Scientifique—Jardins et Sites Historiques*. Madrid: ICOMOS, pp. 312–325.

Araújo, I.A. de (1962). *Arte Paisagista e Arte dos Jardins em Portugal*. 2nd Edn. 2020. Lisbon: CML.

Australia ICOMOS. (2013). Australia ICOMOS. In *Charter for the Conservation of Places of Cultural Significance (The Burra Charter)*. Available at: https://australia.icomos.org/wp-content/uploads/The-Burra-Charter-2013-Adopted-31.10.2013.pdf [Accessed 2023].

Azambuja, S.T. (2002). Real Estate of Necessidades: trendsetter in the art of garden design in 19th-century Portugal. In Castel-Branco, C. (ed.), *Necessidades: The Gardens and Enclosure*. Lisbon: Livros Horizonte, pp. 129–151.

Azambuja, S.T. (2009). *A Linguagem Simbólica da Natureza. A Flora e Fauna na Pintura Seiscentista Portuguesa*. Lisbon: Nova Vega.

Azambuja, S.T. (2021). *Plantas, Animais e Paisagem. Da Iconografia à Iconologia na Pintura dos Séculos XV e XVI em Portugal*. Lisbon: Scribe, ARTIS-IHA da FL/ULisboa, CML.

Azambuja, S.T. (2023). Ações prioritárias na gestão e salvaguarda do Património Paisagístico na próxima década (2023–2033). In Gonçalves, C.V. (Coord. ed.), *Os Próximos 10 Anos do Património Cultural em Portugal: Tendências*. Vila Nova da Baronia: patrimonio.pt, pp. 83–91.

Castel-Branco, C. (ed.). (1999). *Botanic Gardens of Ajuda*. Lisbon: J.B.A./Livros Horizonte.

Castel-Branco, C. (ed.). (2002). *Necessidades: The Gardens and Enclosure*. Lisbon: Livros Horizonte.

Castel-Branco, C. (2008). *The Gardens of the Viceroys—Fronteira*. Alfragide: Oceanos.

Castel-Branco, C. (ed.). (2010). *Water in Portuguese Gardens*. Lisbon: Scribe.

Castel-Branco, C. (2014). *Jardins de Portugal*. Lisbon: Clube do Colecionador dos Correios.

Castel-Branco, C. (2017). *A Índia nos Jardins Portugueses*. Lisbon: Verbo/Babel.

Castel-Branco, C., & Azambuja, S.T. (2020). Restauro de Jardins Históricos: Teoria, Prática e Ensino (1989–2017). In *Actas do 5.° Congresso Internacional Casa Nobre—Um Património para o Futuro*, Vol. II. Arcos de Valdevez: Município de Arcos de Valdevez, pp. 1037–1055.

Cobham, R. (1984). Blenheim: The art and management of landscape restoration. *Landscape Research*, 9(2), 4–14.

Fieldhouse, K., & Woudstra, J. (ed.). (2000). *The Regeneration of Public Parks*. London: Taylor & Francis.

Fowler, P.J. (2003). *World Heritage Cultural Landscapes 1992–2002*. Paris: UNESCO World Heritage Centre.

França, J.A. (2004). *História da Arte em Portugal. O Pombalismo e o Romantismo*. Lisbon: Editorial Presença.

Girot, C. (2016). *The Course of Landscape Architecture*. London: Thames & Hudson.

Gothein, M.L. (1928). *History of Garden Art*. 1st Edn. (in English) Translated by A. Hind. London: J. M. Dent.

Harney, M. (ed.). (2014). *Gardens and Landscapes in Historic Building Conservation*. Oxford: Wiley.

ICOMOS. (1981/1982). *Historic Gardens (The Florence Charter)*. Available at: www.icomos.org/images/DOCUMENTS/Charters/gardens_e.pdf [Accessed 2023].

ICOMOS. (1994). *The Nara Document on Authenticity*. Available at: www.icomos.org/charters/nara-e.pdf [Accessed 2023].

Jellicoe, G., & Jellicoe, S. (1975). *The Landscape of Man: Shaping the Environment from Prehistory to the Present Day*. London: Thames and Hudson.

Jørgensen, K., Karadeniz, N., Mertens, E., & Stiles, R. (ed.). (2019). *The Routledge Handbook of Teaching Landscape*. New York: Routledge.

Jørgensen, K., Stiles, R., Mertens, E., & Karadeniz, N. (2022). Teaching landscape architecture: A discipline comes of age. *Landscape Research*, 47(2), 167–178.

Monteiro, A., Castel-Branco, C., & Fonseca, L. (1999). Restauro e Manutenção: Apogeu e Declínio de um Jardim. In Castel-Branco, C. (ed.), *Botanic Gardens of Ajuda*. Lisbon: J.B.A./Livros Horizonte, p. 143.

Serrão, V. (2003). *História da Arte em Portugal. O Barroco*. Lisbon: Editorial Presença.

Steinitz, C. (1990). A framework for theory applicable to the education of landscape architects (and other environmental design professionals). *Landscape Journal*, 9(2), 136–143.

Steinitz, C. (ed.). (2012). *A Framework for Geodesign: Changing Geography by Design*. Redlands: ESRI Press.

Treib, M. (2011). Must landscapes mean? Approaches to significance in recent landscape architecture. In Treib, M. (ed.), *Meaning in Landscape Architecture and Gardens*. New York: Routledge, pp. 82–125.

Treib, M. (2019). The landscape of landscape history. In Jørgensen, K., Karadeniz, N., Mertens, E., & Stiles, R. (eds.), *The Routledge Handbook of Teaching Landscape*. New York: Routledge, pp. 191–203.

Viterbo, S. (1906). *A Jardinagem em Portugal: Apontamentos para a sua História*. Coimbra: Imprensa da Universidade.

Watkins, J., & Wright, T. (2007). *The Management and Maintenance of Historic Parks, Gardens and Landscapes: The English Heritage Handbook*. London: Frances Lincoln.

10 Nature-based solutions in the teaching of landscape design by Manuel de Sousa da Câmara

Paulo Farinha-Marques

Introduction

Since the introduction of the Landscape Architecture degree in the Portuguese academic circuit by Francisco Caldeira Cabral in 1942, teaching and learning in this area of knowledge have been structured by a combination of ecological, technical, and artistic concepts. Such synthesis has been most relevant in the core subjects of the discipline such as design, planning, and management of the outdoor space and the entire landscape. In these, the guiding principles of the intervention adopt a systemic vision based on the understanding and implementation of the functioning and dynamics of the ecosystems.

Manuel de Sousa da Câmara (1929–1992), agronomist and Landscape Architect, was one of the most remarkable disciples of Cabral. In 1981–1982, he was the person that led the implementation of the five-year full-time programme in Landscape Architecture at the High Institute of Agronomy, University of Lisbon. This was the beginning of an independent degree in the discipline, following the bright steps of the founder, which he conducted for eight years with distinctive ingenuity and innovation.

He advocated that the organisation and design of outdoor spaces should be based on the suitable location and integration of human activities into existing ecosystems, rigorously observing their microclimate, terrain landform, geological substrate, water resources, biological communities, and cultural values.

Notions such as *carrying capacity*, *substitute ecosystem*, *minimised disorder*, *viability*, and *durability* of the intervened spaces oriented the design concepts and the overall detailing. Specifically, these ideas were accomplished in proposals comprising 1) spatial models harmonizing ecological values (diversity), human use (functionality), and aesthetic performance (e.g., mosaic landscapes/clearing-edge-woodland spatial model); 2) in-depth knowledge and handling of vegetation; 3) multi-layered compositions, etc.; 4) adequate water management infrastructure (irrigation, drainage, and impoundment); 5) spatial design that facilitates mechanised maintenance.

The conception of this chapter was based on the reading and selection of information registered in notebooks during the classes of the curricular unit Green

DOI: 10.4324/9781003454007-14

Figure 10.1 Photo of a student's notebook.

Space Construction II (Construção de Zonas Verdes II), taught by Manuel de Sousa da Câmara, during the 4th year of the Landscape Architecture degree, 1985–1986.

Handwritten notes were then organized, interpreted, and rewritten in a new joint text that proposes a single narrative, integrating some of the concepts presented in the classes, all grounded on the interpretation of the functioning of natural-based systems.

Special attention was given to my notebook, which I intensively explored, recalling very relevant and inspiring moments of fundamental learning of Landscape Architecture.

Principles: Carrying capacity and substitute ecosystem

The problem of the carrying capacity of the Portuguese territory is directly linked to the close relationship between rainfall and biomass, in which there are great asymmetries. A rainy northwest (1000–1500 mm/year), temperate and green almost all year round, contrasts with a semi-arid southeast (300–400 mm/year), dry, yellow, progressively closer to a desert situation. An easily communicable ecological indicator can be used in this context, such as a cow. Generally speaking, while in Minho ½ ha of local resources are necessary to sustain a cow, in Alentejo 10 ha are sometimes necessary to achieve this end (20 times more). It is immediately apparent that the viability and even the wealth of a region are directly linked to its ability to generate biomass, and this capacity defines its ability to sustain demanding consumers, such as large herbivores.

As a fundamental principle, the landscape architect plans and designs in harmony with the natural and cultural conditions of the space where he intervenes; for this, he must observe and fully understand its systemic and metabolic context and draw inspiration from it to generate integrated, sustainable, and long-lasting solutions. This knowledge allows us, above all, to take advantage of the functioning of natural systems in favour of human populations and consequently to propose consonant interventions that simultaneously enable human use and maintain the intrinsic qualities and values present.

In this sense, the landscape architect is impelled to adapt the existing situation to a new one that can better support human pressure without degrading it. This adaptation or adjustment can, for example, involve interventions that occasionally influence or modify the microclimate and the topography of the site, the quality of the soil, the circulation of water and the permeability of the substrate, and the type of vegetation cover.

This means that, in most circumstances, humans in high density cannot use intensively and directly the natural ecosystems (or those dominated by living elements); to avoid ecosystem degradation, it is necessary to adapt its spatial organisation and land cover to such pressure in order to guarantee the viability of the living systems.

The pressure of high human density requires the "hardening" of the land, which allows for an increase in the carrying capacity of the spaces; this hardening can be achieved by covering the ground with inert materials (e.g., built structures or paving) or by maintaining a vegetation cover that is resistant to trampling. This requires the practice of irrigation and frequent cutting of the aerial part of the plants, combined with regular fertilisation, reduction of soil compaction, and root pruning (scarification). In addition to these actions, there is also the selection of the most favourable species and the prevention of antagonistic species, the eradication of infesting or invasive species (e.g., some ruderal plants), the removal of waste and general cleaning of the spaces, and the elimination or minimisation of situations that indicate a lack of spatial order.

Carrying capacity is generally a function of the type of ground cover, rainfall and irrigation, and the type of maintenance. It varies as follows: 1) it is higher on paved land than on vegetated land; 2) it is higher on regularly mown and irrigated grassland than on similar non-irrigated land; 3) it is higher on regularly mown grassland than on non-irrigated natural woodland.

This is when the concept of substitute ecosystem arises, which, according to Sousa da Câmara, is an ecosystem modified by human intervention; it adapts the existing ecosystem to increase its carrying capacity so that it can support human use without being degraded, maintaining its most important characteristics such as soil fertility, permeability and aquifer recharge, and long-lasting vegetation cover.

This substitute ecosystem requires a set of characteristics and adaptations to best meet its design as follows: 1) the organisation and zoning of the space, adapting the biophysical suitability of each unit to the proposed activities; 2) a woody and perennial vegetation cover, undemanding in terms of installation and maintenance conditions, easily adaptable, stable, combining fast-growing species (pioneers) and long-lasting species (climax); 3) a space design that facilitates the mechanisation of maintenance operations; and 4) a careful selection of areas to be irrigated to ensure that those of greater intensity of use are the most efficiently irrigated.

Within the scope of this reflection, it is assumed that all ecosystems have a limited carrying capacity; when this is exceeded, their regeneration is compromised, resulting in biological impoverishment and degradation. This physical, ecological, and physiological capacity, especially of green spaces, tends to increase with latitude, with rainfall more evenly distributed throughout the year, and with the possibility of organic matter recharging on the soil surface.

Guidelines for project development

Main principles

Based on functional, aesthetic, and economic objectives, a landscape designer develops clear, simple, and congruent proposals suitable to the site's environmental context. The functional objectives assume the use of vegetation, paving, and equipment is appropriate for human use, always considering the durability of the materials and the problems arising from misuse, such as vandalism. The aesthetic objectives focus on the enhancement of views of special interest, on distinctive aspects of the place, and on philosophical issues that inspire the design. The economic objectives ensure the viability of the idea and its perpetuation, through a correct management strategy that relies mainly on mechanisation and automation.

In terms of the spatial overall composition, the design of parks and gardens should follow an aesthetic philosophy that affirms a simple and uncluttered layout, based on the appropriateness to the topographical conditions and rational use of other natural and cultural resources, which promotes the proper functioning and expression of the natural systems.

Such an approach can be inspired by naturalistic climax landscapes and pre-industrial rural landscapes adapted to human use and optimized management.

Composition and vegetation

The composition defines structuring spatial units whose presence and articulation guarantee a fundamental diversity important for the multifunctionality and viability of the whole. These spatial units are the clearing, the edge, and the woodland. The clearing constitutes the open, illuminated, and sunny space, being essential for extended views and activities of movement. The woodland is the place of intense woody covering, shade, recharge, and conservation of resources. The edge is the transitional space between the two previous ones, the most biodiverse and aesthetically stimulating.

Awareness of the need for sunbathing gave urban green spaces a different opportunity for use than they had had until the 19th century. Shade and paleness were replaced by tanned bodies and a new culture of outdoor health, motivated by the need for vitamin D and physical exercise for all social and age groups. The clearing, wide, unobstructed, and preferably flat becomes an indispensable spatial element in any modern park or garden. The clearing can easily be obtained with the creation of a meadow, a lawn, or a lake, all of which are favourable to mitigate the negative effect of the albedo.

In each case, the recreation opportunities are very different; clearings, with regularly mowed and watered grass cover (e.g., meadows or lawns) create better opportunities for active recreation (movement activities), are more easily accessible to a greater and more diverse number of users and allow for more easily mechanised maintenance (higher carrying capacity).

Clearings with high radiation are complemented by dense woods with a strong shading and soil protection effect; this shading effect is also important in deterring the growth of weeds under their canopies, contributing to greater stability and order in the space. An example of this is the clusters and small stone pine woods used in the urban development of Olivais Sul, northeast of Lisbon.

The transition between the clearing and the woodland is made by a spatial unit of enormous importance, the edge of the woodland. The main advantages of the edge can be summarised as follows: 1) it is an area with favourable solar radiation and a good shadow-sun balance; 2) it is an area of shelter and refuge, where one can see without being seen; 3) it is an area which also offers selective protection against winds, undesirable views, predators, pollution, etc.; 4) it is an area with high biological and sensory diversity (e.g., the area with the highest diversity of blooms). For these reasons, the layout of the paths and the location of the resting areas (sitting areas) or viewpoints should be associated with the edge of the woodland.

The interventions with vegetation should make the best use of the various floristic strata available, proposing multi-stratified compositions that guarantee the simultaneous presence of trees, shrubs, and herbaceous plants. In woodland areas and at the edge (border), three main strata should always be considered, with the perennial flowering herbaceous plants having a special visual and ecological importance, in covering, finishing, and redoing the border, preventing the garden spaces from being invaded by undesirable plants, such as ruderal plants.

The use of isolated trees and shrubs is of little interest due to the difficulty they create for mechanised maintenance, as well as the low ecological and aesthetic performance they confer to the compositions. In the Portuguese climatic situation, with relatively long, dry, and hot summer periods, the use of isolated trees or shrubs does not produce sufficient shade or soil protection. As an alternative, denser multi-stratified woody clumps should be designed, consisting of several individuals and several species, planted at close spacing so that they can quickly intertwine their crowns and create a relevant and conspicuous green mass in the landscape.

Water, landform, and soil

The clearings with permanent water bodies, forming lakes or ponds, are more restrictive to direct human use but greatly contribute to the ecological and aesthetic diversification of the spaces, still maintaining the possibility of certain types of active recreation (e.g., boats, bathing/swimming). Alternatively, if the presence of water bodies is not permanent, the clearings can function as seasonal retention basins of peak flows from torrential rains, which are directed there, constituting "stormwater basins", especially in the autumn-winter period; in the dry months, when precipitation is rare, they can be used as meadows or lawns for recreation.

These temporary water reservoirs are of enormous importance, especially in the hotter and drier areas where the irrigation of green spaces is crucial during the summer. This use of rainwater can also be done more efficiently and for many purposes, through the cistern system linked to the collection of less polluted water from roofs.

In this sense, it is very important to maximise the permeability of the catchments including urban areas, especially in the headwaters, ensuring that natural drainage lines dissipate run-off velocity, increase temporary retention capacity (i.e., increase the time of concentration of flows), and promote infiltration and storage. Low-lying areas must be transformed into green spaces that dissipate and accommodate flood waters.

From the point of view of management and maintenance, the project should integrate these aspects from the initial phases of its design. Thus, it is necessary to overcome constraints that prevent or hinder mechanisation and automation, promoting 1) the definition of homogeneous areas that relate suitability, activity, and maintenance; 2) the creation of gentler slopes, favourable for the operation of machines, irrigation, and drainage; 3) the increase of mechanisable areas (e.g., larger and wider clearings make mechanical cutting and automatic irrigation of

Figure 10.2 Influence of Sousa da Câmara's teaching in one of his students' landscape design interventions (clearing-edge-woodland spatial model): the path along the edge; subtle terrain grading for stormwater retention—Parques da Asprela, Porto.

vegetation more efficient); and 4) the selection of materials more resistant to the weather and human use and less demanding in resources in their maintenance processes.

Sloping areas make it difficult for machinery to operate; this can be significantly improved by grading the land to increase the flatter areas that are more suitable for mechanical maintenance, particularly of the vegetation cover (e.g., lawns, meadows, or grasslands); sloping areas, which are less accessible, can only be made possible with a free-growing shrub and tree cover that is less demanding on maintenance and consequent mechanisation.

In this context, it is also essential to exercise topographic concordance, ensuring continuity of access between all spaces, even adopting sigmoidal slopes, if necessary.

The limitations introduced by the slopes are also reflected in the opportunity for automatic irrigation systems, with the distribution of water under pressure, namely sprinkling and spraying. To optimise automatic irrigation, it is necessary to locate the clearings in areas with low gradients (0–5%) and design them free from obstacles and with dimensions following the range of the irrigation nozzles.

In turn, its flow rate must be in accordance with the permeability of the soil. In general, the water infiltration rate in most Portuguese soils is low, which means that a significant part of the distributed water does not infiltrate, but runs off on the surface, therefore not meeting local needs. This means that the water supply must be slow, making use of low precipitation rates distributed throughout the day, avoiding unnecessary losses by surface run-off, evaporation, or accumulation on the surface (formation of mudflats). Irrigation is the most important factor in the creation and management of green space and a key factor in its viability.

Sustainability in the urban context

It is important to highlight that the effect and success of green spaces in highly artificialized cities and industrial areas depend on urban planning. Fundamentally, this planning foresees the creation of connected green spaces with a homogeneous and balanced distribution in the urban network (*continuum naturale*). This connectivity network should be based primarily on natural drainage lines operating in the open air, which can sustain riparian woodland, an essential element for the existence of robust and efficient green corridors.

The sustainability of the green structure in the urban ecosystem depends on five main aspects: 1) adequate location; 2) precise suitability to the purpose for which they are intended; 3) integration of well-dimensioned irrigation and drainage systems; 4) selection of species well adapted to the ecological and social conditions of the site; and 5) possibility of mechanisation of the main maintenance operations.

This means that, in the medium and long-term, the organisation and conception of green spaces articulate and optimise functional needs, ecological and aesthetic qualities, and socio-economic aspects.

Conclusion

The text presented earlier highlights some of the concepts and solutions professed in Manuel de Sousa da Câmara's course unit Green Space Construction II, 1985–86 (Construção de Zonas Verdes II). It aims at systematizing and disseminating his lucid and sharp approach on the subject of green space design within the scope of Landscape Architecture.

The understanding of the natural systems' metabolism and the necessary change of spaces to accommodate human use through an appropriate spatial organisation and design, in a context of viability and durability (sustainability) underpins the learning of landscape design practice.

For Sousa da Câmara, design solutions must be based on knowledge, rationality, critical thinking, and the aesthetics of simplicity; they are intended for all, universalistic and inclusive, manifesting themselves frontally through objectivity, utility, quantification, promptness, and functionality.

There is still much to explore in Manuel de Sousa da Câmara's class notebooks. At the time, students recorded this information, sometimes without fully understanding its scope, but often feeling its importance and awakening a desire to know more.

Bibliography

Câmara, M.S. (1985). Planeamento das Zonas Verdes no Espaço Urbano. *Arquitetura Paisagista*, 1, 2–6.

Câmara, M.S. (1986). Planeamento da Paisagem na Perspectiva dum Arquitecto Paisagista na Actividade Privada. In *Atas do Simpósio sobre Ordenamento do Território Assistido por Computador—Relatório final do Simpósio*, pp. 30–36.

Castel-Branco, C. (2003). O Vale do Silêncio de Sousa da Câmara: Prioridade à Ecologia. *Jardins*, 1(5), 14–18.

Costa, L.R. (1986). *Caderno de notas das aulas de Construção de Zonas Verdes II proferidas por Manuel Sousa da Câmara*. Unpublished document.

Cunha, A. (2015). *Análise e Interpretação de Obras de Arquitetura Paisagista de Manuel Sousa da Câmara (1929–1992)*. Master Dissertation, Instituto Superior de Agronomia, Universidade de Lisboa, Lisboa.

Farinha-Marques, P. (1986). *Caderno de notas das aulas de Construção de Zonas Verdes II proferidas por Manuel Sousa da Câmara*. Unpublished document.

Ferreira, M.J. (1986). Espaços Verdes, Espaços de Recreio. *Boletim GTH, Gabinete Técnico da Habitação da Câmara Municipal de Lisboa*, 7(50/51), 322–326.

Portela Marques, T. (1986). *Caderno de notas das aulas de Construção de Zonas Verdes II proferidas por Manuel Sousa da Câmara*. Unpublished document.

11 The legacy of three generations of architects-professors

Francisco Caldeira Cabral, Manuel de Sousa da Câmara, and João Ferreira Nunes

Eduardo Costa Pinto

Francisco Caldeira Cabral, his team of "collaborators", and the foundation of the "School"

Landscape Architecture was first taught in Portugal at the *Instituto Superior de Agronomia* (ISA) in Lisbon. The course curriculum was drafted in 1941 by Francisco Caldeira Cabral, founder and head of the "Curso Livre" Landscape Architecture (CLAP). From the beginning, it was a combination of agronomy and fine arts. Like the Agriculture College of Berlin, where Cabral attended the Landscape Architecture programme between 1936 and 1939, the Lisbon course programme essentially repeated the curricular model of the German School (Antunes, 2019). According to Cabral, this duality between agronomy and fine arts had always tended towards agronomy in most courses then taught in Europe because the basis of Landscape Architecture is gardening and horticulture. However, it is still an art, and this is how he explained this duality: "Our trade is gardening, but our art is landscape architecture" (Cabral, 2009).

It was his wish to bring this artform to Portugal that led Cabral to create the CLAP in 1942 and teach it for 33 years. During his professional life, not only did he teach, but he also carried out over a hundred projects in Portugal and abroad, both private and public commissions, some of which with his inner circle of "collaborators" (Cabral, 2009), such as the Belver Dam (1952) and the Montijo Park (1956), in collaboration with Edgar Fontes; the Sugar Cane Irrigation Project in Incomati, Mozambique (1956), with Ribeiro Telles and Viana Barreto; or the remodelling of Avenida da Liberdade (1956–60)[1] and the Lisbon German School gardens (1962)[2], with Ribeiro Telles. Therefore, the singularity and originality of his legacy is not limited to the creation of the first School of Landscape Architects in Portugal. In addition to his work as an architect and professor, we must also credit his scientific activity and research, the construction of a body of theoretical work and the production of knowledge. Examples are the conferences and scientific meetings in which he participated; the creation of the Centre for Landscape Architecture Studies (CEAP) in 1953; or study trips with his CLAP's students, in Portugal and abroad (Antunes, 2019).

DOI: 10.4324/9781003454007-15

Moreover, his published texts constitute an important written legacy (Pinto, 2014). They helped establish notable concepts such as "*continuum naturale-culturale*", which he imported from Germany, or "issues that are still debated today: environment, quality of life, planning, protected areas, zoning, nature reserves, etc." (Barreto, 2009).

In Europe, the CLAP was the fourth course of its kind to be implemented. For Barbara Birli (2016), "Cabral may be seen as first example of knowledge transfer from one country to another and an early form of transnational cooperation in landscape architecture education with one university providing knowledge for the foundation of another university". Despite his extensive work in teaching, project practice or research, Cabral's international recognition reached a new height when he became the fifth president of the International Federation of Landscape Architects (IFLA), between 1962 and 1966.

Domestically, his recognition as "a true Master"[3] emerged in the first generation of Cabral's students at ISA, who were also, in the words of Landscape Architect António Viana Barreto, his "extended family"[4]. This generation became a notorious team of "collaborators" of landscape architects/agronomists, who followed in his footsteps in a country under the rule of a dictatorial regime in which almost everything was needed: landscape planning, environmental law, projects on public green spaces, etc. According to Viana Barreto, with these interpreters and from these new practices an "authentic revolution" began: "in the gardens, nurseries, production of flowers, in the trade of plants with other countries, gardening techniques, gardening training, . . . at Lisbon City Hall".[5] The Master-disciple relationship between Cabral and the first generation of landscape architects in Portugal was essential for the effective recognition of the profession in Portugal and for the creation, decades later, in 1981, of the Degree in Landscape Architecture at ISA.

Manuel de Sousa da Câmara, the atelier GPSC, and the course restructuring of 1981

The number of Landscape Architecture curricula in Europe grew exponentially, from seven courses in the 1940s, to 51 courses in the late 1970s (Birli, 2016). In Portugal, between the 70s and 80s, there were important changes in how Landscape Architecture was being taught. In the aftermath of the Portuguese dictatorship, in 1975, during a troubled political time, Cabral passed the leadership of the CLAP on to Sousa da Câmara, who held the position until 1978, and was later replaced by forestry engineer Luís Soares Barreto (Antunes, 2019). From September 1979 onwards, Professor Carlos Baeta Neves was appointed coordinator, and José Marques Moreira, Joaquim Elias Gonçalves, and Sousa da Câmara were responsible for teaching CLAP's specific subjects (Telles et al., 1992). In 1981, this free course, taught at ISA since 1942, was finally replaced by the degree in Landscape Architecture. That same year, at the University of Évora, the degree in Biophysical Planning and Landscape Architecture was created by Ribeiro Telles, thus extending the teaching of the profession to other regions of the country.

The 1981 the study plan of the degree in Landscape Architecture at ISA continued to be strongly based on agronomy. Indeed, the first 2 years of both courses were identical.[6] As to the course curriculum, it was mainly designed by the figures of Sousa da Câmara and José Marques Moreira (Telles et al., 1992). In Teresa Andresen's words, the "CLAP was greatly influenced by the personality of Sousa da Câmara, who was clearly the leader of the teaching staff". His leadership was reinforced "by his knowledge, intelligence, generosity and pioneering in the use of computers applied to Landscape Architecture" (Antunes, 2019) and to planning.

The importance of Sousa da Câmara for the teaching of Landscape Architecture at ISA can be traced in correspondence from figures such as Julius Fábos and Baeta Neves (Castel-Branco, 2022), or in the words of Cabral: "as a special activity we also want to highlight the original Models for projects, using Automatic Calculation, an activity which he pioneered in our country" (Cunha, 2015). We find this same recognition not only from his colleagues, such as Ilídio de Araújo[7], Ribeiro Telles, or Edgar Fontes (Telles et al., 1992), but also in the testimonies of the new generation of landscape architects that the "Master" taught. In a letter addressed to the president of ISA's Scientific Council in 1985, signed by a group of students, among them Cristina Castel-Branco, Luís Paulo Ribeiro, and João Ferreira Nunes (today architects-professors in the same institute), they wrote: "a unique professor, highly regarded at home and abroad, whose knowledge, experience, pedagogical qualities and scientific prominence we appreciate so much".[8]

Sousa da Câmara's professional career, spanning over 22 years before he joined the CLAP's teaching staff, and especially his work for the Lisbon City Hall, allowed him, according to Teresa Andresen, to know "in depth the problems of the city" at a time when the "Lisbon City Hall was a continuation of the CLAP" (Antunes, 2019). Highlights from this time are the projects for *Vale do Silêncio* Park, with Álvaro Dentinho, and for the new access routes to the *25 de Abril* Bridge, which earned him public praise in the "Municipal Gazette", in 1969 (Cunha, 2015). After 1971, he developed projects as a liberal professional, and then founded his atelier *Gabinete de Projectos Sousa da Câmara* (GPSC) in 1982. Meanwhile, he developed several studies and projects, of different scope and scale, e.g. "Recovery of Gardens and Historic Places", "Environmental Impact Studies and Regional Planning", and "Landscape Planning" in Portugal, Macau, Luanda, and Saudi Arabia (Cunha, 2015).

In a time of great transformation, of great hope, where everything was possible and everything was necessary, just as important as the projects built was his legacy as "teacher, researcher and pioneer in issues such as the association of aesthetic simplicity with sustainability or the introduction of informatics as a planning tool" (Cunha, 2015).

João Ferreira Nunes, the atelier PROAP, and the internationalisation of the Lisbon School

In Europe, the 1990s and the first two decades of the 21st century corresponded to a phase of maturity in Landscape Architecture education. There were over 77 such courses by the end of this period (Birli, 2016). In Portugal, the "Bologna Process" was put in motion in 1999, and it was a reflection of the intergovernmental reform of higher education in Europe. That same year, the teaching of Landscape Architecture extended to the Universities of Trás-os-Montes and Alto Douro (UTAD) and Algarve (UALG), and in 2001, to the University of Porto (UP). This fact significantly contributed to the exponential growth of the annual number of graduates in Landscape Architecture in Portugal, rising from 159, between 1995–2000, to 718, between 2005–2010[9]. In this context, ISA's curricular programmes underwent further restructuring, which culminated in the implementation, in 2009, of a new bachelor's + master's degree (and also a PhD programme) that materialised the Bologna principles. In the new curriculum of the integrated master's degree at ISA, the specific subjects of Landscape Architecture (such as project, landscape planning, and Landscape Architecture theory) gained structural importance, now making up 45% of the total subjects taught over a 5-year (3+2) period[10]. Moreover, the first 2 years of the bachelor's syllabus changed and became different from those of agronomy.

At the beginning of the 21st century, the recognition of the Portuguese School of Landscape Architecture became an effective and transversal reality in the areas of teaching, professional practice, and research. According to Ribeiro Telles, "the awards they have received, the competitions they have won, attest" to the quality and value of Portuguese landscape architects abroad, "and justify the demand for Landscape Architecture courses in our universities by students from schools qualified to attend the "Erasmus" programme and undertake post-graduate and doctoral studies" (Telles, 2009). At the turn of the century, part of that recognition in the particular case of the Lisbon School was due to a new generation of professors-architects who repeated the model of their Masters, namely Cristina Castel-Branco (ACB Paisagem/ISA: 1989–present), João Ferreira Nunes (PROAP/ISA: 1991–present), Teresa Alfaiate (*Margem–Arquitectura Paisagista*/ISA: 1992–present), or Luís Paulo Ribeiro (Topiaris/ISA: 1999–present). Following Cabral or Sousa da Câmara, that model was based on the interdependence between teaching/research and professional practice.

Of this new generation of "ambassadors" of the Lisbon School, João Ferreira Nunes and the atelier "PROAP–Studies and Projects of Landscape Architecture" play a decisive and remarkable role in Portugal and abroad. For Ribeiro Telles, PROAP's projects are not only "significant references of the School of Landscape Architecture" but also "innovative and creative, they develop the art of creating the Landscape, answering society's current problems, coherently valuing the ecological, environmental and cultural circumstances of places, and creating the beautiful in the image of Man" (Telles, 2010).

In an increasingly globalised work environment, João Ferreira Nunes and PROAP's projects, as well as his teaching and research, are becoming a prime example of the internationalisation of the Portuguese Landscape School. During its 35 years of existence, PROAP, which has offices in Lisbon, Luanda, and Treviso, has developed projects on four continents. These projects are the result of a large multidisciplinary team of landscape architects (many of them formed by the

Lisbon School and former students of João Ferreira Nunes), architects, engineers, photographers, artists, etc.[11] The international competitions PROAP has won, with projects such as the Antwerp Waterfront, in Belgium, and the Valdebebas Park, in Madrid, or the five nominations for the renown "Rosa Barba" Award at the Barcelona International Biennial of Landscape Architecture, have consolidated the atelier PROAP as an international reference in the field of project practice and investigation. At the same time, the current academic work of João Ferreira Nunes as scientific director of the University of Cagliari, Italy, as professor at the Academy of Architecture in Mendrisio, Switzerland, or as guest lecturer/speaker in conferences all over the world, continues to influence the new generations of architecture and Landscape Architecture students. Moreover, the publication of books, monographies, articles, and dozens of texts demonstrate that the importance of PROAP is not limited to projects built. An analysis of their theoretical work shows clearly that the art of "building landscape, through a project of landscape architecture", by João Ferreira Nunes and the atelier PROAP, is about "positively manipulating natural metabolic factors, adding to them a poetic, ideological and artistic sense" and "transmitting fertility, productivity and diversity, with the awareness of the cultural importance of this gesture" (Nunes, 2010). The continued practice of engaging in "Lost Competitions" (Nunes et al., 2011) in Landscape Architecture is also one of the trademarks of atelier PROAP since its creation, which transcends the competitions won or the projects built. This represents a unique example of project research within the Portuguese context, with innovative results regarding theory and the way of "communicating" in Landscape Architecture, the representation techniques, or the team and multidisciplinary work.

Conclusions

It is not only an impossible task, but also not the purpose of this chapter, to reduce the "Lisbon School" to three architects. When we talk about each of these architects, we are also invoking their Masters, their students, their ateliers, and their teams of collaborators, without forgetting that each of them has undeniably left a mark and continue to influence a history of longevity, multiple visions, and the singularity of a "School" now celebrating its 80th anniversary. Over these 80 years, while it is true that the teaching and practice of Landscape Architecture in Portugal emerged under a dictatorial regime and a cultural and social model dominated by white men, it is also true that much has changed and continues to change. As major paradigm shifts, we would like to highlight the following: change of political regime; professionalisation of the practice and teaching of Landscape Architecture; environmental, landscape, and landscape planning legislation changes; increase in the number of courses taught at universities in Portugal; the percentage of female graduates[12] and female landscape architects who reconcile teaching with practical work in ateliers; and the scale of the job market, which is increasingly globalised.

We focused our research on these three professors-architects because we wanted to question the existence of a model: as a *modus operandi*, founded in the nature of the relationship "Master-disciple" shared by these three generations of landscape architects; as a possible "seal of quality"; as a factor of differentiation and identity of a School, at a time when teaching Landscape Architecture has become a reality both more necessary and more global.

The results can be divided into three aspects: the interdependence between professional practice and research/teaching at ISA; the relationship between "Master-disciple" and among "collaborators"; and the international recognition achieved by the Masters, who established new horizons for future generations of landscape architects.

In all three case studies, teaching, professional practice, and research are interdependent activities, conducted in parallel and complementarily. This fact was only possible thanks to the "generosity, optimism and enthusiasm" with which Cabral and Sousa da Câmara always regarded their own profession. These qualities about their Masters have always been recognised by the generations of students that followed them, including João Ferreira Nunes, who describes Sousa da Câmara as "my dear Master" (Nunes, 2015). Now João Ferreira Nunes' students and disciples see them in him as well. According to testimonies of students and "collaborators", these qualities helped to connect the classroom with these professors-architects' offices, who "always kept their doors open" (Cunha, 2015). This was often an opportunity for students to get a first look at professional practices and the labour market, as well as to build trust and develop working relationships (Figures 11.1 and 11.2), which today are no longer limited to Portugal. Instead, there is now an extensive network of "collaborators" in foreign universities, research centres, and the private sector in many different countries, whose academic educations started at ISA. The legacy and the international recognition achieved by Cabral, Sousa da

Figure 11.1 Meeting room of atelier GPSC–Sousa da Câmara Projects Studio (Photograph by Cristina Castel-Branco).

Figure 11.2 Meeting room of atelier PROAP–Studies and Projects of Landscape Architecture (Photograph by Duarte Belo).

Câmara, or João Ferreira Nunes is also proof of the quality of Landscape Architecture teaching in Portugal, especially at ISA—an international reference in this area. A lasting legacy that commits the new generations of disciples and "collaborators" to an idea of "School" as a "seal of quality and assurance" (Ramos, 2009) beyond borders. In the words of Manuela Raposo Magalhães, the "professor-student relationship model" that she encountered as a student of Cabral in the 1960s was the same she would find years later, as a professor at ISA, in the Berlin School, and which "with the adaptations that time requires, is the one practiced today in the Landscape Architecture schools in Portugal" (Magalhães, 2009).

Although the genesis of Landscape Architecture in Portugal was directly influenced by the "German School" and Prof. Cabral's Masters, such as Wiepking, Barth, or Migge (Antunes, 2019), the Lisbon School, in its 80 years of history, has built its own distinct identity, rooted in the "intimate relationships" between garden design, the city, and the "essence and form of the Portuguese landscape as fundamental values of our country's cultural identity" (Telles, 2009). This identity is founded on the interchange of concepts between agronomy and fine arts; the close collaboration with entities such as CEAP/Linking Landscape, Environment, Agriculture And Food (LEAF) and the Lisbon City Hall; and the relationships based on generosity and sharing between Masters (professors-architects) and disciples.

In conclusion, if the creation of the Landscape Architecture School in Portugal by Cabral contributes to open "an enormous window over the Portuguese landscape" (Telles, 2009), it is up to the new generations of landscape architects to respect and learn from the work done by the Masters who followed him.

Notes

1 *In* http://proffranciscocaldeiracabral.portaldojardim.com/projectos/decada-de-1950, October 23, 2023.
2 *In* http://proffranciscocaldeiracabral.portaldojardim.com/projectos/decada-de-1960, October 23, 2023.
3 *In* http://proffranciscocaldeiracabral.portaldojardim.com/video/ilidio-de-araujo, October 23, 2023.
4 *In* http://proffranciscocaldeiracabral.portaldojardim.com/video/antonio-viana-barreto, October 23, 2023.
5 *In* http://proffranciscocaldeiracabral.portaldojardim.com/video/antonio-viana-barreto, May 20, 2024.
6 Portaria n.º 1072/81, de 17 de dezembro, Diário da República n.º 289/1981, Série I de 1981–12–17, pp. 3288–3289.
7 *In* http://proffranciscocaldeiracabral.portaldojardim.com/video/ilidio-de-araujo, May 20, 2024.
8 *In* https://80anosap.isa.ulisboa.pt/wp-content/uploads/2022/10/ASS1-01_02.pdf, October 23, 2023.
9 *In* https://apap.pt/wp-content/uploads/2017/08/Caracterizacao-da-Arquitectura-Paisagista-em-Portugal.pdf, October 23, 2023.
10 *In* https://apap.pt/wp-content/uploads/2017/08/Caracterizacao-da-Arquitectura-Paisagista-em-Portugal.pdf, May 20, 2024.
11 *In* www.proap.pt/pt-pt/proap/past-collaborations-2, October 23, 2023.
12 *In* https://apap.pt/wp-content/uploads/2017/08/Caracterizacao-da-Arquitectura-Paisagista-em-Portugal.pdf, May 20, 2024.

Bibliography

Antunes, A. (2019). *A influência alemã na génese da Arquitetura Paisagista em Portugal.* PhD thesis, Faculdade de Ciências da Universidade do Porto, Porto.

Barreto, A. (2009). Professor Francisco Caldeira Cabral. In *Francisco Caldeira Cabral—Memórias do Mestre no Centenário do seu Nascimento.* Lisboa: Argumentum, pp. 10–11.

Birli, B. (2016). *From Professional Training to Academic Discipline: The Role of International Cooperation in the Development of Landscape Architecture at Higher Education Institutions in Europe.* Doctoral dissertation, Technische Universität Wien, Vienna.

Cabral, F. (2009). Transcrição de Entrevista ao Professor Francisco Caldeira Cabral. In *Francisco Caldeira Cabral—Memórias do Mestre no Centenário do seu Nascimento.* Lisboa: Argumentum, pp. 67–97.

Castel-Branco, C. (2022), Conference "MANUEL SOUSA DA CÂMARA, INOVATION IN LANDSCAPE ARCHITECTURE TEACHING (1977–1987)", held on September 30, 2022.

Cunha, A. (2015). *Análise e Interpretação de obras de arquitectura paisagista de Manuel de Sousa da Câmara (1929–1992).* Master thesis, ISA/UL, Lisboa.

Magalhães, M. (2009). Sobre o Professor Caldeira Cabral. In *Francisco Caldeira Cabral—Memórias do Mestre no Centenário do seu Nascimento.* Lisboa: Argumentum, pp. 34–36.

Nunes, J. (2010). Construir Paisagem. In *PROAP—Arquitectura Paisagista.* Lisboa: Note, pp. 9–11.

Nunes, J. (2015). Os Círculos da Memória. In *Sete círculos/Seven Circles.* Porto: Circo de Ideias, pp. 67–73.

Nunes, J., Ribas, C., Zoilo, I, & Jacinto, N. (2011). *Concursos Perdidos/Lost Competitions.* Lisboa: PROAP.

Pinto, M. (2014). *O legado escrito de Francisco Caldeira Cabral-Construção do pensamento teórico em arquitectura paisagista.* Master thesis, Faculdade de Ciências da Universidade do Porto, Porto.

Ramos, M. (2009). Recordando um Mestre Invulgar. In *Francisco Caldeira Cabral—Memórias do Mestre no Centenário do seu Nascimento*. Lisboa: Argumentum, pp. 39–48.

Telles, G. (2009). Gonçalo Ribeiro Telles. In *Francisco Caldeira Cabral—Memórias do Mestre no Centenário do seu Nascimento*. Lisboa: Argumentum, pp. 14–16.

Telles, G. (2010). Introducción. In *João Ferreira Nunes [PROAP]. Monográficos del Paisaje*. Barcelona: Asflor Ediciones, p. 10.

Telles, G., Castel-Branco, C., Fontes, E., Ribeiro, L., Magalhães, M., & Saraiva, M. (1992). Cinquenta anos de ensino de arquitectura paisagista no Instituto Superior de Agronomia. *Agros—Revista da Associação de Estudantes do Instituto Superior de Agronomia*, A.75(1), Janeiro-Junho, 42–44.

12 Landscape Architecture and urban ecology research as means for knowledge transfer on urban resilience in Maputo, Mozambique

Ana Beja da Costa

Introduction

Addressing fast-growing urban areas in Africa is urgent as the continent faces a fast urban expansion in ways that cities cannot cope with or efficiently react to. In parallel, the loss of traditional lifestyles and economic opportunities has led to a drastic cut between fast-growing settlements, their population, and the landscape systems in which they lie. In the PhD research "Mangroves of Maputo. Towards urban resilience through green infrastructure" (Beja da Costa, 2020), the aim was to find alternatives to the production of a generic city (Koolhaas et al., 1998) in the specific context of Maputo, in Mozambique, grounded in a sustainable perspective towards its coastal ecosystems and the local culture. The ultimate goal was to create a methodology incorporating local landscape values into the city's planning culture and promote knowledge sharing in the field of Landscape Architecture between Portugal and Mozambique, to plan toward sustainability. The research aimed at enhancing the cooperation and intellectual capacity exchange between Portuguese and Mozambican Universities—University of Lisbon (UL) and Universidade Eduardo Mondlane (UEM) on urban planning, Landscape Architecture, and ecology, taking advantage of the inevitably shared planning culture between the two countries. It also intended to allow the application of Landscape Architecture and urban ecology know-how developed in the past 80 years in Portugal in a distinct cultural and environmental context, validating knowledge in a geographical setting to enrich what is currently being practiced in the Portuguese context. Thus this represented an opportunity to generate new cooperation within the scope of knowledge development, sustainable management of natural resources, and global partnerships within the Millennium Development Goals frame (UNCSD, 2012). Fortunately, to this aim, synergies arose throughout the research process: the fieldwork trips and interviews were only possible with the support, contacts, and advice of the Faculty of Architecture and physical planning of the Eduardo Mondlane University (FAPF–UEM) teaching staff. In sequence, several lectures were given at the FAPF–UEM at different moments of the bachelor's course in architecture and physical planning. There was also the opportunity to join the colloquium "Landscape and Urbanism" organizing committee, organized by KU Leuven at the FAPF–UEM. Finally, the participation in conferences in Maputo was

DOI: 10.4324/9781003454007-16

also a means for dissemination of the research contents and results. This chapter will firstly contextualize the research aims of the PhD research that were carried out and summarize the contents of the most relevant pedagogic experiences and knowledge exchange moments. A reflection on what these added to the pedagogic experience at the FAPF–UEM are described, contemplating the outputs related to Landscape Architecture and urban ecology that were disseminated.

Contextualizing the PhD research: Common points between Maputo and Lisbon (1942–1974) and the post-independence split

In Portugal, Landscape Architecture teaching and practice was introduced by Cabral in 1942. In defining the role of the landscape architect in the Portuguese context, he envisioned interventions in the urban context but that the profession should also play an active role in rural planning, highlighting the relevance of urban green areas not as "green spots" but as a "continuous and coherent system" through means of the interpenetration of town and countryside that should be recovered (Cabral, 1993). Later on, Ribeiro Telles (1994, 1997) stressed the relevance of the *continuum naturale* that takes the existing landscape characteristics as a condition for human activities. At this point, landscape planning was based on Cabral's doctrine, and the planning process followed three steps: analysis, diagnosis, and proposal, which was progressively reflected in the city planning carried out for many cities, both in Portugal, and in Portugal's former colonies (Nunes Silva, 2015). An example is the 1969 *Lourenço Marques* masterplan (*Plano Director da Cidade de Lourenço Marques*)[1] developed following these steps, and it included extensive site analysis, a synthesis of servitudes and constraints, and based on that, the general masterplan proposal. This document strongly involved landscape architect Álvaro Dentinho, and it was the first time that a broad reading of the metropolitan area was made. It also included a piece on "Landscape planning: Development according to suitability" and a partial plan for a (never implemented) central linear park (Azevedo et al., 1969) along a riverbed (Figure 12.1), coincident with the current drainage canal along the *Joaquim Chissano* Avenue. A parallel can be made, on the one hand, between the masterplan's suitability studies that conditioned the design proposals and McHarg's ecological planning method (McHarg, 1969), published in the same year; and, on the other hand, between the central linear park design, which included the road integration along the riverbed, with leisure and sports areas and a continuous pedestrian trail along the park, and the greenway design principles (Fabos & Ryan, 2006).

After Mozambique's independence in 1975, there was a clear shift in Maputo's development paradigm, associated with the radical international support for Mozambique's liberation and the change from colonialism to a new government with Marxist-Leninist inspiration (Cahen, 1994). The built infrastructure (for a limited period) and the land (until current times) became state property through the implementation of nationalization policies. Migration, especially during the civil war in the 1980s, and population growth afterward (from 1992 onwards), has deeply changed the city and its immediate periphery (Oppenheimer & Raposo, 2007). In its peri-urban areas, horizontal urban sprawl was the means for urban

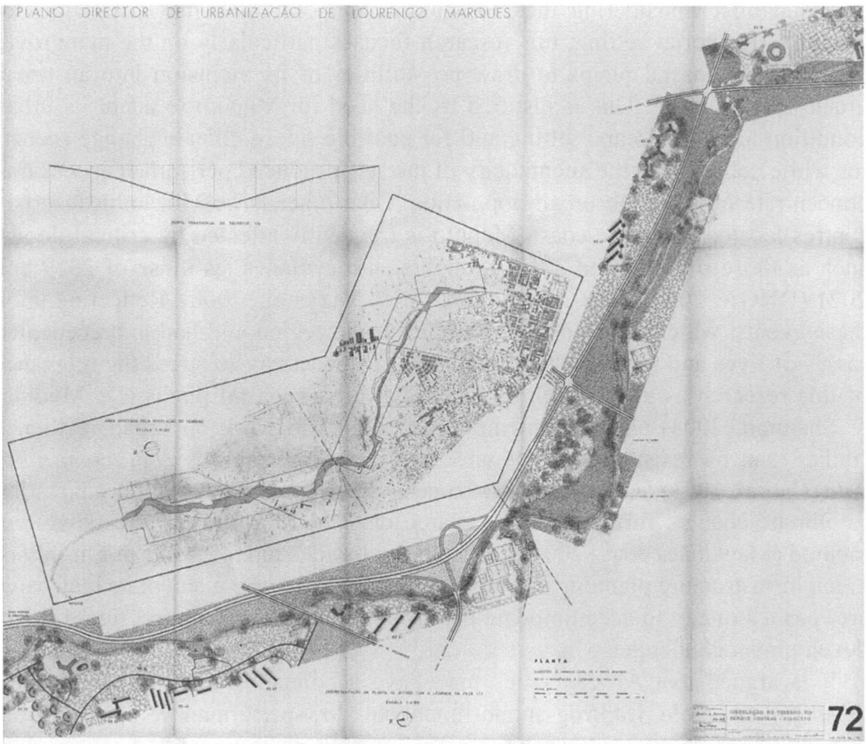

Figure 12.1 Central Park Grading Plan (Modelação do Terreno do Parque Central), Plano Director da Cidade de Lourenço Marques (Azevedo et al., 1969).

Source: Plan by A. Dentinho in Azevedo et al., 1969.

expansion. Nowadays, Maputo is Mozambique's largest and economically most important city, accounting for about 30% of the country's gross domestic product. The surrounding 'semi-rural' areas are undergoing a fast process of densification and Greater Maputo, which includes the administratively separate city of Matola, has a population of over two million (Jenkins & Eskemose, 2011). These processes are broadly impacting Maputo Bay coastal areas, placing ecosystems under several threats related to urbanization. In the city's coastal plains, the recently constructed Maputo ring road and the Katembe bridge have further pushed urban development towards the municipality's more or less vacant land, compromising the mangrove ecosystems and flood plains of this territory.

Green infrastructures as a means to respond to current urban development challenges

As stated by Forman (2014), "it has become the time when coastal cities across the globe, regardless of their scale and development state, cannot ignore the often dramatic ongoing climate events," such as droughts, cyclones, monsoonal storms, and tsunamis. Considering the aforementioned points and the specificities of Maputo's territorial setting, this research focuses particularly on the mangroves and adjacent coastal plains to draw possibilities of its inclusion into an urban green infrastructure. This is justified by the need for Maputo to adapt its urban condition to its landscape setting and for possible future climate change scenarios while considering the spontaneity of the self-produced peri-urban spaces that concentrate most of the urban population. Like other large Mozambican urban centres located along the coast, Maputo is frequently affected by critical floods, such as those of 2000 and 2013, droughts, and cyclones, as those of 2019 and 2021 (DNGA, 2005; INGC, 2009; RM, 2017; Vaz et al., 2000). Reflecting upon these events, which led to drastic changes in the region and had unprecedented losses of lives and infrastructural damage, it is necessary to stress the relevance of this research as an opportunity to rethink urban coastal plains (De Meulder & Shannon, 2008) and to contemplate strategies to work on urban resilience (rather than on resistance) to climate events. Taking Cabral's *continuum naturale* (Cabral, 1993) and by adopting an ecosystem-based approach for adaptation to climate change, furthering the research on Ribeiro Telles' landscape-system method (Magalhães et al., 2007), research discussions reflect on the possibility of green infrastructure planning in Maputo having as a central element its mangrove areas as a strategy to accommodate the aforementioned current and future urban development challenges.

It is argued that African cities may have a competitive advantage as their development could leapfrog mono-functional, low-performance, unsustainable infrastructures of conventional urban development paths towards greener urban economies (Pellegrino et al., 2015). Focusing on the urban realm, green infrastructures represent an opportunity for SSA cities. These are functionally integrated landscape networks supported by hybrid/built infrastructures that provide complementary and multifunctional landscape functions to communities (Ahern, 2007). In this sense, green infrastructures are based on a multifunctional foundation and act as multiservice providers and therefore have the potential to redefine the aesthetics, cultural identity, and place attachment in relation to urban landscapes (Ribeiro & Barão, 2006). The concept's application has shifted towards proactive planning and design strategies that recognize the complexities of social-ecological interactions (Cunha, 2017) to improve urban resilience.

The PhD research as an opportunity for knowledge exchange

The research has a strong multidisciplinary character, and the scientific background of Landscape Architecture and human settlements was the motivation to study possibilities for implementing urban green infrastructure at Maputo's coastal plains. Parallel to this, one of the goals of the thesis was to demonstrate how the results and research milestones might be shared in Maputo. Thus, in order to incorporate local concepts and values on the landscape, specifically on well-functioning natural systems within the planning processes, socio-anthropological research was carried out, strongly based on fieldwork observations and interviews.

In addition to regularly published peer-reviewed articles in international journals, an exchange platform with the FAPF–UEM was planned to be organized since the outline of the research process. This was meant to bring together stakeholders, planners, municipality staff and development companies, and experts, as well as planners of both countries, to promote the sharing of knowledge acquired throughout the research, contributing to validating the research tracks on the local governance.

Results and discussion

Landscape Architecture and urban ecology lectures at the FAPF–UEM

The lectures were given to the FAPF–UEM architecture undergraduate students. In 2017, the lecture "Ecological Structure—Principles and Application" included two parts: a) urban ecology principles and b) its application in Landscape Architecture design. In 2018, the lecture "Landscape Architecture design process" focused on the drawings that commonly constitute the sketch design and final design in Landscape Architecture, accompanied by design references and examples for its application.

FAPF–UEM architecture undergraduate thesis co-supervision

Taking on the momentum of the given lectures, there was the opportunity to co-promote three undergraduate theses on landscape-related research topics. These presented new perspectives on urban development and supported parts of the PhD research through a research-by-design approach (Figure 12.2):

- "Infulene Valley revitalization: A green corridor as urban sustainability strategy" by Ivan Muchanga (2018) researched possibilities for the environmental recovery of the Infulene river, which is the border between Maputo and Matola municipalities, proposing a new (intermunicipal) urban linear park, providing new public areas that support ecological functions.
- "Coastal wetlands and climate change: An environmental requalification of Língamo" by Dinshi Cangy (2018) focuses on the Língamo salt pans area in Matola, which is being targeted for urban expansion. The thesis explored means to solve the problem of degradation and unsustainable conversion of mangroves in the estuarine system along the coast of Matola, aiming at the transformation of the wetland into an urban park accessible to all, improvement of the existing neighbourhood, construction of an ecological village, and mangrove recovery.
- "Landscape Architecture as spatial revitalization tool: The UEM Campus case study" by Leonardo Neves (2019) focused on the UEM campus site, in Maputo, as an essential part of the city's green infrastructure. Its location on a plateau, low construction density, and vast extension of permeable soils led to research on possibilities for re-designing the campus open spaces as a large-scale water retention area, with defined areas and functions related to the university.

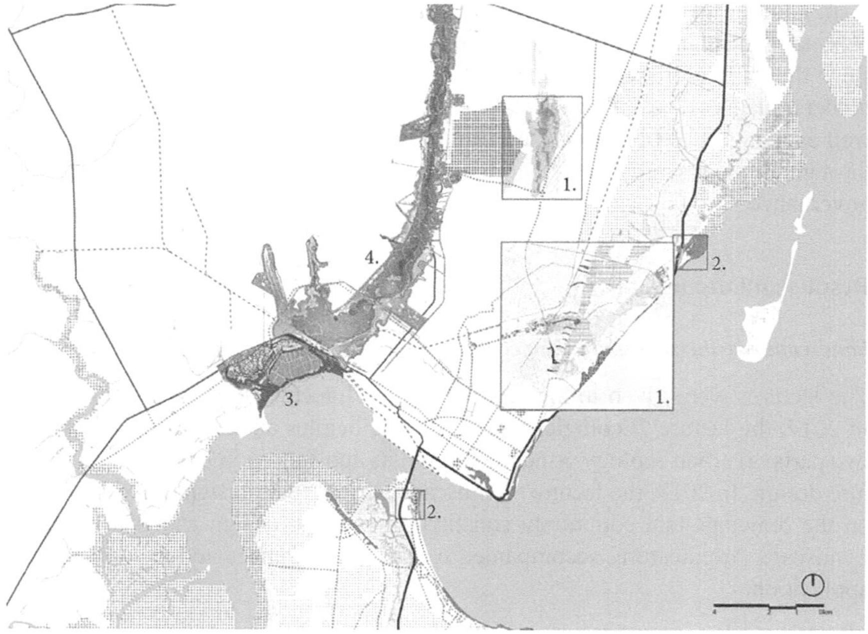

Figure 12.2 Compilation collage of recent academic design thesis projects that contribute to Maputo's green infrastructure (KUL, FAPF–UEM, and ISA).

Source: Map by the author.

Landscape and urbanism seminar

The Landscape and Urbanism Colloquium was organized by the KUL in partnership with the University of Zambeze, held at the FAPF–UEM as a closing event of the "Landscape urbanism strategies for the Zambeze River Basin, Mozambique" research project. The project integrated one of the three design studios of the KUL Spring studios that shared a common research theme, "Urbanized Deltas," that included the Tete region in the Zambeze river basin (Mozambique). The colloquium presented design and planning scenarios and urban-rural interventions with the landscape as the driver of change, fitting rural development sustainably into current and future challenges such as climate change, water management, and natural resources in the Tete region. By bridging the rural and urban characters of Tete and Maputo, the debate on context-based designs related to climate change through new settlement typologies and strategies in both regions had positive feedback from the participants (Figure 12.3).

Conferences in Maputo

Both the "Middle Class Urbanism in the Global South" conference, organized by the Aarhus University at the FAPF–UEM, in Maputo in February 2018, and the "Right to the City and the challenges in the implementation of the New Urban

Figure 12.3 Landscape and Urbanism Colloquium at the FAPF–UEM, organized by KUL with the support of ISA–ULisboa.

Source: Photograph by the author.

Agenda in Mozambique" conference, co-organized by the Centre for Environment, Biodiversity and Quality of Life Law of the College of Law of UEM and the UN-HABITAT Mozambique in April 2021 were an opportunity to showcase findings on landscape-related issues. They allowed for the debate to bring a new perspective on preserving the existing green infrastructure, improved as public space networks that add value to the real estate investments being made in Maputo's coastal plains.

Conclusions

Bearing in mind that no accredited Landscape Architecture course is taught in Mozambique, the lectures aimed at teaching fundamental concepts and giving simple examples for their application, showcasing reference projects. Topics on landscape planning and design possibilities to achieve resilient urban forms when facing the current environmental and development challenges triggered a shared interest among architecture and physical planning students and academic staff. Lecture contents on fundamental urban ecology concepts, green infrastructure, and landscape design helped to raise questions on possible means for application in the Mozambican context. A relevant follow-up of these were the three undergraduate theses on landscape-related research topics, such as green infrastructure, coastal resilience, and urban agriculture in Maputo, that presented new perspectives for urban development. The Landscape and Urbanism Colloquium brought to light design possibilities for sustainable development in a de-centralized rural area of Mozambique in Africa's fourth-largest river basin. It promoted knowledge exchange among several academic institutions on design experimentations for landscape and urbanism, raising awareness on how architects and related professions have a role in tackling climate change processes in the region's vulnerable settlements. Finally, the conferences' participation in Maputo was relevant, concerning the oral dissemination of findings to a broader local audience. The debate they generated among peers, concerning the ongoing process of urbanization in the coastal plains, both in the socio-anthropological perspective on forms of space appropriation, as in terms of what that implies in the intrinsic relation of urban resilience and coastal ecosystems, was very positive, contributing to the debate of themes related to both the Landscape Architecture and urban ecology fields in Mozambique. It is clear that the post-independence academic, theoretical, and practice frameworks in Mozambique were directed in a clear rupture with theory and practice as developed in Portugal. Landscape Architecture, as advocated at that time in Portugal, and that is continued up to today, was no exception. In this context, it is important to highlight the pertinence and relevance of the pedagogic format and practical methodologies applied throughout the events that took place in Maputo between 2017 and 2021 that generated a genuine interest and effective understanding on ecology concepts, landscape methods, and design outputs. Hopefully, it was an opportunity for a possible reconciliation between theory and methods in Landscape Architecture in Mozambique and for its potential establishment as a relevant discipline in face of contemporary development challenges. In turn, testing landscape theory and methods as they have been developed in the past 80 years in Portugal, in a different geographical, climate, and cultural context brings valuable insights on the methodologies' efficiency and validity, showing a potential to be continued as a valuable academic and professional exchange.

Note

1 Lourenço Marques was the name of the city of Maputo during Mozambique's colonial period.

Reference list

Ahern, J. (2007). Green infrastructure for cities: the spatial dimension. In Novotny, V., & Brown, P. (eds.), *Cities of the Future: Towards Integrated Sustainable Water and Landscape Management*. London: IWA, pp. 267–283.

Azevedo, M., Dentinho, A.P., & Bruschy, J. (1969). *Relatório Geral do Plano Director de Urbanização de Lourenço Marques*. Lisboa: Câmara Municipal de Lourenço Marques.

Beja da Costa, A. (2020). *Mangroves of Maputo. Towards Urban Resilience through Green Infrastructure*. Lisboa: PhD Universidade de Lisboa.

Cahen, M. (1994) *Les bandits. Un historien au Mozambique*. Lisboa: Foundation Calouste Gulbenkian.

Caldeira Cabral, F. (1993). *Fundamentos da Arquitectura Paisagista (Landscape Architecture Fundamentals)*. 2nd Edn. Lisbon: Instituto da Conservação da Natureza.

Cangy, D. (2018). *Coastal Wetlands and Climate Change: An Environmental Requalification of Língamo*. Maputo: Bachelor in Architecture and Physical Planning Universidade Eduardo Mondlane.

Cunha, N.S. (2017). *The National Ecological Network and a Land Morphology Model*. Lisboa: PhD Universidade de Lisboa.

De Meulder, B., & Shannon, K. (2008). Water and the city. The 'Great Stink' and Clean Urbanism. In Shannon, K., De Meulder, B., d'Auria, V., & Gosseye, J. (eds.), *Water Urbanisms*: *Vol. 1 Urbanism Fascicles OSA*. Amsterdam: SUN Publishers, pp. 5–9.

DNGA. (2005). *Avaliação da vulnerabilidade às mudanças climáticas e estratégias de adaptação*. Maputo: Ministério para a Coordenação da Acção Ambiental.

Fabos, J., & Ryan, R.L. (2006) An introduction to greenway planning around the world. *Landscape and Urban Planning*, 76, 1–6. https://doi.org/10.1016/j.landurbplan.2004.09.028.

Forman, R.T.T. (2014) *Urban Ecology: Science of Cities*. Cambridge, NY: Cambridge University Press.

INGC. (2009). *Synthesis Report. INGC Climate Change Report: Study on the Impact of Climate Change on Disaster Risk in Mozambique*. INGC. Available at: www.mz.undp.org/content/dam/mozambique/docs/Environment_and_Energy/INGC_Synthesis_Report_ClimateChange_Low-1.pdf.

Jenkins, P., & Eskemose, J. (2011). Developing cities in between the formal and informal. In *ECAS 2011–4th European Conference on African Studies: African Engagements: On Whose Terms? 15–18 June 2011, Sweden*. The Nordic Africa Institute, Upsala.

Koolhaas, R.A., O.M., Mau, B., & Sigler, J. (1998). *S,M,L,XL: Small, Medium, Large, Extra-Large*. 2nd Edn; first published: 1995 Edn. New York, NY: The Monacelli Press.

Magalhães, M.R., Abreu, M.M., Cortez, N., & Lousã, M. (2007). *Estrutura Ecologica da Paisagem, conceitos e delimitação—escalas regional e municipal (Ecological Network—Landscape, concepts and delimitation—regional and municipal scales)*. Lisbon: ISAPress.

McHarg, I.L. (1969). *Design with Nature*. 2nd Edn. Hoboken, NJ: John Wiley & Sons, Inc.

Muchanga, I. (2018). *Infulene Valley Revitalization: A Green Corridor as Urban Sustainability Strategy*. Bachelor in Architecture and Physical Planning. Maputo: Universidade Eduardo Mondlane.

Neves, L. (2019). *Landscape Architecture as spatial revitalization tool: The UEM Campus case study*. Bachelor in Architecture and Physical Planning. Maputo: Universidade Eduardo Mondlane.

Nunes Silva, C. (ed.). (2015). *Urban Planning in Lusophone African Countries*. 1st Edn. London and New York: Routledge.

Oppenheimer, J., & Raposo, I. (eds.). (2007). *Subúrbios de Luanda e Maputo [Suburbs of Luanda and Maputo]*. Lisboa: Edições Colibri.

Pellegrino, P., Ahern, J., & Becker, N. (2015) Green infrastructure: Performance, appearance, economy and working method. In Czechowski, D., Hauck, T., & Hausladen, G. (eds.), *Revising Green Infrastructure Concepts Between Nature and Design*. Boca Raton: Taylor & Francis Group, pp. 385–403.

Ribeiro, L., & Barão, T. (2006). Greenways for recreation and maintenance of landscape quality: Five case studies in Portugal. *Landscape and Urban Planning*m, (76), 79–97. https://doi.org/10.1016/j.landurbplan.2004.09.042.

Ribeiro Telles, G. (1994). Paisagem Global. Um conceito para o Futuro (Global Landscape. A concept for the future), *Iniciativa*, (no. Special Issue).

Ribeiro Telles, G. (1997). *Plano Verde de Lisboa (The Green Plan of Lisbon)*. Lisboa: Edições Colibri.

RM. (2017). *Plano Director para a Redução do Risco de Desastres 2017–2030 (2017–2030 Disaster Risk Reduction Masterplan)*. Maputo: República de Moçambique.

UNCSD. (2012). *The Future We Want*. Available at: www.uncsd2012.org/content/documents/727The%20Future%20We%20Want%2019%20June%201230pm.pdf.

Vaz, E.A., Remane, A.I., Vaz, A.C., Quadros, C., Mazoio, J., Lemos de Brito, J.E., Ronda, J., Chembeze, J., Cumbe, M., Sacartal, N., Brito, R., Nhassengo, S., & Cumba, T. (2000). *Levantamento da situação criada na cidade de Maputo após as fortes chuvadas caídas em princípios de Fevereiro. Identificação dos principais problemas existentes e propostas para acções imediates e a médio/longo prazo (Survey of the situation created in Maputo City after the heavy rains that fell at the beginning of February. Identification of the main existing problems and proposals for immediate and medium/long term actions)*. Maputo: Conselho Municipal de Maputo.

13 Breaking siloes and embracing the future

Western and Indigenous lenses in the education of landscape architects

Bruno Marques and Jacqueline McIntosh

Introduction

As both a profession and a field of inquiry, Landscape Architecture has changed significantly over the past four decades. Not limited to simply bridging the traditional divide between arts and sciences, it has extended to integrate concepts and approaches from various disciplines such as the humanities, technologies and more (Herlin, 2004). The assimilation of perspectives from different disciplines and their translation into practical guidelines for effective design, planning and management of our landscapes across the globe has made it crucial to address the contemporary issues that shape our environment across all scales of development (Deming & Swaffield, 2011; Fry, 2001; Gobster et al., 2010; Herlin, 2004; Holling et al., 1998; LaGro, 1999; Lenzholzer et al., 2013; Meijering et al., 2015; Milburn & Brown, 2003; Naveh, 2001; Redman, 1999; Stokols, 2011; Tai, 2003; Tress et al., 2001). Of necessity, landscape architects have had to become experts in providing design responses to climate, cultural, environmental, economic and social change issues.

Designing and preparing a suitable education to address these professional challenges has generated an imperative change to education, research and practice. Tertiary providers continue to focus on developing pedagogic knowledge and skills to address the multifaceted nature of Landscape Architecture and establish stronger research agendas that enable the preservation of the discipline. Professional degree programmes, including Landscape Architecture, are typically accredited by a professional rganization or agency, bringing educators and practitioners together. Jointly concerned with the scope of the profession, they introduce accreditation standards and recognition procedures to tertiary providers aimed at developing curricula and improving core knowledge (Imbert, 2007). However, although major progress has been made concerning the expanded role of Landscape Architecture, many significant issues remain unresolved.

With the widening divide between developed and emerging economies and growing environmental pressures, cultural values concerning landscape have been lost. Traditional wisdom and Indigenous knowledge, which embrace a more holistic approach to resource management, introduce fresh perspectives and leadership as the current Western-dominated ecological, economic and social models fall short. The long-standing connection with land through forests, wetlands, rivers,

DOI: 10.4324/9781003454007-17

coastal areas and mountains provides Indigenous cultures with a sense of identity, belonging and well-being. This is cultivated by engaging with the human-nature relationship balance as part of daily life and wellness, experiencing the natural environment as 'home'. Nevertheless, very little has penetrated curriculum development and practice in Landscape Architecture worldwide.

This chapter explores the changing nature of Landscape Architecture by examining Indigenous approaches. It does this by looking critically at the opportunities afforded by collective strategies to ensure that culturally and socially driven landscape can be understood. This can diversify the experience of Landscape Architecture students at the formal academic and professional career-building stages of their lifelong learning. Critically, this chapter offers new ways of approaching landscape and contributes to scaffolding a transformation in intercultural Landscape Architecture education and professional educational policies.

Indigenous knowledge

Indigenous knowledge, also known as local knowledge, traditional science or wisdom, is context-specific knowledge that has helped Indigenous communities survive and thrive throughout time (Inglis et al., 1993). Academic sources recognise that Indigenous knowledge is situated within broader cultural traditions, making it empirical and local as it is orally transmitted through imitation and demonstration, generally the consequence of practical engagement in everyday life (Ellen & Harris, 2000). Consequently, many practices, beliefs and values, such as oral narratives, tales, songs, customs and approaches to birth, healing, and death, are shared by almost all Indigenous communities worldwide (Bear, 2000). One fundamental aspect of Indigenous knowledge is the recognition of the interrelation between the spiritual, the natural and the self (Greenwood & de Leeuw, 2007), which connect all forms of life, the environment and the cosmos (Figure 13.1) (Lee & Armstrong, 2016; McIntosh et al., 2021; Russell & Shedd-Steele, 2003; Young & Koopsen, 2010). This universally shared Indigenous view creates a connection between cultures.

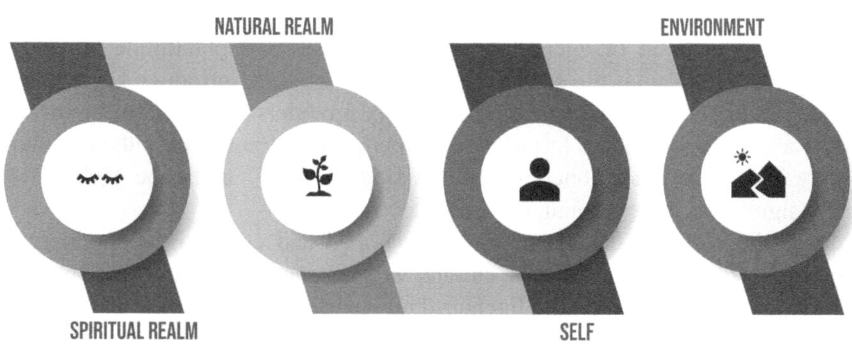

Figure 13.1 Fundamental aspects of Indigenous knowledge.

Source: Author's own.

For many Indigenous peoples, the disconnection from ancestral land has disrupted how they connect to place and has limited their ability to adapt to the endemic landscape. The severing of the link to the place where they sought protection, guidance, understanding, knowledge and identity, as well as elemental health and well-being, has ultimately led to adverse health and well-being outcomes (Marques et al., 2020, 2021, 2022). Some explain this disconnection as a result of colonial appropriation or commodification and privatisation of land; others speak to the forceful expulsion of local populations from their ancestral lands and the suppression of alternative forms of production and consumption (Harvey, 2006; Sassen, 2010). Many studies also report that access to physical and non-physical properties of the land is a strong indicator of Indigenous emotional well-being (Cajete, 2011; Panelli & Tipa, 2007; Robertson & Ljubicic, 2019). The risk of Indigenous knowledge being lost and as a result the loss of valuable perspectives and strategies for living sustainably, both ecologically and socially, is a problem identified by many (Carter, 2005; Harmsworth & Awatere, 2013; Mark & Lyons, 2014; Senanayake, 2006).

Climate change presents an urgent imperative for a changed relationship with the environment. For Indigenous people, under-represented minorities and deprived communities, climate change may result in the need to leave ancestral lands, increasing the risk of losing traditions and, ultimately, a sense of identity. Working with local communities and developing place-specific initiatives can help achieve climate change adaptation goals, increase biodiversity by focusing on native planting initiatives, protect traditional knowledge, address land grief and trauma, foster community participation, develop local socio-economic capacity and create a supportive context for intergenerational knowledge and culture to thrive.

Although intergenerational knowledge transfer can happen in any context, proximity to ancestral lands and natural entities frequently featured in Indigenous narratives makes it easier for the recipients to grasp the spirit of the stories. Whereas a modern family home may be sold at any time without bringing about feelings of a guilty conscience, ancestral lands hold a sacred status among most Indigenous people, and any damage done to these lands will negatively influence the health and well-being of the entire affected group. Similarly, a decontextualised narration would require much imagination on the listener's part as narrating the stories of creation and development of the world in contextualised settings involves all of the senses, again contributing to better knowledge transmission and a more immersive and site-specific experience (Figure 13.2).

In more urban contexts, common issues are generally reported as associated with noise and light. These have the ability to overwhelm users, removing the predictability and legibility essential for maintaining a sense of control outdoors. High-paced places can be demanding, forcing those with lower physical and mental health to choose places of refuge with appropriately paced rhythms. Access to nature and outdoor spaces promotes physical activity and stress reduction, particularly improving quality of life. Strategies such as planting naturalised wild-like vegetation, sheltering walkways and access to water in ways that encourage curiosity or playfulness can promote physical activity or enable foraging for food or medicinal plants (Figure 13.3). The feel and ambience of a site has great significance.

Figure 13.2 Learning from drawing in the sand with the Murujuga peoples, Western Australia.

Source: Author's own.

Most Indigenous cultures uphold responsibility for guardianship and stewardship of the land. As many ecosystems and associated places provide a sense of place and belonging, ecological restoration of waterways and the education of traditional cultural practices are fundamental (Figure 13.4). For example, waterways can provide access to traditional food sources and protect sacred sites. Access to open

Figure 13.3 Gubinge, an Australian superfood, collected by the Murujuga peoples.

Source: Author's own.

Figure 13.4 Place of guardianship is important for Māori culture such as Lake Wairarapa, the third largest in Aotearoa New Zealand's North Island.

Source: Author's own.

spaces for cultural interaction can bring different age groups of people together while creating the potential to develop ecological corridors, restore waterways, re-introduce wetlands for restoring riparian edges and adapt re-naturalised landscapes for well-being promotion. There seems to be a strong relationship between vegetation and other landscape components as it permits the creation of spatial variability and sensory engagement while improving ecological health and delivering healing narratives.

Guardianship extends beyond just the land but to an active role as guardians of the life force in each other and nature, elevating the physical, spiritual, mental and communal well-being (Mark & Lyons, 2014). This complex intersection of people, place and 'nature-culture' relations maintains the balance between mind, body, spirit, family, genealogy and land. Traditional healing thereby is about the link between spirit, mind, body and land. Such beliefs demonstrate how Indigenous peoples conceptualise the body as an arbiter of interaction with the environment enacted through both speech and naming. This includes naming places, which impresses ancestors and deities into the landscape so that a place and its knowledge cannot be separated (Carter, 2005).

International studies explain the connection to land as place attachment, often defined as genus loci, which attaches people to the essence or spirit of a certain place (Tuan, 1990). The desire for harmony with nature and interconnectedness between the natural and the spiritual realms are recurring themes in literature (Greenwood & de Leeuw, 2007; Kirmayer et al., 2003; Mark & Lyons, 2014). However, unity and connection with the land are fundamental characteristics that define the existence of Indigenous communities and are therefore reflected in all aspects of Indigenous life (Carter, 2005).

Discussion

We live in a time of unprecedented change, which threatens our existence. Landscape Architecture, both as a profession and a field of inquiry, can no longer make tweaks to curriculum or add branches to enquiry, rearranging the deck chairs in the Titanic. A fresh approach is proposed for consideration. We not only need to change what we do, but we need to change how we think about what we do. Indigenous knowledge offers insights into how this might be achieved.

The move to a more holistic way of thinking will require changes to how we practice and structure our curricula in Landscape Architecture. One of the key necessary changes is the reformation of a deep and personal connection to the landscape that can ground scholars, students and practitioners to establish a more holistic and respectful framework. From the literature reviewed, together with extensive engagement with local Indigenous groups, a set of recommendations to address Landscape Architecture curricula is proffered to offer insights into how we can better cater to the future needs of our students and profession.

Address environmental changes with empathy: An important aspect is the impact of climate change in the near future and how much of that will put some of our ecosystems under pressure, which may threaten food and water security.

Uphold local knowledge: Landscape Architecture has the tools to work collaboratively with local communities to create meaningful spaces closer to sacred and significant sites that enable intergenerational transmission of knowledge.

Restore ecological and human health as a priority in resource management: For students and professionals, identity in relation to land is highly important as it distinguishes us from any other built environment disciplines.

Re-introduce the body to the landscape: By re-introducing the body to the landscape, explorative or playful path options with aspects of discovery could encourage interaction, including exercise through horticultural activity and cultivation of plants for medicinal or craft purposes while enabling social interaction. Sitting, walking, playing, picking up fruit or gardening or through intergenerational activities to encourage collective participation.

Support mental restoration: Evidence is clear that the environment can have devastating effects on mental and spiritual health, leading to loss of social capital and sense of community and diminished mental health while helplessly observing the degradation of local ecosystems. Looking at Landscape Architectural solutions, mental restoration can be provided through landscape components that viscerally engage users and create a safe and healthy place for connecting with self and others.

Promote participatory design strategies: The adoption of participatory design strategies as a means of increasing education of the population at large and using Indigenous worldviews is something to consider when looking at curriculum development. Participatory or co-design processes foster a human-centred design approach that commonly brings together users and stakeholders to address an issue.

Invest in holistic infrastructure: Infrastructure renewal offers opportunities to rethink how heavy solutions might be made more inclusive. Solutions can be crafted as the heart of the landscape, where people meet and forge connections with more consistent relationship building and better communication to foster mutual respect and trust.

Protect our landscapes: Landscape remains critical to cultural identity and maintaining a sense of place and belonging. Landscapes are inseparable from ancestors, events, occupations and cultural practices. Loss of land, traditions and landscape degradation profoundly affect humanity. Design strategies that promote cultural healing, connection with healthy landscapes and group and intergenerational activities can balance apparent economic imperatives.

These recommendations follow a national trend in New Zealand and many other countries with Indigenous populations to acknowledge different ways of knowing and incorporate local knowledge. Coincidentally, local and Indigenous knowledge and its inclusion in global climate adaptation, biodiversity protection and policy processes are being addressed by various countries worldwide.

Conclusion

Landscape Architecture in Portugal is in a pivotal position. It can either remain contained within its current silo or seek a 'sea change' for both the profession and academia. Rather than limit its potential agenda to bridging the gap between arts and sciences, it can embrace the opportunity to assimilate perspectives from both different disciplines and also different ways of knowing. This chapter explores the recent trends in Landscape Architecture within New Zealand that incorporate Indigenous knowledge to obtain an entirely new way of thinking which can better address the social and cultural aspects of sustainability. Indirectly, it posits that a shift of perspective might offer opportunities to Landscape Architecture education and practice in Portugal that could be otherwise be overlooked.

Curricula that engage with the different professional realms can be seen as a positive tool to aid in developing and implementing better design and have more holistic outcomes as part of modern living. However, this chapter calls for more than that. It proposes that directly incorporating local knowledge and Indigenous values and approaches can offer new opportunities for living holistically with nature and supporting the health and well-being of our communities. To do this, we collectively need to better understand those unique characteristics that make up our natural and built environments and how they affect our physical, mental, spiritual and communal health and well-being. This new perspective could become the tipping point for Landscape Architecture.

References

Bear, L.L. (2000). Jagged worldviews colliding. In Battiste, M.A., & Barman, J. (eds.), *Reclaiming Indigenous Voice and Vision*. Vancouver: UBC Press, pp. 77–85.

Cajete, G. (2011). Indigenous knowledge: The Pueblo metaphor of indigenous education. In Battiste, M. (ed.), *Reclaiming Indigenous Voice and Vision*. Vancouver: UBC Press, pp. 181–191.

Carter, L. (2005). Naming to own place names as indicators of human interaction with the environment. *AlterNative: An International Journal of Indigenous Peoples*, 1(1), 6–24. https://doi.org/10.1177/117718010500100102.

Deming, M.E., Swaffield, S. (2011). *Landscape Architectural Research: Inquiry, Strategy, Design*. Hoboken, NJ: John Wiley & Sons.

Ellen, R., & Harris, H. (2000). Concepts of indigenous technical knowledge in scientific and developmental studies literature: A critical assessment. In Ellen, R., Parkes, P., & Bicker, A. (eds.), *Indigenous Environmental Knowledge and Its Transformations: Critical Anthropological Perspectives, Studies in Environmental Anthropology*. Amsterdam: Harwood Academic, pp. 1–34.

Fry, G.L.A. (2001). Multifunctional landscapes—towards transdisciplinary research. *Landscape and Urban Planning*, 57(3–4), 159–168. https://doi.org/10.1016/S0169-2046(01)00201-8.

Gobster, P.H., Nassauer, J.I., & Nadenicek, D.J. (2010). Landscape journal and scholarship in landscape architecture: The next 25 years. *Landscape Journal*, 29(1), 52–70. https://doi.org/10.3368/lj.29.1.52.

Greenwood, M., & de Leeuw, S. (2007). Teachings from the land: Indigenous people, our health, our land, and our children. *Canadian Journal of Native Education*, 30(1), 48–53.

Harmsworth, G.R., & Awatere, S. (2013). Indigenous Māori knowledge and perspectives of ecosystems. In Dymond, J. (ed.), *Ecosystem Services in New Zealand—Conditions and Trends*. Lincoln: Manaaki Whenua Press, pp. 274–286.

Harvey, D. (2006). Neo-liberalism as creative destruction. *Geografiska Annaler: Series B, Human Geography*, 88(2), 145–158.

Herlin, I.S. (2004). New challenges in the field of spatial planning: Landscapes. *Landscape Research*, 29(4), 399–411. https://doi.org/10.1080/0142639042000289037.

Holling, C.S., Berkes, F., & Folke, C. (1998). Science, sustainability and resource management. In Berkes, F., Folke, C., & Colding, J. (eds.), *Linking Social and Ecological Systems: Management Practices and Social Mechanisms for Building Resilience*. Cambridge: Cambridge University Press, pp. 342–362.

Imbert, D., (2007). Landscape architects of the world, Unite! Professional organizations, practice, and politics, 1935–1948. *Journal of Landscape Architecture*, 2(1), 6–19. https://doi.org/10.1080/18626033.2007.9723376.

Inglis, J.T., International Program on Traditional Ecological Knowledge, International Development Research Centre (Canada), International Association for the Study of Common Property. Meeting. (1993). *Traditional Ecological Knowledge: Concepts and Cases. Canadian Museum of Nature, International Program on Traditional Ecological Knowledge*. Ottawa, Canada: International Development Research Centre.

Kirmayer, L., Simpson, C., & Cargo, M. (2003). Healing traditions: Culture, community and mental health promotion with Canadian aboriginal peoples. *Australas Psychiatry*, 11(1), S15–S23. https://doi.org/10.1046/j.1038-5282.2003.02010.x.

LaGro, J.A. (1999). Research capacity: A matter of semantics? *Landscape Journal*, 18(2), 179–186. https://doi.org/10.3368/lj.18.2.179.

Lee, C.C., & Armstrong, K.L. (2016). Indigenous models of mental health intervention: Lessons from traditional healers. In Casas, J.M., Suzuki, L.A., Alexander, C.M., & Jackson, M.A. (eds.), *Handbook of Multicultural Counseling*. Singapore: SAGE Publications, pp. 441–456.

Lenzholzer, S., Duchhart, I., & Koh, J. (2013). 'Research through designing' in landscape architecture. *Landscape and Urban Planning*, 113, 120–127. https://doi.org/10.1016/j.landurbplan.2013.02.003.

Mark, G., & Lyons, A. (2014). Conceptualizing mind, body, spirit interconnections through, and beyond, spiritual healing practices. *EXPLORE*, 10(5), 294–299. https://doi.org/10.1016/j.explore.2014.06.003.

Marques, B., Freeman, C., & Carter, L. (2022). Adapting traditional healing values and beliefs into therapeutic cultural environments for health and well-being. *IJERPH*, 19(1), 426–445. https://doi.org/10.3390/ijerph19010426.

Marques, B., Freeman, C., Carter, L., & Pedersen Zari, M. (2020). Sense of place and belonging in developing culturally appropriate therapeutic environments: A review. *Societies*, 10(4), 83–101. https://doi.org/10.3390/soc10040083.

Marques, B., Freeman, C., Carter, L., & Pedersen Zari, M. (2021). Conceptualising therapeutic environments through culture, indigenous knowledge and landscape for health and well-being. *Sustainability*, 13(16), 9125–9147. https://doi.org/10.3390/su13169125.

McIntosh, J., Marques, B., & Mwipiko, R. (2021). Therapeutic landscapes and indigenous culture: Māori health models in Aotearoa/New Zealand. In Spee, J.C., McMurray, A., & McMillan, M. (eds.), *Clan and Tribal Perspectives on Social, Economic and Environmental Sustainability*. Bingley: Emerald Publishing Limited, pp. 143–158. https://doi.org/10.1108/978-1-78973-365-520211016.

Meijering, J.V., Tobi, H., van den Brink, A., Morris, F., Bruns, D. (2015). Exploring research priorities in landscape architecture: An international Delphi study. *Landscape and Urban Planning*, 137, 85–94. https://doi.org/10.1016/j.landurbplan.2015.01.002.

Milburn, L.-A.S., & Brown, R.D. (2003). The relationship between research and design in landscape architecture. *Landscape and Urban Planning*, 64(1–2), 47–66. https://doi.org/10.1016/S0169-2046(02)00200-1.

Naveh, Z. (2001). Ten major premises for a holistic conception of multifunctional landscapes. *Landscape and Urban Planning*, 57(3–4), 269–284. https://doi.org/10.1016/S0169-2046(01)00209-2.

Panelli, R., & Tipa, G. (2007). Placing well-being: A maori case study of cultural and environmental specificity. *EcoHealth*, 4, 445–460. https://doi.org/10.1007/s10393-007-0133-1.

Redman, C.L. (1999). Human dimensions of ecosystem studies. *Ecosystems*, 2, 296–298.

Robertson, S., Ljubicic, G. (2019). Nunamii'luni quvianaqtuq (It is a happy moment to be on the land): Feelings, freedom and the spatial political ontology of well-being in Gjoa Haven and Tikiranajuk, Nunavut. *Environment and Planning D: Society and Space*, 37(3), 542–560. https://doi.org/10.1177/0263775818821129.

Russell, K., & Shedd-Steele, R. (2003). The meaning of health in mammography screening for African American women. *Health Care for Women International*, 24(1), 27–39. https://doi.org/10.1080/07399330390169981.

Sassen, S. (2010). A savage sorting of winners and losers: Contemporary versions of primitive accumulation. *Globalizations*, 7(1–2), 23–50.

Senanayake, S.G.J.N. (2006). Indigenous knowledge as a key to sustainable development. *Journal of Agricultural Sciences—Sri Lanka*, 2(1), 87–94. https://doi.org/10.4038/jas.v2i1.8117.

Stokols, D. (2011). Transdisciplinary action research in landscape architecture and planning: Prospects and challenges. *Landscape Journal: Design, Planning, and Management of the Land*, 30(1), 1–5. https://muse.jhu.edu/article/415278.

Tai, L. (2003). Doctoring the profession. *Landscape Architecture*, 93(11), 64–73.

Tress, B., Tress, G., Décamps, H., & d'Hauteserre, A.-M. (2001). Bridging human and natural sciences in landscape research. *Landscape and Urban Planning*, 57(3–4), 137–141. https://doi.org/10.1016/S0169-2046(01)00199-2.

Tuan, Y.-F. (1990). *Topophilia: A study of Environmental Perception, Attitudes, and Values*. New York: Columbia University Press.

Young, C., & Koopsen, C. (2010). *Spirituality, Health, and Healing: An Integrative Approach*. 2nd Edn. G—Reference, Information and Interdisciplinary Subjects Series. Burlington, MA: Jones & Bartlett Learning.

14 From the concept of cultural landscape to its application in conservation policies and higher education in Portugal

Raquel Carvalho, Teresa Portela Marques and Cristina Castel-Branco

Introduction

In 2022, we celebrated 80 years of Landscape Architecture teaching in Portugal. We also celebrated 50 years of the United Nations Educational, Scientific and Cultural Organization (UNESCO) "Convention Concerning the Protection of World Cultural and Natural Heritage" (1972) and 30 years of the inclusion of Cultural Landscapes as a category of World Cultural Heritage (1992).

The term cultural landscape emerged in German geography (*kulturlandschaft*) at the end of the 19th century (Potthoff, 2013; Martin & James, 1993; Jones, 2003) and was developed and disseminated by Carl Sauer, in the 1920s, in the study of cultural geography at the Berkeley School, University of California. It evolved to mean landscapes encapsulating historical, aesthetic, ecological, social, spiritual, technical, economic and other tangible and intangible values that are especially cherished by societies.

Presently, although widespread in the academic community and among landscape conservation professionals, the term reveals different understandings, and it remains vague to broader society. The concept embraces the idea of the higher significance of a humanized landscape that is distinguished from others by conveying specific values and socio-ecological, economic and cultural values (Jacques, 1995; Antrop, 2005; Taylor, 2008). The cultural landscape is an anthropogenic product generated by human relations with the environment, an intentional material and/or immaterial appropriation of the territory reflecting a specific way of life, a vision or an ideology. They are representative and distinctive landscapes that give us a feeling of belonging and rootedness (Carvalho & Marques, 2019).

The concept has been continuously debated and supported by policies and operational guidelines to promote cultural landscapes' identification, characterization, management and conservation.[1]

The integration of the cultural landscape concept into the practice of landscape conservation: a brief international overview

After the Second World War, a growing concern about threats to heritage was initially focused on built heritage. The Athens Charter for the Restoration of Historic

DOI: 10.4324/9781003454007-18

Monuments (1931) opened the door to conservationist practices focusing on built monuments. Nevertheless, measures were also developed to protect the landscape, highlighting beauty, character and cultural identity, as witnessed in the "European Cultural Convention" (1954) and in the "Recommendation concerning the Safeguarding of the Beauty and Character of Landscapes and Sites" (1962). The nature conservation movements, which were already taking place in various parts of the world in the late 19th and early 20th centuries, led to the creation of parks and nature reserves. The concept of a 'nature park', with less restrictive measures than a 'national park', was developed in Europe at the beginning of the 20th century to engage rural planning with the promotion of economic and recreational areas (Pessoa, 1978). The protection of agricultural resources and other modified ecosystems in the context of biodiversity and nature conservation has fostered an interest in cultural landscapes as an expression of sustainable land use (Farina, 2000; Phillips, 1998; Brown et al., 2004).

Cultural landscapes bring together 'culture' and 'nature', and their approach requires adapting the guidelines developed for both built and natural heritage. This highlights the perception of meanings, the relationship between humans and nature and the necessary state of balance, establishing an order in which both participate.

In 1972, the "Convention Concerning the Protection of World Cultural and Natural Heritage" (UNESCO) defined the 'Site' category as "works of man or the combined works of nature and man" (UNESCO, 1972, para. 1) that can be seen as including cultural, humanized landscapes.

Although the term cultural landscape only appears later, its meaning is expressed in The International Union for Conservation of Nature (IUCN) "Protected Landscapes": "distinctive landscape patterns are created by the integration of specific natural and cultural features that present aesthetically attractive land and water settings. These may result through traditional land use practices of a cultural group in response to their environment" (Commission on National Parks and Protected Areas, 1978, p. 17).

The recognition of Indigenous places and values early stressed in Australia brought the intangible heritage component into debate, impacting the idea of cultural significance as expressed in the "Burra Charter", adopted by the International Council on Monuments and Sites (ICOMOS) Australia in 1979. The intangible values of the landscape are reflected in representations, oral expressions, traditional practices, knowledge, artefacts and spaces that communities or individuals recognize and transmit from generation to generation (UNESCO, 2003).

In the late 1980s and early 1990s, Ken Taylor[2] (2017) explained that momentum as an opportunity to develop theory and practice on the idea of cultural landscape, as a result of a growing awareness that built heritage does not function as an 'isolated island' but rather in interdependence with people, social structures and ecosystems.

The concept was quickly established in management and conservation landscape plans in the United States. The National Park Service (NPS) adopted the term cultural landscape as a cultural property type and developed pioneering guidelines to identify, assess and manage cultural landscapes of historic and rural character (Melnick et al., 1984; Birnbaum, 1994).

The Cultural Landscape category was considered from 1992 for the inclusion of properties in the UNESCO World Heritage List. The first meeting of experts from different fields was held at La Petite Pierre in France (Luengo & Rossler, 2012). Three main categories of cultural landscapes and two subcategories were then included in the "Operational Guidelines for the implementation of the Convention": "defined landscape designed and created intentionally by man . . . ; 'organically evolved landscape . . .' (a) relict (or fossil) landscape; b) continuing landscape . . . ; 'associative cultural landscape'" (World Heritage Committee, 1994, p. 14). According to Añón (2020), a strategic move impacting the visibility of the cultural landscape concept was taken in 1999, with the renaming of the ICOMOS–IFLA International Scientific Committee on Historic Gardens and Sites (established in 1971) to International Scientific Committee on Cultural Landscapes (ISCCL).

The "Conservation of Cultural Landscape Areas' Recommendation" (Council of Europe, 1995) emphasized theoretical and operational principles for conserving and managing cultural landscape areas. It became a reference document focusing on cultural landscapes, an important advancement, proposing concrete actions. It also recognized the vital role of training and research, suggesting educational measures: "Specific programmes should be provided for students intending to work, and for practitioners already engaged, in the conservation of cultural landscape areas and general landscape policies" (Council of Europe, 1995, p. 9). Besides the growing interest in cultural landscapes, the study and conservation of landscape, overall, reached a milestone with the "European Landscape Convention" (ELC) (Council of Europe) in 2000, promoting the identification, management, planning and protection of landscapes, raising awareness of civil society to participate in the implementation of landscape policies.

Sustainable development, cultural diversity and recognition of cultural heritage are current paradigms of the 21st century (Council of Europe, 2005; UNESCO, 2005; EUROPA NOSTRA et al., 2018), strengthening the importance of identifying and conserving cultural landscapes.

The cultural landscape in Portuguese higher education and conservation policies

The 'humanized landscapes' as considered by the first generation of Portuguese landscape architects (1942–1975) are products of a particular human culture, testimonies of the experience and knowledge of humans over a particular place enabling the creation of ecosystems that, through new forms of sustainable management, provide food, shelter, recreation opportunities and other primary needs ('replacement ecosystems'). These landscapes are expressions of the work of successive generations in the humanization of the land (Cabral, 1967; Araújo, 2008; Telles, 2016).

As an academic topic, the now internationally called cultural landscapes were handled by the first Portuguese landscape architect and professor Cabral at ISA.[3]

Since the 1950s, the discipline of "History of Garden Art" and subjects such as nature conservation, protection of rural landscapes, historic sites and gardens were taught and discussed in his texts.[4] Cabral left us a legacy cherished by the following generations and developed in the contemporary context.[5] From 1981 on, the curricula included disciplines of landscape evolution and landscape planning, among others like history of art and landscape aesthetics. Nowadays, the cultural landscapes subject is addressed in the title of subjects from three of the five Portuguese Landscape Architecture master's degrees (Universities of Lisbon, Porto and Algarve).

Concerning landscape conservation practice, European policies and orientations were followed. The creation of the first nature parks in the 1970s[6] allowed landscape management and conservation plans to tackle ethnographic resources, the traditional economy and natural resources. The idea of preserving intangible values and balanced landscapes was already expressed in the Portuguese Constitution:

> b) Plan the territorial space to construct biologically balanced landscapes; c) Create and develop natural and recreational reserves and parks, as well as list and protect landscapes and sites, to guarantee nature conservation and the preservation of cultural values of historical or artistic interest.
>
> (Constituição da República Portuguesa, 1976, p. 66)

In 1987, the environment framework law presented several references to the landscape, including the need for management instruments as

> important for the maintenance of landscape and cultural plurality. . . . A development strategy that engages the populations in the protection of these values. . .. The inventory and evaluation of types of rural and urban landscapes, comprising abiotic and cultural elements. . .. The identification and mapping of visual and aesthetic values of natural and artificial landscapes.
>
> (Assembleia da República Portuguesa, 1987, p. 19)

In 1988 a joint programme of the Portuguese Association of Landscape Architects (APAP) and the Portuguese Institute of Cultural Heritage (IPPC) launched the first historic gardens restoration programme, involving many young landscape architects who got scientific support from Professor Cabral and other academics and senior landscape architects.[7] Following the new interest in this subject, several historic gardens started to be studied and restored. In 1995, the 'Cultural Landscape of Sintra' was included in the World Heritage List (UNESCO), one of the first world cultural landscapes to be listed. The following entries were the 'Alto Douro Wine Region' (2001), the 'Landscape of the Pico Island Vineyard Culture' (2004) and the 'Sanctuary of Bom Jesus do Monte in Braga' (2019), raising the recognition of the theme at a national level.[8] These experiences were introduced in academic educational practices, serving particularly as case studies for the development of methodologies for the characterization of cultural landscapes and to assessing their value (Figures 14.1 and 14.2).

Figure 14.1 Quinta da Bizelga, Tomar: The wheat fields in the valley and the main house at the top of the hill.

Source: Raquel Carvalho, 2022

In 2003, the Portuguese Association of Historic Gardens (AJH) was founded[9] and since then that has been very dynamic in relation with academia in studying and promoting landscape heritage through practical restoration actions, technical support to gardens' owners and managers, production of studies on cultural landscapes, organisation of courses, conferences and visits to gardens. Its initiatives are open to the academic community, namely the online inventory of historic gardens and other designed landscapes that also acts as an educational resource. In 2004, the work "Contributions to the identification and characterization of the landscape in mainland Portugal"[10] was published responding to ELC's call for greater awareness of landscapes, opening several lines of research and ways to operationalize concepts such as identity and character, central issues to the study of cultural landscapes. The recently launched National Policy for Architecture and Landscape (Conselho de Ministros, 2015) recognizes the role of cultural landscapes in enhancing diversity and local and regional identities and in promoting the development of programmes that foster awareness, information, and legislative measures. The legal protection of cultural landscapes continues to be referred to as the Framework Law for the Safeguarding of Cultural Heritage (2001), according to the categories of 'monuments', 'groups of buildings' and 'sites', following the international Granada Convention.[11]

Figure 14.2 Field trip to the case study area as part of the methodology for the characterization of cultural landscapes. Quinta da Bizelga, Tomar, 2022 (Conservation and Management of Cultural Landscapes—master's degree in Landscape Architecture at ISA).

Source: Raquel Carvalho, 2022

There is still work to be done for the recognition, by the general public and government entities, of cultural landscapes as a very strategic value for society and a very significant part of a common heritage. Yet the work of landscape architects, particularly in academia, has been quite diverse and intense (thesis, publications, scientific courses and conferences, etc.) sharing the same vision of the pioneers and applying it to actions that comply with the needs and the challenges of the contemporary societies.

Concluding remarks

The cultural landscape concept allowed us to rethink how to work with the natural, cultural, tangible and intangible elements and processes that reinforce landscape identity and promote the protection and durability of landscapes. This recognition was progressively included in strategies to face the challenges resulting from globalization that could lead to the disappearance of these landscapes of identity. Such strategies included protecting natural areas and historic gardens, followed by acknowledging landscapes of other cultural significance.

The gradual increase in the study and development of conservation methodologies on cultural landscapes, tested in Landscape Architecture courses, enriches the theoretical body on the subject. Cabral and the first generation of landscape architects were great connoisseurs of cultural landscapes, understanding and describing elements and natural processes and applying this knowledge in their projects while passing it on to their students. The inclusion of disciplines related to the restoration of gardens and historical landscapes in academic curricula, the production of master's and doctoral theses, the organization of meetings and scientific conferences, scientific articles, exhibitions and catalogs allowed for the improvement of technical and scientific knowledge.

Beyond education, the concept is applied in professional practice. The inscription of Portuguese cultural landscapes in the UNESCO Heritage List challenged the study and debate on landscape heritage. Nevertheless, the adequacy of the existing national legislation regarding the protection of cultural landscapes is currently being debated to seek solutions that promote an adequate framework for safeguarding these landscapes with social, ecological, economic and historical dimensions. Cultural landscape conservation would highly benefit from the development of guidelines to designate them as a separate category in heritage protection in Portugal.

Notes

1 The doctoral thesis "Cultural Landscape: from concept to professional practice. The transhumance route between Serra da Estrela and Campinas de Idanha" (authors translation) by Raquel Carvalho, was an opportunity to research and develop strategies towards cultural landscapes conservation.
2 Expert of the ICOMOS–IFLA International Scientific Committee on Cultural Landscapes (ISCCL) (IFLA–International Federation of Landscape Architects).
3 Cabral graduated as a landscape architect during the 1930s in Berlin. The teaching of Landscape Architecture started at ISA in 1942 as a 'free course'. From 1981, the course was recognized as a bachelor's degree.
4 The master thesis by Mariana Abranches Pinto, "Francisco Caldeira Cabral's written legacy. Construction of theoretical thought in Landscape Architecture" (2014) (authors translation) reveals Cabral's awareness to nature conservation and landscape protection in a universal and humanist vision (Pinto, 2014). This legacy from the 1940s and 1950s conveys the knowledge acquired in Berlin and also in field trips and visits. It also expresses his professional expertise related to garden design and landscape planning (Pinto, 2014, Antunes, 2019).
5 That is the case of the doctoral thesis "The cultural landscape and the uniqueness of place—a greenway heritage network for landscape conservation of Lisbon metropolitan", developed in 1998 at the UMass, Amherst, by Luís Paulo Ribeiro, who employed the term and concept cultural landscape, then well-established in the United States.
6 In 1963, Cabral proposed a national network of parks and reserves that lead to the creation of the first national park (Peneda-Gerês) in 1971.
7 Besides Cabral, Teresa Andresen, Ribeiro Telles, Edgar Fontes, Ilídio de Araújo and Aurora Carapinha, among others, set up a group of senior landscape architects who fostered, in the Portuguese academic and professional practice, the understanding of cultural significance and values in historic gardens.
8 Entries involved landscape architects, professors in Landscape Architecture degree courses (namely Teresa Andresen, co-coordinator in the 'Alto Douro Wine Region' and

Teresa Andresen and Teresa Portela Marques, coordinators of the entry 'Sanctuary of Bom Jesus do Monte in Braga'), among others landscape architects.

9 By landscape architects Cristina Castel-Branco, Ana Luísa Soares, Teresa Chambel and finance manager Iole Sala.

10 Contributos para a Identificação e Caracterização da Paisagem em Portugal Continental (Cancela d'Abreu et al., 2004).

11 Adopted in 1985.

Bibliography and references

Añón, C. (2020). Interview by Raquel Carvalho [in person]. 17th February.

Antrop, M. (2005). Why landscapes of the past are important for the future. *Landscape and Urban Planning*, 70(1–2), 21–34.

Antunes, A.C.D.S. (2019). *A influência alemã na génese da Arquitetura Paisagista em Portugal*. Unpublished PhD thesis, Faculdade de Ciências da Universidade do Porto.

Araújo, I.A. (2008). Sobre a gestão do património agrário e a sustentabilidade das paisagens culturais. In *Património Paisagístico: os caminhos da transversalidade*. Lisboa: Associação Portuguesa Arquitectos Paisagistas, pp. 63–76.

Assembleia da República Portuguesa. (1987). *Lei de Bases do Ambiente*. Lei n.11/87—Diário da República n.º 81/1987, Série I de 1987–04–07, pp. 1386–1397.

Birnbaum, C. (1994). *Protecting Cultural Landscapes: Planning, Treatment and Management of Historic Landscapes*. Washington D.C.: U.S. Department of the Interior, National Park Service.

Brown, J., Mitchell, N., & Beresford, M. (eds.). (2004). *The Protected Landscape Approach Linking Nature, Culture and Community*. Gland, Switzerland and Cambridge, UK: IUCN.

Caldeira Cabral, F. (1967). A Paisagem como Expressão de Cultura. *Brotéria*, LXXXIV(4), 442–452.

Cancela d'Abreu, A., Correia, T.P., & Oliveira, R. (eds.). (2004). *Contributos para a Identificação e Caracterização da Paisagem em Portugal Continental* (5 vol). Lisboa: Direcção-Geral do Ordenamento do Território e Desenvolvimento Urbano.

Carvalho, R., & Marques, T.P. (2019). A evolução do conceito de paisagem cultural. *Revista de Geografia e Ordenamento do Território*, (16), 81–98.

Comission on National Parks and Protected Areas. (1978). *Categories, Objectives and Criteria for Protected Areas*. Switzerland: IUCN/Rockefeller Brothers Fund.

Conselho de Ministros. (2015). *Política Nacional de Arquitectura e Paisagem*. 1.a série—nº 130.

Constituição da República Portuguesa. (1976). Diário da República n.º 86/1976, Série I de 1976–04–10.

Council of Europe. (1995). *Recommendation No R(95) 9 on the Integrated Conservation of Cultural Landscape Areas as Part of Landscape Policies*. https://search.coe.int/cm/Pages/result_details.aspx?ObjectID=09000016804fd025

Council of Europe. (2005). *Faro Convention on the Value of Cultural Heritage for Society*. Treaty series n.º 199.

EUROPA NOSTRA, German Cultural Heritage Committee (DNK), and Prussian Cultural Heritage Foundation (SPK) (eds.). (2018). *The Berlin Call to Action. Cultural Heritage for the Future of Europe*. EUROPA NOSTRA.

Farina, A. (2000). The cultural landscape as a model for the integration of ecology and economics. *BioScience*, 50(4), 313.

Jacques, D. (1995). The rise of cultural landscapes. *International Journal of Heritage Studies*, 1(2), 91–101.

Jones, M. (2003). The concept of cultural landscape: Discourse and narratives (Chapter 3). In Palang, H., & Fry, G. (eds.), *Landscape Interfaces* (1). Dordrecht: Springer, pp. 21–51.

Luengo, A., & Rossler, M. (eds.). (2012). *World Heritage Cultural Landscapes*. Elche: Ayuntamiento de Elche.

Martin, G.J., & James, P.E. (1993). *All Possible Worlds: A History of Geographical Ideas*. 3rd Edn. London: John Wiley & Sons.
Melnick, R.Z., Elkinton, S., Sponn, D., & Saxe, E.J. (1984). Cultural Landscapes: Rural Historic Districts in the National Park System. *Bulletin of the Association for Preservation Technology*, 16(3/4), 80.
Pessoa, F. (1978). *Parques Naturais*. Lisboa: Serviço Nacional de Parques, Reservas e Conservação da Natureza.
Phillips, A. (1998). The nature of cultural landscapes – A nature conservation perspective. *Landscape Research*, 23(1), 21–38.
Pinto, M.A. (2014). *O legado escrito de Francisco Caldeira Cabral Construção do pensamento teórico em arquitetura paisagista*. Unpublished master's thesis, Faculdade de Ciências da Universidade do Porto.
Potthoff, K. (2013). The use of 'cultural landscape' in 19th century German geographical literature. *Norsk Geografisk Tidsskrift – Norwegian Journal of Geography*, 67, 49–54.
Taylor, K. (2008). Landscape and memory: Cultural landscapes, intangible values and some thoughts on Asia. In *16th ICOMOS General Assembly and International Symposium: 'Finding the spirit of place—between the tangible and the intangible'*. Quebec: ICOMOS, pp. 1–14.
Taylor, K. (2017). *Landscape, culture and heritage. Changing perspectives in an Asian context*. Doctoral dissertation, School of Architecture and Built Environment, Deakin University.
Telles, G.R. (2016). *Gonçalo Ribeiro Telles, Textos Escolhidos*. Lisboa: Argumentum.
UNESCO. (1972). *Convention concerning the Protection of the World Cultural and Natural Heritage*. Paris: UNESCO World Heritage Centre.
UNESCO. (2003). *Convention for the Safeguarding of the Intangible Cultural Heritage*. Paris: UNESCO World Heritage Centre.
UNESCO. (2005). *Convention on the Protection and Promotion of the Diversity of Cultural Expressions*. Paris: UNESCO World Heritage Centre.
World Heritage Committee. (1994). *Operational Guidelines for the Implementation of the World Heritage Convention*. Paris: UNESCO World Heritage Centre.

Part III

Portuguese perspectives on theory and methods in Landscape Architecture. Theory versus circumstances

The help of methods in Landscape Architecture

Cristina Castel-Branco

Introduction

Is Landscape Architecture an art? More specifically, can we consider Landscape Architecture to be a fine art whereby the material characteristics of the land, the climate, the local inhabitants, and the culture of the artist coalesce into a million possibilities that then limit the possibility of a universal theory guiding practitioners? Can a creative act involving the manipulation of the Earth and the flora via the landscape architect's vision be enhanced by methodologies that assist the design process and techniques that support representation models? How much do landscape architects borrow from other domains to develop their practice and further their research? Where are we at the start of the 21st century?

When Maria Matos Silva, as editor, asked me to introduce this chapter addressing the topic of theory and methods in Landscape Architecture, I turned to Ian Thomson's article reviewing methods, theories, and research advancements in Landscape Architecture where he tells us that his background was philosophy, which he credits with altering his approach to Landscape Architecture:

> Prior to becoming a landscape architect, I studied philosophy as an undergraduate. . . . As a result I carry around with me certain ways of thinking. . . . The most useful of these is the idea that the meaning of a word is to be found by considering its use.
>
> (Thompson, 2017)

DOI: 10.4324/9781003454007-19

Following Ian Thomson, my background prior to joining the profession was music, and this seems to have affected my Landscape Architecture practice as a designer; it also influences my capacities as an academic researcher and teacher when students evidence elements of artistic effusion.

Francisco Caldeira Cabral was a musician and agricultural engineer prior to introducing Landscape Architecture in Portugal in the early 1940s. Cabral embraced the profession and soon expanded his talents to both teaching and research, later describing the field to an audience at the London Institute of Landscape Architecture in 1966 as follows:

> Landscape Architecture is an art, very subtle, with a technique which is very elaborate and based on a science which is very broad. However, art and science are, of course, related. Perhaps one can explain the relation between art and technique and science by describing the playing of Beethoven sonata—the art is the artistic appreciation of the music; the playing of the piano is a technique and the mechanism of the piano is the result of a construction of science.
>
> (Cabral, 1964, as cited in Andresen, 2001, p. 20)

In search of theories and methods in Landscape Architecture, both in this chapter and more generally, we are guided by Cabral's assertion that Landscape Architecture is an art, or rather a fine art based on science.

Cabral also describes the use of Landscape Architecture, the "materials" of this practice, and its complete dependence on Nature. As such, we can consider Landscape Architecture exceptional in that it has its genesis at the hand of humans but is left to time and natural processes to develop it to its mature aesthetic stage. For Cabral, we only "write the overture" in creating a landscape, leaving the final act to Nature:

> I would say that it [Landscape architecture] is the art of designing the living environment of man. As such it is a fine art, . . . Landscape Architecture works with living space, is dynamic and four-dimensional. . . . the last of the arts in time, . . . the art of our time; because, I think, the recognition of time as a dimension in which we live was never so acute and so truly felt as in our day. . . . music is also an art in which time is important, but there is a great difference since music is entirely commanded by man, while Landscape Architecture is not. We just induce and try to convince nature to collaborate with us. That is what the ancients called ars cooperative naturae.
>
> (Cabral, as cited in Andresen, 2001, p. 20)

This expression was first coined by the Romans, revealing centuries ago the wisdom of a profession capable of cooperating with nature and its laws to create an art product.

Ian Thomson recalls Elisabeth Meyer, who asserted that design and theory must be based on observation and experience, coupled with the sensory (all the senses, not just vision):

> Theoretical work . . . Grounding in the immediate, the particular, and the circumstantial . . . is an essential characteristic of landscape architectural design and theory. Landscape theory must rely on the specific and not the general (1997).
>
> (as cited in Thompson, 2017, p. 46)

And he criticizes Meyer's contradiction; how can the contingent and situational issue become an essence (i.e., something immutable)? Let's embrace Thomson's evocative term "circumstantial" to reinforce the idea of an endless combination of possible circumstances that are dependent on the environment, the local people, and the professional involved in changing the landscape. The epistemology of the word "circumstantial" is clear "occasion and situation", depending on time and place. The spirit of this term implies the inherent fluidity of each situation we encounter to design and plan a landscape, also changing through time, and thus impeding the establishment of some long-desired Landscape Architecture theory or conforming to a position that Thomson refers to as "uncomfortably perhaps", but always at the borders of other research domains he refers to as "neighbouring empires": science, arts, and humanities. At its most authentic, the art of Landscape Architecture is nurtured by an inspired synthesis of the different sciences and requires a multifaceted capacity to invent new spaces grounded on the wisdom of scientific laws.

The book "Research in Landscape Architecture, Methods and Methodology", by Adri Van Den Brink, Diedrich Bruns, Hilde Tobi, and Simon Bell (Van Den Brink et al., 2017), was seminal, and it presents a thorough review of the different methods—notably those of Ian MacHarg, Simon Swafield, and Carl Steinitz—which were instrumental for the analysis of this theme. Landscape research domains identified in the same book were also analysed to compare with the five chapters presented in this third part. To measure the Portuguese contribution to research in Landscape Architecture, a survey of 102 PhD theses clustered by decade was conducted, resulting in a chart showing an exponential curve from two PhDs granted in the 80s to 56 during the 2010–2020 decade.

How much of an art is Landscape Architecture?

Similar to how Cabral views Landscape Architecture as an art, Claude Monet, the great painter of light and water, said about his own garden at Giverny: "*mon plus grand chef-d'oeuvre est mon jardin* [my greatest *chef-d'oeuvre* is my garden]".[1] Monet places the design and building of his garden above all other of his immortal artistic achievements—notably, the many incomparable paintings hanging in the world's finest museums. In referring to Giverny as he did, he catapults his landscape work to aesthetic preeminence, supporting Cabral's equivalency of Landscape Architecture with art.

In the same line of reasoning, the conductor Seiji Osawa wrote:

> The impact of music on man is something very direct and natural, a bit like the effect that poetry produces. The sounds of music penetrate immediately and directly into the listener's heart and instantly trigger feelings of joy or sadness. The harmonies, sounds and rhythms have an immediate impact with almost no use of the brain.[2]

The analogy is so perfect as to be irresistible in its application to the aesthetic perception of landscapes: "The impact of landscape on man is something very direct and natural, . . . the vision of landscape penetrates immediately and directly into the viewers heart". While not all music and landscapes will have that piercing effect on the listener or viewer, I would argue that we have all been moved by landscapes achieved through the cooperation between humans and Nature, that are products of the referred *as co-operative Naturae.*

To complexify the perception of the beholder in the landscape, "Landscape is something mental as well as something physical. Cultural geographers, such as Denis Cosgrove and Stephen Daniels . . . pointed out that 'landscape' is not a neutral term, but an ideologically charged 'way of seeing'"(Thompson, 2017, p. 38). Like a Monet painting hanging in a museum, the product of Landscape Architecture is perceived and interpreted by the person who sees it and depends on his/her background that influences its reception.

When considering Landscape Architecture as an art, one cannot help but turn to Brazilian artist and landscape architect, Roberto Burle Marx (1909–1994). This great master of landscape modernism, with extensive experience in designing and building gardens, also correlated the two:

> I am an artist and I approach the design of gardens as so, not as a painter. . .because the garden involves the dimensions of time and space which must be fully and literally included in the project and not as an illusion in a canvas.[3]

Jacques Leenhardt, a critique d'art writes about Burle Marx: "The way he introduces into space the double aesthetic perception in which both the body and visual perception are called is a central element and deep personal mark of Burle Marx garden design" (Jacques et al, 1994, p. 31). Vera B. Sequeira comments on Burle Marx's artistic compositions

> on the one hand the spatial composition is previously organized, remaining a central element for the legibility of the garden, on the other hand the paths have a form of spread and distribute that do not aim at a point but allow the body, when walking, to alter each of the images it perceives.
>
> (Vera Beatriz, 2001, p. 50)

In other words, the "artistic frames" of each point of view are given for continuous but differentiated perception at each moment.

Adding to this design complexity, instead of relying on one's vision alone for an aesthetic appreciation of a landscape, there are four other senses one can use to perceive a space more fully: smell, touch, hearing, and taste. The Japanese professor Yoshio Nakamura has written about this phenomenological aspect of landscape creation:

> Could we not say that a landscape (fûkei in Japanese) appears to us in a relationship of co-suscitation (engi kankei) established between our body and a place, perceived as a matrix? If the arrangement of the body in the place changes, then new co-stimulation relationships are formed. The place and the body, mediated by the landscape, are firmly chained by these co-stimulating relationships. . . . our body is projected into a field space . . . then walks while tracing lines of landscape embroidery . . . thus the expansion of the body by making the amalgamation of the five senses, walks in the field.[4]

For Burle Marx, the immediate circumstances, the visitor's perception played the pivotal role in creating a designed place. This giant of the profession has left us with a seminal statement that captures this ideology: "*O tempo completa as nossas ideias*" (Time completes our ideas; author's translation).

While essential for our profession, art alone is not sufficient for landscape design composition, thus the influence of science and technology for Landscape Architecture are presented in the next section.

Scientific research domains used in Landscape Architecture methods

Gathering opinions from well-known practitioners to convey and support our conviction that artistic creativity must serve as the genesis for the professional practice of Landscape Architecture is but a first step. The synthesis of other bodies of knowledge and scientific approaches are needed to develop artistic solutions to the endless challenges this multidisciplinary avocation presents. What seems evident is the need for a congregation of knowledge that can "cooperate" with Nature. This recognition was illustrated at Barcelona's Landscape Architecture Biennale in 2012,[5] with an 8 points star diagram[6] (Figure PIII.1).

As we are at the crossroad of other professions, these eight distinct sources of knowledge and inspiration can be successfully integrated in most Landscape Architecture design and planning products. The influence of three well-known and inspiring books in our field are here highlighted, all published in the latter half of the 20th century, to help define Landscape Architecture's professional scope and its hybrid nature from science to art. The first is "Design With Nature" by Ian McHarg,[7] one of last century's most influence Landscape Architecture books that conveys how McHarg advocates for blending science and philosophy in the creation of ecological and productive landscape design. The second is "Design With Climate: Bioclimatic Approach to Architectural Regionalism" by Victor Olgyay,[8]

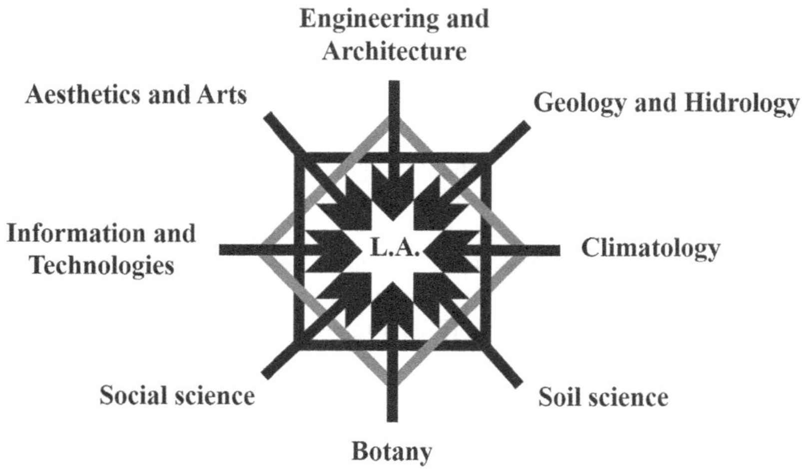

Figure PIII.1 *8-points star diagram* presented by Cristina Castel-Branco at Barcelona's Landscape Architecture Biennale in 2012. The item "Information and Technologies" was not on the original 8 points star diagram and was added after the publication of Carl Steinitz's Geodesign diagram in 2012.

wherein he presciently emphasises the impact of climate on site, materials, and even the movement of air. Deriving principles from engineering, physics, biology, and meteorology, Olgyay demonstrates how science and aesthetics can coalesce in producing sound design concepts. Finally, there is "Design for Human Ecosystems" by John Lyle,[9] a wise landscape architectural researcher, gone too soon, who included in his thoughts the influence of time, regenerative design, and the human dimension in landscape ecology.

Landscape Architecture's evolving methods

During the 1980s, the Landscape Architecture programme in Lisbon (Higher Institute of Agronomy *(Instituto Superior de Agronomia, ISA)*, Technical University of Lisbon) was headed by one of Cabral's first students and leading academician and professional, Sousa da Câmara,[10] who added computer-aided technologies and geographic information systems (GIS). Using an early WANG computer, Sousa da Câmara introduced his students (now the senior teaching staff at the Universities of Porto, Lisbon, Algarve, and Trás-os-Montes e Alto Douro)[11] to the new digital era in Landscape Architecture. This expansion of technological knowledge and skills resulted in a new teaching domain that further increased the multidisciplinary nature of Landscape Architecture. GIS, digital cartography, and AutoCad became required courses in ISA's 5-year Landscape Architecture degree as early as 1990.

Also, Sousa da Câmara's teaching and practice followed "Patrick Geddes (1854–1932) S-A-D, Survey, Analysis and Design, which was used for town planning and adopted by landscape architects" (Thompson, 2017, p. 39) but incorporated an intermediate step of synthesis prior to proposing any changes to the landscape.

1 Analysis of the landscape (topography, geology, water lines and ridges, slopes, etc.).
2 Synthesis/diagnosis of the assets and constraints of a site.
3 Proposal/project presenting the changes to be introduced in the landscape.

Sousa da Câmara's method remains a cornerstone of teaching and practice for those who learned Landscape Architecture in Lisbon's University at the School of Agriculture (ISA, 1977–1987) and now teach at other universities. The interesting concept here is that the analysis step relies on scientific data collected during the survey (and identified in the 8-points-star diagram) and precedes the synthesis step that necessitates the comprehensive and thoughtful incorporation of all data. The proposal/project phase is therefore nurtured by the preceding storage and processing of information (in our brain) about the site and can be followed by an intuitive and creative process, giving way to art. In short, a sequential, thoughtful, and balanced design approach, responding both to the functional and the aesthetic requirements of a site, is essential for both the training and practice of Landscape Architecture.

In 2012 when Carl Steinitz published his seminal work, "A Framework for Geodesign" (Steinitz, 2012), he chose as its cover a four-part diagram that stressed in equal measures the importance of the people of the place, design professions, geographic sciences, and information technologies. Steinitz also advocated for a "six-step-design method", the basis of Geodesign for approaching a large Landscape Architecture problem in complex areas, and proposed six models in sequence:[12] i) representation; ii) process, iii) evaluation, iv) change, v) impact, and vi) decision. Ian Thomson refers this design process as "a flowchart which organised the landscape planning process around a series of models each associated with a question . . . In the full version of the process each of these steps must be gone through three times"[13] (Thompson, 2017, p. 39) and comments that Steinitz's systematic method "appeals to those who wish to bring rigour to their practice, though it probably has little appeal to designers and those who place themselves at the more artistic pole of the landscape profession" (Thompson, 2017, p. 39).

A compared analysis of Steinitz's "six-step method" with Sousa da Câmara's three-step process to approach Landscape Architecture yielded interesting parallels. One of the three Steinitz extra steps is a consequence of a split in Steinitz's framework of the Sousa da Câmara's analysis step into representation models and process models. This separation proposed by Steinitz is useful for two reasons. The first is the fact that some factors in the landscape are static (or nearly) and can be mapped as a fixed consideration rather than dynamic; such a division helps to organise landscape systems in two distinct categories: immutable or mutable. During the representation step we can assess exposure, slope, hydrography geology, soils, annual sunlight hours, and many other factors for which static existing cartography is easy to obtain using GIS. Processes analysis include more dynamic factors that change easily by themselves or are induced by external factors such as population growth, underground water salinization, tides, sediments contamination, animal dynamics, tree pests, fire hazards, transportation systems, and many others. This division between representation and process models encourages a distinct consideration of each factor in terms of how the landscape can be described and how it operates and for whom. Addressing these questions is essential for the next stage: evaluation. The *evaluation model* corresponds to Sousa da Câmara's synthesis/diagnosis step, which is augmented by identifying both assets and liabilities as well as the spatial strategies that every site in and of itself will inspire.

The *change model*, the fourth step, correlates well with the proposal/project phase in Sousa da Câmara's sequence. Steinitz's *impact model* is very useful in that it collects the stakeholders related to the project opinion with a critical question: *What differences might the changes cause*? Although this essential phase could limit the creative design process, the design solution will inevitably be more robust as it reflects the people's opinion and the "pruning" of some potentially negative elements with emphasis on others that will serve the site and the people more faithfully.

The final phase in the Steinitz sequence, the *decision model*, does not exist in the so-called Sousa da Câmara method. This knowledge-based step enables the landscape design team to answer the question "*Should the landscape be changed?*" and choose among different alternatives arriving at a viable and thoughtful solution. The correlation between the Sousa da Câmara and Steinitz methods also helps to explain why Carl Steinitz's six-step process was so easily integrated and adapted into the LINK PhD programme with three associated universities: Lisbon, Porto and Coimbra.

Dissertations, thesis and publications: How is research doing in Portugal?

To approach the quest for theory and methods in Landscape Architecture in Portugal, the last three decades of scholarly research was considered, and the data reflects an analysis of 102 doctoral theses from Portuguese students of Landscape Architecture who graduated in Portugal, having as a background a 5-year degree, or the equivalent post Bologna bachelor's (3 years) and master's degrees (2 years). Also considered was the work of foreign students who came and earned their Landscape Architecture PhDs in Portugal.

The raw results of this data reflect exponential programme growth since the first two PhDs in Landscape Architecture were granted in Portugal in 1987: one woman and one man, both earning their final degree from the School of Agriculture at the University of Lisbon. Both were agronomists with a specialization in Landscape Architecture and were students of Cabral in his Autonomous Course in Landscape Architecture *(Curso Livre de Arquitectura Paisagista, CLAP)*.[14]

During the 1990s, 13 doctoral degrees were registered (six women, six men), from the Universities of Lisbon and Évora, and one each from Aveiro University, Sheffield University in the United Kingdom, and the University of Massachusetts at Amherst in the United States. During the first decade of the 21st century, the number of PhDs was 20 (14 women and six men)—from the University of Lisbon (5), the University of Évora (6), and the rest from universities with younger Landscape Architecture programs: University of Porto (2) and University of Trás-os-Montes e Alto Douro (UTAD) (3). The other three PhD theses were successfully defended at foreign universities and one at the Lisbon Institute of Engineering (IST).

During the second decade of our century the number of PhD graduates jumped to 56 (35 women and 21 men); this dramatic increase can be linked to the creation of an associated PhD programme run by the Universities of Lisbon, Porto and Coimbra, guided by the supervisory and teaching orientation of Carl Steinitz. This associated doctoral degree in Landscape Architecture and urban ecology—the so-called LINK program—reflects the joint decision among the three universities to collaborate instead of competing for both students and teaching faculty, thereby creating a stronger body of knowledge in Landscape Architecture. In a rough analysis, these 56 theses were granted by the Universities of Porto (10), Évora (10), Lisbon (11), foreign universities (5), and a total of five from other Portuguese universities in other domains such as art history. The other 15 doctorates were earned by LINK programme graduates.

Despite this encouraging scenario of expanding PhD research and accomplishment, the scientific papers from Portugal are still relatively scarce, causing us to agree with Maggie Roe in her critique of the profession: "[We] do not assure enough research in Landscape architecture: One of the common cries is that Landscape Architecture does not have a good quantitative research basis and therefore is generally regarded as 'wishy-washy'" (Roe, 2017, p. 68). This assessment may indeed be accurate within the Portuguese academic community—but for what I argue are legitimate reasons. Because we as landscape architects and researchers borrow from other sciences and professions; we are by comparison generalists in many fields and thus comparatively less well equipped than those who research only in their restricted field. Moreover, landscape scholars publish their research articles in several outlets, but compared with other disciplines the choices can be relatively narrow. Maggie Roe also emphasises this difficulty: "A key issue in crossover research areas is that there are often limited outlets for publication because many journals still tend to focus upon disciplinary lines" (Roe, 2017, p. 68). When we use ecological principles and practices to design an area of conservation, we are less knowledgeable than the ecologists; and when we use art history to prepare a master plan for the restoration of a historical garden, we are less familiar with the historic documentation and materials. Consider also the research in health and landscape conducted by Catharine Ward Thompson,[15] which is based on the deep medical research of doctors who know so much more about public health. Botanists are always more knowledgeable about the plants, trees and shrubbery we use in our garden design, so we will be hampered in that output of scholarship compared to those specialists. In contrast, however, landscape architects excel in synthetising information and using it for designing and implementing robust solutions that cooperate with Nature. In essence, then, one cannot be a specialist in all the fields of science that interact in the practice and study of Landscape Architecture; thus, we must know a bit about each "Empire", in the language of Ian Thomson. Nevertheless, this multi-language of ours represents an essential tool that enables us to create landscapes that others can enjoy using the "borrowed knowledge" acquired from other domains.

I note with optimism that although the 99 doctoral dissertations awarded in Portugal have yet to be published as scientific papers, this trend is changing. Since 2015[16] research is being published, and before PhD candidates are awarded their terminal degree, they usually have published four or five peer-reviewed papers. The assertion that "all PhD-level work ought to make a contribution to knowledge" (Thompson, 2017, p. 58) is finally respected in Portugal, and the emphasis on the critical documentation of Landscape Architecture and the theories, models and research that support it is taken seriously.

Research domains in Landscape Architecture

Using the Delphi method, Jurian Meijering and colleagues compiled what panels of experts considered to be important Landscape Architecture research domains and classified them according to percentages from most (#1 at 37%) to least important (#15 at 4%) (Meijering et al., 2017, p. 89). This approach was applied to the 1990s 12 PhD dissertations completed in Portugal. As shown in Table PIII.1, their

Table PIII.1 Summary of the research domains in Landscape Architecture presented by Jurian Meijering et al. The second column presents the number of Landscape Architecture PhD theses in Portugal during the 20th century. The third column presents the distribution of research domains when applied to the 8-point star diagram.

	Research domains	*(% most important)*	*#PT PhD theses (20th century)*	*8-point star correlation*
1	Human dimension of planning and design	37	2	Social sciences
2	Green urban development	33	1	Sciences
3	Built environments and infrastructure	30	1	Arch. Eng.
4	Global landscape issues	30	0	Sciences
5	Theories	24	0	Hybrid
6	Measuring Landscape Architecture performance and impact	24	0	Hybrid
7	Rural and natural environments	22	2	Sciences
8	Historic dimension of planning and design	20	2	Social Sc.
9	Biophysical dimensions of planning and design	20	3	Sciences
10	Values and ethics	17	0	Hybrid
11	Artistic creativity	13	1	Aesthe. Art
12	Policy and governance	9	0	Social Sc.
13	Aquatic environments	7	0	Sciences
14	Tools and technologies	7	0	Info. Techno
15	Landscape Architecture education	4	0	Hybrid

research areas were associated each to one of the 15 research domains in the chart. The resulting list of themes is presented with the number of theses belonging to a domain in the last column.

Note that the three most significant research domains served as dissertation topics for 33% (four total) of the surveyed 12 doctoral theses. Even though the next three research areas (global landscape issues, theories, and measuring Landscape Architecture performance and impact) were identified as important research domains, those topics were not investigated by any of the 12 doctoral candidates. Rather, the following three topical areas (rural and natural environments, historical dimension of planning and design, and biophysical dimensions of planning and design) were most represented (59%) with seven students selecting these research areas and just one student researching artistic creativity. From this simple analysis it may be stated that the diversity of research domains is rich and represents a mature effort in Landscape Architecture research.

Nonetheless, the goal of publishing "evidence-based scholarly papers" appears to have been elusive for these first 12 PhDs students since very little of their research was published; nor did any report a single guiding theory in the vast research domain of Landscape Architecture. We are content, however, that the aggregate research output from Portugal's early doctoral students has helped to better define the scope of the profession, supporting the advancement of its body of knowledge.

Regarding the current profile of Landscape Architecture research in Portugal, the importance of the "research by design" concept cannot be overlooked. Cabral, as well as his teaching successors Ribeiro Telles, Viana Barreto, Edgar Fontes and Sousa da Câmara, were role models and educated the next generation of practitioners and researchers, while also maintaining an active professional practice. As a result through their 35 years of teaching (1989–2024), each of the current four senior professors[17] of Landscape Architecture at the University in Lisbon have taught and advised students, while also creating hundreds of projects for gardens, parks and landscapes within their private offices. These efforts have enabled them to support a thriving "laboratory" of "research for design" and "research by design", thus advancing within their practice the technologies of the digital world, the evolution of materials in outdoor space, the learning through experience that the real world offers and the enrichment of the visual archive that each circumstance of a project can provide.

Despite this rich background of practical experience in the realm of Landscape Architecture, far too little has been published in the academic world of papers that reflect the experiences and assessments of these design professionals. Unfortunately, just a small number of non-peer-reviewed books have been published that fully convey Landscape Architecture projects with their strategies, outdoor solutions, historical inventories and adaptations from other models and procedures. Moreover, there are very few researchers in our field who are not also practitioners, with little time left to stop, collect and report their cumulative process of knowledge and experience.

Beyond Portugal's borders, Nick Dines from the University of Massachusetts must be singled out as particularly adept at teaching, practicing and publishing.

Doctorates completed by Portuguese landscape architects or completed in Portugal in the field of landscape architecture

Date	*Name*	*Gender*	*Country of origin*	*University*	*Faculty / Department*	*PhD programme*	*Thesis title*	*Research domain*	*On-line repository*	*Supervisors*
1987	Maria Antónia de Sá L. C. Castro e Almeida	Female	Portugal	Universidade Técnica de Lisboa	Instituto Superior de Agronomia	Engenharia Agronómica	Aplicação de Métodos de Engenharia Biofísica na Estabilização do Sistema Dunar – Península de Tróia	Agricultura, Silvicultura e Pescas		
1987	Sidónio da Costa Pardal	Male	Portugal	Universidade Técnica de Lisboa	Instituto Superior de Agronomia	Engenharia Florestal	Planeamento do Território – Instrumentos para a Análise Física	Engenharia do Ambiente		
1990	Alexandre D´Orey Cancela D´Abreu	Male	Portugal	Universidade de Évora	Universidade de Évora	Artes e Técnicas da Paisagem	Caracterização do Sistema Biofísico com Vista ao Ordenamento do Território	Artes		Gonçalo Ribeiro Telles
1990	Nuno José de Noronha Mendonça	Male	Portugal	Universidade de Évora	Universidade de Évora	Artes e Técnicas da Paisagem	Para uma Poética da Paisagem	Artes		
1992	Maria Teresa Lencastre de Melo Breiner Andresen	Female	Portugal	Universidade de Aveiro	Universidade de Aveiro	Ciências Aplicadas ao Ambiente	Para a Crítica da Paisagem	Ciências da Terra e Ciências do Ambiente		
1993	Maria Cristina F. Ataíde Castel Branco	Female	Portugal	Universidade Técnica de Lisboa	Instituto Superior de Agronomia	Arquitectura Paisagista	O Lugar e o Significado: os Jardins dos Vice-reis	Engenharia Civil		
1994	José Maria da Silva Pinto Barbosa	Male	Portugal	Universidade de Évora	Universidade de Évora	Artes e Técnicas da Paisagem	Da Praça Pública em Portugal	Artes		

(*Continued*)

(Continued)

Date	*Name*	*Gender*	*Country of origin*	*University*	*Faculty / Department*	*PhD programme*	*Thesis title*	*Research domain*	*On-line repository*	*Supervisors*
1995	Leonel de Sousa Fadigas	Male	Portugal	Universidade Técnica de Lisboa	Faculdade de Arquitectura	Planeamento Urbanístico	A Natureza na Cidade. uma Perspectiva para a Sua Integração no Tecido Urbano	Geografia Económica e Social		
1995	Maria Adalgisa Alves Palmeiro Cruz de Carvalho	Female	Portugal	Universidade de Évora	Universidade de Évora	Artes e Técnicas da Paisagem	Variação da Diversidade Ecológica numa Paisagem Rural em Transformação – o Caso do Vale da Ribeira de Seda-raia	Artes		
1995	Maria da Graça Magalhães do Amaral Neto Lopes Saraiva	Female	Portugal	Universidade Técnica de Lisboa	Instituto Superior de Agronomia	Arquitectura Paisagista	O Rio Como Paisagem – Gestão de Corredores Fluviais no Quadro do Ordenamento do Território	Engenharia Civil		
1996	Aurora da Conceição Parreira Carapinha	Female	Portugal	Universidade de Évora	Universidade de Évora	Artes e Técnicas da Paisagem	Da Essência do Jardim Português	Artes		Gonçalo Ribeiro Telles
1997	Maria Manuela Cordes Cabêdo Sanches Raposo de Magalhães	Female	Portugal	Universidade Técnica de Lisboa	Instituto Superior de Agronomia	Arquitectura Paisagista	Morfologia da Paisagem	Engenharia Civil		Gonçalo Pereira Ribeiro Telles

(*Continued*)

(Continued)

Date	*Name*	*Gender*	*Country of origin*	*University*	*Faculty / Department*	*PhD programme*	*Thesis title*	*Research domain*	*On-line repository*	*Supervisors*
1998	Luís Paulo Ribeiro	Male	Portugal			Landscape Architecture	The cultural landscape and the uniqueness of place: A greenway heritage network for landscape conservation of Lisbon Metropolitan Area	Urban planning| Area planning & development| Landscaping	https://scholar-works.umass.edu/dissertations/AAI9823768/	
1998	Maria da Conceição Martins Lopes de Castro	Female	Portugal	Universidade de Évora	Universidade de Évora	Artes e Técnicas da Paisagem	Estudo Eco-morfológico de Duas Herbáceas Ruderais da Flora Portuguesa – Salvia Verbenaca I. e Scabiosa Atropurpurea I.	Artes		
1999	Paulo Jorge Rodrigues Farinha Marques	Male	Portugal	University of Shefield	Faculty of Architectural Studies University of Sheffield (UK)		The integration of reservoir margins in northern Portugal	Landscape Architecture		
2000	Robert Manners Moura	Male	Moçambique	Universidade de Trás-os-Montes e Alto Douro			Para o Desenvolvimento de Sistema Integrado de Conservação da Natureza em Portugal	Ciências da Terra e Ciências do Ambiente		Maria Teresa Lencastre de Melo Breiner Andersen

(Continued)

(Continued)

Date	Name	Gender	Country of origin	University	Faculty / Department	PhD programme	Thesis title	Research domain	On-line repository	Supervisors
2001	Isabel Maria Henriques Martinho da Silva	Female	Portugal				Historic Anthropogenic Factors Shaping the Rural Landscape of Portugal's Interior Alentejo			
2002	Maria Teresa Amaro Alfaiate	Female	Portugal	Universidade Técnica de Lisboa	Instituto Superior de Agronomia	Arquitectura Paisagista	Expressão dos Valores do Sítio na Paisagem	Engenharia Civil		Gonçalo Ribeiro Telles
2003	Maria José Dias Curado	Female	Portugal	Universidade de Aveiro	Universidade de Aveiro	Ciências Aplicadas ao Ambiente	O Planeamento e a Gestão das Paisagens Culturais – Alto Douro Vinhateiro: Contributos e Aplicação	Ciências da Terra e Ciências do Ambiente		Maria Teresa Lencastre Melo Breiner Andresen
2006	Ana Luengo Añon	Female	Espanha	Universidade de Évora	Universidade de Évora	Artes e Técnicas da Paisagem	Aranjuez – Utopia y Realidad	Artes		Aurora Conceição Parreira Carapinha
2006	Domingos Manuel Mendes Lopes	Male	Portugal	Universidade de Trás-os-Montes e Alto Douro	Universidade de Trás-os-Montes e Alto Douro		Estimating Net Primary Production in Eucalyptus Globulus and Pinus Pinaster Ecosystems in Portugal	Agricultura, Silvicultura e Pescas		
2006	Luísa Maria Lopes Pires Genésio Simeão Versos	Female	Portugal	Universidade Técnica de Lisboa	Instituto Superior de Agronomia	Arquitectura Paisagista	Desenho da Paisagem	Engenharia Civil	http://hdl.handle.net/10198/6191	Sidónio da Costa Pardal

(*Continued*)

(Continued)

Date	*Name*	*Gender*	*Country of origin*	*University*	*Faculty / Department*	*PhD programme*	*Thesis title*	*Research domain*	*On-line repository*	*Supervisors*
2007	Ana Isabel Costa Febrero de Queiroz	Female	Portugal	Universidade do Porto	Faculdade de Ciências	Arquitectura Paisagista	A Paisagem de Terras do Demo. Análise Integrada da Sua Evolução ao Longo do Século XX e Valorização Através da Literatura	Engenharia Civil		Maria Teresa Lencastre de Melo Breiner Andresen; Nuno Miguel dos Santos Ferrand de Almeida
2007	Ana Luísa Brito dos Santos de Sousa Soares Ló de Almeida	Female	Portugal	Universidade Técnica de Lisboa	Instituto Superior de Agronomia	Arquitectura Paisagista	O Valor das Árvores e Floresta Urbana de Lisboa	Engenharia Civil		Maria Cristina da Fonseca Ataíde Castelo-Branco; Francisco Manuel Cardoso de Castro Rego
2008	Ana Luísa Figueiredo Lavrador da Silva	Female	Portugal	Universidade de Évora	Universidade de Évora	Artes e Técnicas da Paisagem	Paisagens de Baco – Identidade, Mercado e Desenvolvimento – Estudo de Percepção e de Representação Aplicado Às Regiões Demarcadas: Vinhos Verdes, Douro, Dão, Bairrada e Alentejo	Artes		Maria Teresa Pinto Correia
2008	Anne Marije Van Doorn	Female	Países Baixos	Universidade de Évora	Universidade de Évora	Artes e Técnicas da Paisagem	Reconstruction, Understanding and Evaluation of Rural Landscape Changes in Peripheral Areas	Artes		Maria Teresa Pinto Correia

(*Continued*)

(Continued)

Date	Name	Gender	Country of origin	University	Faculty / Department	PhD programme	Thesis title	Research domain	On-line repository	Supervisors
2008	Helena Cristina Fernandes Ferreira Madureira	Female	Portugal	Universidade do Porto	Faculdade de Ciências	Arquitectura Paisagista	A Infra-estrutura Verde da Bacia do Leça- uma Estratégia para o Desenvolvimento Sustentável na Região Metropolitana do Porto	Engenharia Civil		Maria Teresa Lencastre de Melo Breiner Andresen; Ana Maria Rodrigues Monteiro de Sousa
2008	Marcelo de Almeida Oliveira	Male	Brasil	Universidade de Évora	Universidade de Évora	Artes e Técnicas da Paisagem	A Preservação dos Espaços Verdes em Cidades Classificadas no Brasil. Ouro Preto e Olinda, um Olhar Sobre Duas Realidades	Artes		Aurora Conceição Parreira Carapinha
2008	Maria do Rosário Gaspar de Oliveira	Female	Portugal	Universidade de Évora	Universidade de Évora	Artes e Técnicas da Paisagem	Gestão Integrada da Paisagem no Concelho de Mértola: Novos Parâmetros para a Implementação da Política Agrícola Comum	Artes		Maria Teresa Pinto Correia
2009	Ana Isabel Loupa Ramos	Female	Portugal	Universidade Técnica de Lisboa	Instituto Superior Técnico	Engenharia do Ambiente	As Dinâmicas da Paisagem Rural – a Reformulação de Cenários como Instrumento de Apoio ao Planeamento			Maria da Graça Magalhães do Amaral Neto Lopes Saraiva

(Continued)

(Continued)

Date	*Name*	*Gender*	*Country of origin*	*University*	*Faculty / Department*	*PhD programme*	*Thesis title*	*Research domain*	*On-line repository*	*Supervisors*
2009	Cláudia Patrícia Oliveira Fernandes	Female	Moçambique	Universidade de Trás-os-Montes e Alto Douro	Universidade de Trás-os-Montes e Alto Douro	Ciências do Ambiente	Caracterização e Previsão do Comportamento Florístico-estrutural da Vegetação da Região Demarcada do Douro	Ciências da Terra e Ciências do Ambiente		António Luis Crespi
2009	Desidério Luís Sares Batista	Male	Portugal	Universidade de Évora	Universidade de Évora	Artes e Técnicas da Paisagem	Paisagem, cidade e Património. O Sistema Urbano Olhão-Faro-Loulé. Propostas para uma Estratégia de Intervenções Integradas de Requalificação Urbana e Valorização Ambiental	Artes		Virgolino Ferreira Jorge
2009	Mário Luís Soares Fortes	Male	Portugal				A xestión da auga na paisaxe romana do occidente peninsular		https://minerva.usc.es/xmlui/handle/10347/2532	Fernando Acuña Castroviejo
2010	Teresa Dulce Portela Marques	Female	Portugal	Universidade Técnica de Lisboa	Instituto Superior de Agronomia	Arquitectura Paisagista	Dos Jardineiros Paisagistas e Horticultores do Porto de Oitocentos ao Modernismo na Arquitectura Paisagista em Portugal	Engenharia Civil		Luís Paulo Almeida Faria Ribeiro

(Continued)

(Continued)

Date	*Name*	*Gender*	*Country of origin*	*University*	*Faculty / Department*	*PhD programme*	*Thesis title*	*Research domain*	*On-line repository*	*Supervisors*
2010	Tiago Monteiro Henriques dos Santos	Male	Portugal	Universidade Técnica de Lisboa	Instituto Superior de Agronomia	Arquitectura Paisagista	Landscape and Phytosociology of the Paiva Rivers Hydrographical Basin	Engenharia Civil		José Carlos Augusta da Costa
2011	Diana Surová	Female	República Checa	Universidade de Évora	Instituto de Investigação e Formação Avançada	Artes e Técnicas da Paisagem	Identification of Landscape Preferences for Integration in the Decision Support System: Contributions for Management of Montado Like Multifuncional System	Artes		Maria Teresa Pinto Correia
2011	Luís Carlos Loures	Male	Portugal	Universidade do Algarve	Faculdade de Ciências e Tecnologia	Ciências e Tecnologias do Ambiente	Planning and Design in Postindustrial land transformation: East bank Arade river, Lagoa – Case study	Ciências da Terra e Ciências do Ambiente	http://hdl.handle.net/10400.1/2382	Thomas Panagopoulos
2011	Maria da Conceição Marques Freire	Female	Portugal	Universidade de Évora	Instituto de Investigação e Formação Avançada	Artes e Técnicas da Paisagem	Para uma Diferente Aproximação ao Ensino do Projecto de Arquitectura Paisagísta	Artes	http://hdl.handle.net/10174/11089	Alexandre D´Orey Cancela D´Abreu
2011	Rute Sousa Matos	Female	Portugal	Universidade de Évora	Instituto de Investigação e Formação Avançada	Artes e Técnicas da Paisagem	A Reinvenção da Multifuncionalidade da Paisagem em Espaço Urbano-Reflexões	Artes	http://hdl.handle.net/10174/4304	Aurora Conceição Parreira Carapinha

(*Continued*)

(Continued)

Date	*Name*	*Gender*	*Country of origin*	*University*	*Faculty / Department*	*PhD programme*	*Thesis title*	*Research domain*	*On-line repository*	*Supervisors*
2012	Cláudia Margarida Macedo Pereira Ávila Gomes	Female	Portugal	Universidade Técnica de Lisboa	Instituto Superior de Agronomia	Arquitectura Paisagista	O conceito de carácter da paisagem e a sua aplicação na gestão de áreas protegidas: caso de estudo dos Açores.	Engenharia Civil		Luís Paulo Ribeiro
2012	Eva Maria Silveirinha de Oliveira	Female	Portugal	Edinburgh College of Art at the University of Edinburgh			Immigrants and Public Open Spaces: attitudes, prefer-ences and uses			Simon Bell, Penny Travlou
2012	Pedro Miguel Ramos Arsénio	Male	Portugal	Universidade Técnica de Lisboa	Instituto Superior de Agronomia	Arquitectura Paisagista	Qualidade da Paisagem e Fitodiversidade. Contributo para o Ordenamento e Gestão de Áreas Costeiras de Elevado Valor Natural	Engenharia Civil		Javier José Loidi Arregui
2013	Andreia Verónica Cor-reia Quintas	Female	Portugal	Universidade do Porto	Faculdade de Ciências	Arquitectura Paisagista	Desenho e Avaliação da Estrutura Verde Urbana. Modelo de implementação para a promoção da qualidade de vida e valorização da paisagem urbana.	Engenharia Civil		Maria José Dias Curado

(Continued)

(Continued)

Date	Name	Gender	Country of origin	University	Faculty / Department	PhD programme	Thesis title	Research domain	On-line repository	Supervisors
2013	Jorge Manuel Frazão Cancela	Male	Portugal	Universidade de Lisboa	Faculdade de Arquitectura	Urbanismo	Agricultura Urbana em Estrutura Ecológica Municipal.	Geografia Económica e Social	http://hdl.handle.net/10400.5/7183	Leonel de Sousa Fadigas
2014	Bibiana Maria Barracosa Rodrigues Ramos	Female	Portugal	Universidade do Algarve	Faculdade de Economia	Gestão da Inovação e do Território	VALORIZAÇÃO DOS PROCESSOS DE RECUPERAÇÃO DA PAISAGEM ATRAVÉS DA PARTICIPAÇÃO PÚBLICA COM ÊNFASE NA QUALIDADE VISUAL E ESTÉTICA DA PAISAGEM. UM CONTRIBUTO PARA A GESTÃO, PLANEAMENTO E DESIGN DA PAISAGEM	Economia e Gestão		Thomas Panagopoulos; Teresa Anderson; Jon Burley
2014	Cati Oliveira Dinis	Female	Portugal	Universidade de Évora	Instituto de Investigação e Formação Avançada	Gestão Interdisciplinar da Paisagem	Cork Oak (Quercus suber L.): A structural-functional 3D Approach	Agricultura, Silvicultura e Pescas	http://hdl.handle.net/10174/17314	Maria do Rosário Gamito de Oliveira; Nuno Manuel Cabral de Almeida Ribeiro; Peter Sorovy

(*Continued*)

(Continued)

Date	*Name*	*Gender*	*Country of origin*	*University*	*Faculty / Department*	*PhD programme*	*Thesis title*	*Research domain*	*On-line repository*	*Supervisors*
2014	João Rui Dias Pinto Ribeiro	Male	Portugal	Universidade de Évora	Instituto de Investigação e Formação Avançada	Gestão Interdisciplinar da Paisagem	Modelação do crescimento do Pinheiro-manso e sua aplicabilidade a nível da paisagem: aplicação na região da margem esquerda da Bacia do Guadiana.	Agricultura, Silvicultura e Pescas	http://hdl.handle.net/10174/11751	Nuno Manuel Cabral Almeida Ribeiro
2014	Maria Verónica Conte de Morais Fernandes	Female	Portugal	Universidade de Lisboa	Faculdade de Arquitectura	Design	Co – Design em Acções de Pintura em Fachadas Residenciais: Expressão das Identidades Individuais e Colectivas na Construção do Espaço Público	Artes	http://hdl.handle.net/10400.5/8706	Maria João Durão dos Santos, Professora Auxiliar da FA – ULisboa
2014	Nuno Joaquim Costa Cara de Anjo Lecoq	Male	Portugal	Universidade de Évora	Instituto de Investigação e Formação Avançada	Gestão Interdisciplinar da Paisagem	A Herdade da Contenda. Uma paisagem única como paisagem de referência	Agricultura, Silvicultura e Pescas	http://hdl.handle.net/10174/11412	Maria Teresa Amado Pinto Correia; Luís Paulo Almeida Faria Ribeiro

(*Continued*)

(Continued)

Date	*Name*	*Gender*	*Country of origin*	*University*	*Faculty / Department*	*PhD programme*	*Thesis title*	*Research domain*	*On-line repository*	*Supervisors*
2015	Carlos António Bastos de Morais Guerra	Male	Portugal	Universidade de Évora	Instituto de Investigação e Formação Avançada	Gestão Inter-disciplinar da Paisagem	A spatially explicit framework to assess soil erosion prevention vulnerability: contributions ecosystem service mapping and management in Mediterranean land use systems	Outras Ciências Naturais	http://hdl.handle.net/10174/16432	Maria Teresa Amado Pinto Correia
2015	Frederico Meireles Alves Rodrigues	Male	Portugal	Universidade do Porto	Faculdade de Ciências	Arquitectura Paisagista	Da Especificidade do Parque Português Contemporâneo	Outras Ciências Naturais	http://hdl.handle.net/10216/94540	Paulo Jorge Rodrigues Farinha Marques
2015	Gerald Luckhurst	Male	Reino Unido	University of Bristol		Archaeology & Anthropology	Monserrate, an English landscape garden in Portugal (1790-1901)		https://ethos.bl.uk/OrderDetails.do?uin=uk.bl.ethos.684645	
2015	Henrique de Menezes de Almeida Pereira dos Santos	Male	Portugal	Universidade do Porto	Faculdade de Ciências	Arquitectura Paisagista e Ecologia Urbana	Evolução da paisagem rural do continente português no século XX	Engenharia Civil	http://hdl.handle.net/10216/80814	Paulo Jorge Rodrigues Farinha Marques
2015	Laura Cristina Roldão Costa	Female	Portugal	Universidade do Porto	Faculdade de Ciências	Arquitectura Paisagista	A vegetação na Implementação de Projetos de Execução em Arquitetura Paisagista. Caracterização e Definição de Critérios de Avaliação	Engenharia Civil	http://hdl.handle.net/10216/95584	Luis paulo Almeida Faria Ribeiro

(Continued)

(Continued)

Date	*Name*	*Gender*	*Country of origin*	*University*	*Faculty / Department*	*PhD programme*	*Thesis title*	*Research domain*	*On-line repository*	*Supervisors*
2015	Maria Manuela Tavares da Silva Lopes	Female	Portugal	Universidade de Coimbra	Faculdade de Ciências e Tecnologia	Arquitectura Paisagista e Ecologia Urbana	Qualificação da Paisagem de Parques urbanos ribeirinhos através da valorização da sua função educativa. Estudo de caso: O Parque Oriental da cidade do Porto como Laboratório escolar da Paisagem em meio urbano	Engenharia Civil	http://hdl.handle.net/10316/27019	Paula Cristina de Oliveira Castro
2015	Maria Teresa Ferraz Lúcio de Sales	Female	Portugal	Universidade de Évora	Instituto de Investigação e Formação Avançada	Gestão Inter-disciplinar da Paisagem	Percursos na Paisagem	Outras Ciências Naturais	http://hdl.handle.net/10174/14522	Maria Teresa Pinto Correia
2015	Maria Teresa Pinto de Mes-quita Cabral de Moncada Bettencourt da Câmara	Female	Portugal	Universidade do Porto	Faculdade de Ciências	Arquitectura Paisagista e Ecologia Urbana	Contributos da Arquitetura Paisagista para o espaço público de Lisboa (1940 – 1970)	Outras Ciências Naturais	http://hdl.handle.net/10216/85668	Teresa Dulce Portela Marques
2015	Paula Maria Correia da Silva Simões	Female	Portugal	Universidade de Évora	Instituto de Investigação e Formação Avançada	Artes e Técnicas da Paisagem	Guardiões da Paisagem – Os Montes Alentejanos, Lugares de Memória	Artes	http://hdl.handle.net/10174/17365	Aurora Conceição Parreira Carapinha
2015	Ricardo Jorge Quinto Canas	Male	Portugal	Universidade de Évora				Doutoramento em Análise e Gestão de Ecossistemas		

(*Continued*)

(Continued)

Date	Name	Gender	Country of origin	University	Faculty / Department	PhD programme	Thesis title	Research domain	On-line repository	Supervisors
2015	Sérgio Rui Borreicho Coelho Godinho	Male	Portugal	Universidade de Évora	Instituto de Investigação e Formação Avançada	Gestão Interdisciplinar da Paisagem	Remote sensing geographical modelling to assess montado change patterns: causes, impacts and biogeophisical process	Agricultura, Silvicultura e Pescas	http://hdl.handle.net/10174/17738	Artur José Freire Gil; Maria Teresa Amado Pinto Correia; Nuno Alexandre Gouveia de Sousa Neves
2015	Sónia Maria Loução Martins Talhé Azambuja	Female	Portugal	Universidade de Lisboa	Faculdade de Letras	História da Arte, e de Arte, Património e Restauro	A ICONOGRAFIA DA NATUREZA E DA PAISAGEM NA PINTURA PORTUGUESA DOS SÉCULOS XV E XVI. IMAGENS E SIGNIFICADOS	História e Arqueologia	http://hdl.handle.net/10451/22524	Vítor Manuel Guimarães Verissimo Serrão, Teresa Andresen
2016	Agnieszka Anna Olszewska	Female	Polónia	Universidade do Porto	Faculdade de Ciências	Arquitectura Paisagista e Ecologia Urbana	Contemplative Values of Urban Parks and Gardens: Applying Neuroscience to Landscape Architecture	Outras Ciências Naturais	http://hdl.handle.net/10216/85632	Paulo Jorge Rodrigues Farinha Marques
2016	Ana Margarida Pinto da Fonseca	Female	Portugal	Universidade de Évora	Instituto de Investigação e Formação Avançada	Gestão Interdisciplinar da Paisagem	"EMERGETIC EVALUATION OF CATTLE REARING IN A MONTADO FARM"	Ciências da Terra e Ciências do Ambiente	http://hdl.handle.net/10174/19944	Carlos Alberto Falcão Marques; Maria Teresa Amado Pinto Correia

(Continued)

(Continued)

Date	*Name*	*Gender*	*Country of origin*	*University*	*Faculty / Department*	*PhD programme*	*Thesis title*	*Research domain*	*On-line repository*	*Supervisors*
2016	Ana Paula Pinto Gomes da Silva	Female	Portugal	Universidade de Lisboa	Instituto Superior de Agronomia	Arquitectura Paisagista	O PENSAMENTO ECOSSISTÉMICO NO PROJETO DA PAISAGEM: UMA REFLEXÃO SOBRE A INTEGRAÇÃO DE CONCEITOS ECOLÓGICOS NA PRÁTICA PROJETUAL DA ARQUITETURA PAISAGISTA CONTEMPORÂNEA	Artes		Maria Manuela Raposo Magalhães
2016	André Gomes Nascimento	Male	Portugal	Universidade do Porto	Faculdade de Ciências	Arquitectura Paisagista e Ecologia Urbana	Recuperação e integração paisagística de áreas aluvionares degradadas pela extração de inertes	Outras Ciências Naturais	http://hdl.handle.net/10216/89780	Paulo Jorge Rodrigues Farinha Marques
2016	Clara Martins Caldeira da Ponte e Sousa	Female	Portugal	Universidade de Évora	Instituto de Investigação e Formação Avançada	Artes e Técnicas da Paisagem	A condução de espécies herbáceas espontâneas em Portugal, na sua utilização em prados de flor	Outras Ciências Agrárias	http://hdl.handle.net/10174/17756	Maria da Conceição Martins Lopes Castro; Mário José Gouveia Pinto Rodrigues de Carvalho

(*Continued*)

(Continued)

Date	*Name*	*Gender*	*Country of origin*	*University*	*Faculty / Department*	*PhD programme*	*Thesis title*	*Research domain*	*On-line repository*	*Supervisors*
2016	Francisca Lima	Female	Portugal				The Role of Open Spaces in the Future of Depopulated Urban Environments,			
2016	Inês Guedelha Rebelo Marques Duarte	Female	Portugal	Universidade de Lisboa	Instituto Superior de Agronomia	Engenharia Florestal e dos Recursos Naturais	Alteração na diversidade das paisagens em Portugal Continental e do seu valor para a Conservação – entre 1970 e 2007	Agricultura, Silvicultura e Pescas		Francisco Manuel Cardoso de Castro Rego
2016	Maria João Pereira Monteiro Gomes Mendes Ferreira	Female	Portugal	Universidade Nova de Lisboa	Faculdade de Ciências Soci-ais e Humanas	Estudos Urbanos	A Cidade Caminhada: Ambiência Urbana experienciada – duas visitas guiadas no Centro Histórico de Lisboa.	Geografia Económica e Social	http://hdl.handle.net/10362/20276	Luís António Vicente Baptista
2016	Maria Matos Silva	Female	Portugal	Universidade de Barcelona			Public space design for flooding. Facing the challenges presented by climate change adaptation	Public Space and Urban Regeneration		
2016	Miguel António Navas Cândido	Male	Portugal	Universidade de Lisboa	Instituto Superior de Agronomia	Arquitectura Paisagista	ENTRE O SILÊNCIO E A PAISAGEM. Lógica e Imaginação na Construção de um Pensamento Devir Paisagem	Artes		Luis Paulo Almeida Faria Ribeiro

(Continued)

(Continued)

Date	*Name*	*Gender*	*Country of origin*	*University*	*Faculty / Department*	*PhD programme*	*Thesis title*	*Research domain*	*On-line repository*	*Supervisors*
2016	Pedro José Calaza Martínez	Male	Espanha	Universidade de Lisboa	Instituto Superior de Agronomia	Arquitectura Paisagista	INFRAES-TRUCTURA VERDE, SALUD PÚBLICA Y ACTIVIDAD FÍSICA. EVIDENCIAS DE SU RELACIÓN. CASO DE ESTUDIO: LA CORUÑA, ESPAÑA	Artes		Luis Paulo Faria Ribeiro
2016	Selma Beatriz de Almeida Nunes da Pena Baldaia	Female	Portugal	Universidade de Lisboa	Instituto Superior de Agronomia	Arquitectura Paisagista	Reserva Ecológica Nacional – Delimitação a Nível Nacional	Artes	http://hdl.handle.net/10400.5/13499	Maria Manuela Cordes Cabêdo Sanches Raposo de Magalhães
2017	Andreia de Sousa Saavedra Cardoso	Female	Portugal	Universidade de Lisboa	Instituto Superior de Agronomia	Arquitectura Paisagista	PLANEAMENTO AGRO-ALIMEN-TAR E AGRO-URBANISMO NAS REGIÕES METROPOLI-TANAS	Artes	http://hdl.handle.net/10400.5/14960	Maria Manuela Cordes Cabêdo Sanches Raposo de Magalhães
2017	João Miguel de Sampaio e Paiva Marques da Cruz	Male	Portugal				Coltivare la Natura, Costruire il Paisaggio			

(Continued)

(Continued)

Date	*Name*	*Gender*	*Country of origin*	*University*	*Faculty / Department*	*PhD programme*	*Thesis title*	*Research domain*	*On-line repository*	*Supervisors*
2017	Maria Amélia da Fonseca dos Santos	Female	Portugal	Universidade de Évora	Instituto de Investigação e Formação Avançada	Artes e Técnicas da Paisagem	Teoria e Prática de Projecto em Arquitectura Paisagista	Outras Ciências da Engenharia e Tecnologias	http://hdl.handle.net/10174/21344	Aurora da Conceição Parreira Carapinha, Desidério Luís Sares Batista
2017	Natália Sofia Canelas da Cunha	Female	Portugal	Universidade de Lisboa	Instituto Superior de Agronomia	Arquitectura Paisagista	THE NATIONAL ECOLOGICAL NETWORK AND A LAND MORPHOLOGY MODEL. An application to Portugal	Agricultura, Silvi-cultura e Pescas	http://hdl.handle.net/10400.5/14947	Maria Manuela Cordes Cabêdo Sanches Raposo de Magalhães
2018	Artur Miguel Dias Gonçalves da Costa	Male	Portugal	Universidade de Lisboa	Instituto Superior de Agronomia	Arquitectura Paisagista e Ecologia Urbana	WAYS OF ENGAGING. Re-assessing effects of the relationship between landscape architecture and art in community involvement and design practice	Artes	http://hdl.handle.net/10400.5/17517	Luís Paulo Almeida Faria Ribeiro
2018	Gisela Mourão Coelho dos Santos	Female	Portugal	Universidade de Lisboa	Instituto Superior de Agronomia	Arquitectura Paisagista e Ecologia Urbana	GESTÃO SUSTEN-TÁVEL DA ÁGUA NO PROJETO DE ARQUITETURA PAISAGISTA EM CLIMA MEDITERRÂNICO	Artes	http://hdl.handle.net/10400.5/15831	Ana Luísa Brito dos Santos de Sousa Soares

(*Continued*)

(Continued)

Date	*Name*	*Gender*	*Country of origin*	*University*	*Faculty / Department*	*PhD programme*	*Thesis title*	*Research domain*	*On-line repository*	*Supervisors*
2018	José Miguel Esteves Lameiras	Male	Portugal	Universidade do Porto	Faculdade de Ciências	Arquitectura Paisagista	Modelação Digital do Terreno em Arquitetura Paisagista: Interação e Visualização	Outras Ciências Naturais	http://hdl.handle.net/10216/111566	Paulo Jorge Rodrigues Farinha Marques
2018	Marisa da Silva Graça	Female	Portugal	Universidade do Porto	Faculdade de Ciências	Arquitectura Paisagista	Performance of Urban Green Areas in Ecosystem Services Proficiency: A case study in Porto, Portugal	Outras Ciências Naturais	http://hdl.handle.net/10216/112026	Mário Manuel de Miranda Furtado Campos Cunha
2019	Ana Catarina Dias Santos Antunes	Female	Portugal	Universidade do Porto	Faculdade de Ciências	Arquitectura Paisagista e Ecologia Urbana	A influência alemã na génese da Arquitetura Paisagista em Portugal	Outras Ciências Naturais	https://hdl.handle.net/10216/121946	Teresa Dulce Portela Marques
2019	Ana Cristina Bento Lourenço	Female	Portugal	Universidade de Lisboa	Instituto Superior de Agronomia	Arquitectura Paisagista	A PAISAGEM DA CIDADE. CONVERGÊNCIA ENTRE ECOLOGIA E MORFOLOGIA URBANA. O CASO DE LISBOA	Artes	http://hdl.handle.net/10400.5/21225	Maria Manuela Cordes Cabêdo Sanches Raposa de Magalhães
2019	Luís Filipe Lucas Barroso	Male	Portugal	Universidade de Évora	Instituto de Investigação e Formação Avançada	Gestão Interdisciplinar da Paisagem	Transition pathways: Assessing land management strategies in Mediterranean areas	Ciências da Terra e Ciências do Ambiente		Gianluca Brunori; Maria Teresa Amado Pinto Correia

(Continued)

(Continued)

Date	*Name*	*Gender*	*Country of origin*	*University*	*Faculty / Department*	*PhD programme*	*Thesis title*	*Research domain*	*On-line repository*	*Supervisors*
2019	Maria Luísa Monteiro Franco	Female	Portugal	Universidade de Lisboa	Instituto Superior de Agronomia	Arquitectura Paisagista	Ordenamento e recuperação da paisagem do Baixo Guadiana. Um contributo da Arquitectura Paisagista	Artes	http://hdl.handle.net/10400.5/18322	Manuel Fernando Belo Moreira; Maria Manuela Cordes Cabêdo Sanches Raposo Magalhães
2019	Rute Correia Martins Cegonho	Female	Portugal	Universidade de Lisboa	Instituto Superior Técnico	Engenharia do Ambiente	THE ROLE OF SOCIAL-ECOLOGICAL SYSTEMS AND INHERENT ECOSYSTEM SERVICES IN ATTRACTING SKILLED IN-MIGRANTS, INDUCING TRANSITIONS IN RURAL AREAS	Engenharia do Ambiente		Maria do Rosário Sintra de Almeida Partidário
2020	Ana Luísa Arrais Falcão Beja da Costa	Female	Portugal	Universidade de Lisboa	Instituto Superior de Agronomia	Arquitectura Paisagista e Ecologia Urbana	MANGROVES OF MAPUTO. TOWARDS URBAN RESILIENCE THROUGH GREEN INFRA-STRUCTURE	Artes	http://hdl.handle.net/10400.5/21196	Luis Paulo Almeida Faria Ribeiro; Sílvia Manuela Branco Jorge

(Continued)

(Continued)

Date	*Name*	*Gender*	*Country of origin*	*University*	*Faculty / Department*	*PhD programme*	*Thesis title*	*Research domain*	*On-line repository*	*Supervisors*
2020	Paulo Jorge Ávila da Silveira	Male	Portugal	Universidade de Évora	Instituto de Investigação e Formação Avançada	Gestão Interdisciplinar da Paisagem	Modelo de Interacção Espacial: Um instrumento para o planeamento e análise do território	Ciências da Terra e Ciências do Ambiente	http://hdl.handle.net/10174/30100	Eduardo Manuel Vieira de Brito de Azevedo; Tomaz Lopes Cavalheiro Ponce Dentinho, Eduardo Manuel Vieira de Brito de Azevedo
2021	André Filipe Samora Arvela	Male	Portugal	Universidade Nova de Lisboa	Faculdade de Ciências Sociais e Humanas	Alterações Climáticas e Políticas de Desenvolvimento Sustentável	Estratégias para a implementação de uma infraestrutura verde como instrumento de reforço da resiliência do território turístico num quadro de alterações climáticas: a região do Algarve no caso de Silves-Albufeira-Loulé	Ciências da Terra e Ciências do Ambiente	http://hdl.handle.net/10362/119537	Jorge Ricardo Ferreira

(*Continued*)

(Continued)

Date	*Name*	*Gender*	*Country of origin*	*University*	*Faculty / Department*	*PhD programme*	*Thesis title*	*Research domain*	*On-line repository*	*Supervisors*
2021	Diana Filipa Borges Teixeira Fernandes	Female	Portugal	Universidade do Porto	Faculdade de Ciências	Arquitectura Paisagista e Ecologia Urbana	A paisagem ribeirinha em contexto urbano. Um modelo de planeamaneto e governação para a bacia hidrográfia do Rio Tinto	Outras Ciências Naturais	https://hdl.handle.net/10216/135785	Maria José Dias Curado
2022	Ana Catarina Patoilo Teixeira	Female	Portugal	Universidade do Porto	Faculdade de Ciências	Arquitectura Paisagista	Towards a New Nature in Cities: Understanding Novel Urban Ecosystems in the Anthropocene	Outras Ciências Naturais	https://hdl.handle.net/10216/141602	Cláudia Patrícia Oliveira Fernandes
2022	Maria de Fátima Guedes de Andrade de Oliveira Bacharel	Female	Portugal	Universidade de Évora	Instituto de Investigação e Formação Avançada	Artes e Técnicas da Paisagem	A Paisagem Como Centro De Decisão – reflexão sobre a relação ecologia e cultura na paisagem de defesa da Raia Seca da Fronteira Leste de Portugal	Artes	http://hdl.handle.net/10174/32109	Aurora da Conceição Parreira Carapinha; Isabel Alexandra Joaquina Ramos
2022	Sandra Cristina Paul Fernandes Mesquita	Female	Portugal	Universidade de Lisboa	Instituto Superior de Agronomia	Arquitectura Paisagista e Ecologia Urbana	Illustrations, gardens, and herbaria. The discovery and dissemination of Madeira's plants from its early days to the work of Richard Thomas Lowe		https://www.repository.utl.pt/handle/10400.5/27453	Maria Cristina da Fonseca Ataíde Castel-Branco Alarcão Júdice Miguel Pinto da Silva Menezes de Sequeira Nuno de Carvalho Conde Senos

(*Continued*)

(Continued)

Date	*Name*	*Gender*	*Country of origin*	*University*	*Faculty / Department*	*PhD programme*	*Thesis title*	*Research domain*	*On-line repository*	*Supervisors*
2022	Bruno Marques	Male	Portugal	University of Otago, Dunedin, New Zealand		Landscape Architecture and Planning	Māori health, wellbeing and landscape architecture: Therapeutic environments and the Wairarapa region	Geography	http://hdl.handle.net/10523/13588	Claire Freeman; Lynette Carter
2022	Lígia Maria Ferreira Vaz de Figueiredo	Female	Portugal	Universidade de Lisboa	Faculdade de Arquitectura	Restauro e Gestão Fluvial	Metodologia de avaliação da paisagem fluvial : perspetivar o rio no espaço urbano		https://www.repository.utl.pt/handle/10400.5/28037	Ana Isabel Loupa Ramos Maria da Graça Neto Lopes Saraiva Fátima Bernardo
2023	Maria Raquel Farinha Jorge de Carvalho	Female	Portugal	Universidade do Porto	Faculdade de Ciências	Arquitectura Paisagista e Ecologia Urbana	Paisagem Cultural: do conceito à prática profissional. A rota da transumância entre a Serra da Estrela e as Campinas de Idanha	Outras Ciências Naturais	https://hdl.handle.net/10216/149100	Teresa Dulce Portela Marques; Cristina Castel-Branco

(*Continued*)

(Continued)

Date	*Name*	*Gender*	*Country of origin*	*University*	*Faculty / Department*	*PhD programme*	*Thesis title*	*Research domain*	*On-line repository*	*Supervisors*
2023	Ricardo Jorge de Almeida Dias Ribeiro	Male	Portugal	Universidade de Lisboa	Faculdade de Arquitectura	Arquitetura	A Urbanidade das Áreas Costeiras. O Lugar Arquitectónico na Paisagem. Contributo para uma interpretação integrada da ocupação urbana do Parque Natural da Ria Formosa	Artes	http://hdl.handle.net/10400.5/28005	José Duarte Centeno Gorjão Jorge; Maria Manuel Raposo Magalhães
2022?	Cláudia Taborda	Female	Portugal			Landscape Architecture				

Sources		
RENATES – Registo Nacional de Teses e Dissertações	https://renates2.dgeec.mec.pt/ppsq.asp	
CIÊNCIAVITAE	https://www.cienciavitae.pt/	Sónia Talhé Azambuja, 13/08/2023
ORCID, which stands for Open Researcher and Contributor ID	https://orcid.org/	revisão Raquel Carvalho 04/09/2023

Co-authored with C.W. Harris, Dine's "Time-Saver Standards for Landscape Architecture" has long represented an invaluable book for the profession and is the result of teaching and private practice data collecting. This falls in the category of Murphy's "'procedural theories', which originate from design practice and the academic development and technical application of knowledge in a social setting'"([2005], as cited in Thompson, 2017, p. 38).

Ian Thomson has called attention to this double insight of experiencing and publishing:

> What particularly topics might a researcher who is a landscape architect be better place to tackle than someone from a different discipline? Perhaps these topics include such things as the landscape design process in general, the design practices of particular landscape designers and offices, the history of designed landscapes, the education of landscape architects, the aesthetic, social and ecological values underlying landscape interventions, evaluations of the effectiveness of design intervention and so on.
>
> (Thompson, 2017, p. 38)

These themes should be considered an inspiration for all those who consider sharing their empirical recollection required for Landscape Architecture practice.

Concluding thoughts

The challenges associated with the search for theories and methods in Landscape Architecture, coupled with the role of Portuguese professionals in generating knowledge in these areas, can be synthesised in seven main statements:

1 Art continues to have a strong role in Landscape Architecture and should not be considered distinct from the scientific approach required of a Landscape Architecture professional.
2 Landscape Architecture as a transdisciplinary profession has long owed much to science, art and the social sciences, with the more recent 21st-century addition of the information technologies to the range of "borrowed" domains that enrich this profession.
3 An 8-point-star diagram, representing eight distinct sources of knowledge and inspiration for changing a landscape, is presented, where Landscape Architecture is centred as a cross-disciplines profession.
4 Research areas that contribute to theories and methods in Landscape Architecture have been organised in a chart (Meijering et al., 2017); these 15 research domains relate directly to the 8-point-star diagram, combining subjects and advancing the scholarship associated with Landscape Architecture.
5 To measure the Portuguese contribution to research in Landscape Architecture, a survey of 102 theses clustered by decade was conducted, resulting in a chart showing an exponential curve from two PhDs granted in the 80s to 56 in 2010–2020. A sampling of a dozen 20th-century doctoral dissertations awarded in

Portugal was closely analysed and fit well in the 15 diverse domains of interest identified in Jurian's chart.

6 No real theory synthetised in an equation has been reached, and Landscape Architecture only provides methods helped by technology. This grants landscape architects a certain freedom and fulfillment to proceed in their long-lasting dialogue with the land and the Earth.
7 Compared to other disciplines, the profession of Landscape Architecture lacks a rich history and prevalence of published academic peer-review books and papers. Rather, Landscape Architecture is advanced through professional practice and its relation to teaching and nontraditional research, which has long been valued in Portugal—though only scantily published. In this vein, Ian Thomson encourages us to embrace the diversity of the field as rationale enough for continued research and investigation:

> Murphy's chapter on procedural theory is much more straightforward. Under this rubric come such matters as the design process, design programming, data gathering and analysis, landscape planning and landscape suitability analysis. . . . All of these topics are closely related to the profession of Landscape Architecture and if one were looking solely for theory that has developed out of the practice of landscape architects themselves, this would probably be the best places to look
>
> (Thompson, 2017, p. 39).

Overall, it is here argued that Landscape Architecture evidences a long heritage of encapsulating a "science-in-art-out" process of creation. A statement that is defended in this chapter on "Portuguese perspectives on theory and methods in Landscape Architecture". In contributing to this book on "Portuguese Landscape Architecture Education, Heritage and Research: 80 Years of History", this chapter highlights the contemporary practice of Landscape Architecture as having deep roots in multiple disciplines, while remaining a coherent body of knowledge and practice.

Notes

1 Giverny Guidebook, 2010.
2 Castel-Branco, C. (2002). *Jardins com história*. Ed. Inapa, p. 27, quoting Osawa, Seiji. (2000) *The might of Music*, Ulisses, Alitalia Italia.
3 Author's translation.
4 Castel-Branco, C. *in Autour de Nakamura*, Yoshio, coord. Cyrille Marlin, Presses universitaires de Bordeaux, 2021, p. 121, quoting *La raison-coeur des co-suscitations paysagères—les fluctuations du paysage entre corps, lieu et language—Ebisu*, colloque de Shin Hirayu "De la chose au fait dans les milieux humains" 2013
5 Though information and technologies was not in the original 2012 diagram, it was added after the publication of Carl Steinitz's diagram in 2013 as an essential and rich domain required to work in Landscape Architecture.
6 The item information and technologies was not in the original 8-points-start diagram and was added after the publication of Carl Steinitz's Geodesign diagram in 2012.
7 McHarg, I. (1969). *Design With Nature*. New York: Natural History Press.

8 Olgyay, V. (1963). *Design With Climate: Bioclimatic Approach to Architectural Regionalism*. Princeton: Princeton University Press.
9 Lyle, J. (1985). *Design for Human Ecosystems: Landscape, Land use, and Natural Resources*. New York: Van Nostrand Reinhold.
10 For further information: https://80anosap.isa.ulisboa.pt/manuel-sousa-da-camara/.
11 University of Porto (Teresa Andresen, Paulo Marques, Teresa Marques, Isabel Martinho da Silva) University of Lisbon (Luís Paulo Ribeiro, Cristina Castel-Branco, João Ferreira Nunes), University of Algarve (Paula Farrajota) UTAD (Robert Moura, Laura Costa).
12 Since 2009, with the help of Carl Steinitz, the LINK PhD and the master's programs in Lisbon have incorporated Steinitz's six-step method with good results in terms of both student motivation and in project planning and completion.
13 More acurate information about Steinitz's Geodesign method in "A Framework for Geodesign—Changing Geography by Design" (2012).
14 *Curso livre* was a postgraduate course offered to agronomists and forestry engineers at the School of Agriculture, and its subjects are discussed in the introductory text of this book.
15 Catharine Ward Thompson is Professor of Landscape Architecture at the Edinburgh School of Architecture and Landscape Architecture (ESALA). Her research tackles 21st century demands of open space, exploring the environmental-behaviour interactions and emphasising the evidence that access to natural environments can offer health benefits, particulary related do mental well-being. She explores landscape design, stemming from her research on salutogenic environments, to improve community well-being and quality of life.
16 This dates corresponds to the enforcement of a Portuguese law allowing PhD theses to be completed by a number of refereed scientific articles thus promoting publication.
17 Teresa Alfaiate, Luís Paulo Ribeiro, Cristina Castel-Branco and João Ferreira Nunes.

Reference list

Andresen, T. (2001). *Francisco Caldeira Cabral*. Surrey, UK: Landscape Design Trust Monograph Series.

Jacques, L. *et al.* (1994). *Dans les jardins de Burle Marx*. Actes Sud: Crestet Centre d'Arts.

Meijering, J., Tobi, H., Van Den Brink, A., & Bruns, D. (2017). Assessing research priorities and quality. In *Research in Landscape Architecture. Methods and Methodology*. London: Routledge, Taylor & Francis Group.

Roe, M. (2017). The challenge of publication. In Adri van den Brink, Diedrich Bruns, Hilde Tobi, Simon Bell (eds.), *Research in Landscape Architecture. Methods and Methodology*. London: Routledge, Taylor & Francis Group.

Steinitz, C. (2012). *A Framework for Geodesign—Changing Geography by Design*. 1st Edn. Redlands, CA: ESRI Press.

Thompson, I. (2017). The role of theory. In *Research in Landscape Architecture. Methods and Methodology*. London: Routledge, Taylor & Francis Group.

Van Den Brink, A., Bruns, D., Tobi, H., & Bell, S. (eds.). (2017). *Research in Landscape Architecture. Methods and Methodology*. London: Routledge, Taylor & Francis Group.

Vera Beatriz, S. (2001). *Burle Marx: Transverse Landscapes*. Cosac Naify.

15 In between revolutions

Drivers of landscape change in Interior Alentejo from the Liberal Revolution to the April Revolution

Isabel Martinho da Silva

Introduction

Research on drivers of landscape change has been a central topic in the last decades. It allows us to gain insight about past and present drivers and relevant actors and to use this knowledge to inform future decisions and lead landscape evolution towards a more sustainable and resilient path (Bürgi et al., 2022). Drivers of landscape change are often classified as political, socio-economic, cultural, technological, and environmental (Bürgi et al., 2005; Van der Sluis et al., 2019).

Ancient landscapes, landscapes shaped by the economic ancient regime,[1] were substantially shaped by environmental constraints. In most Europe countries, the transition to the so-called modern landscapes or post-liberal landscapes occurred around the 1850s with the introduction of technical innovations (Bürgi et al., 2005). In Portugal, we can claim that the 1820 Liberal Revolution created the conditions to the emergence of these landscapes (Silbert, 1966).

According to Van der Sluis et al. (2019), recent landscape changes in EU countries are perceived to be driven mainly by the EU policy and the global economy. The EU policy is perceived as the main driver of landscape changes at local level, with common agricultural policy (CAP) being acknowledge as the main evolution driver in rural and/or agricultural areas.

The goal of this research is to describe the evolution of Interior Alentejo (IA) landscape and identify its main drivers of change between two important historical marks: the 1820 Liberal Revolution and the 1974 April Revolution. The Liberal Revolution substituted a constitutional regime for absolutism, introducing major legal, juridical, fiscal, and administrative reforms. These reforms resulted in major changes in the country's rural landscapes. The April Revolution inaugurated a democracy in Portugal, creating the necessary conditions to its integration in the European Economic Community (later renamed as European Union) in 1986. This integration made CAP the main driver of rural landscape changes from there on.

Data and methods

For purposes of this research, IA, composed of 30 counties, is defined as the region located south of High Alentejo and east of Litoral Alentejo (Figure 15.1). It has a

DOI: 10.4324/9781003454007-20

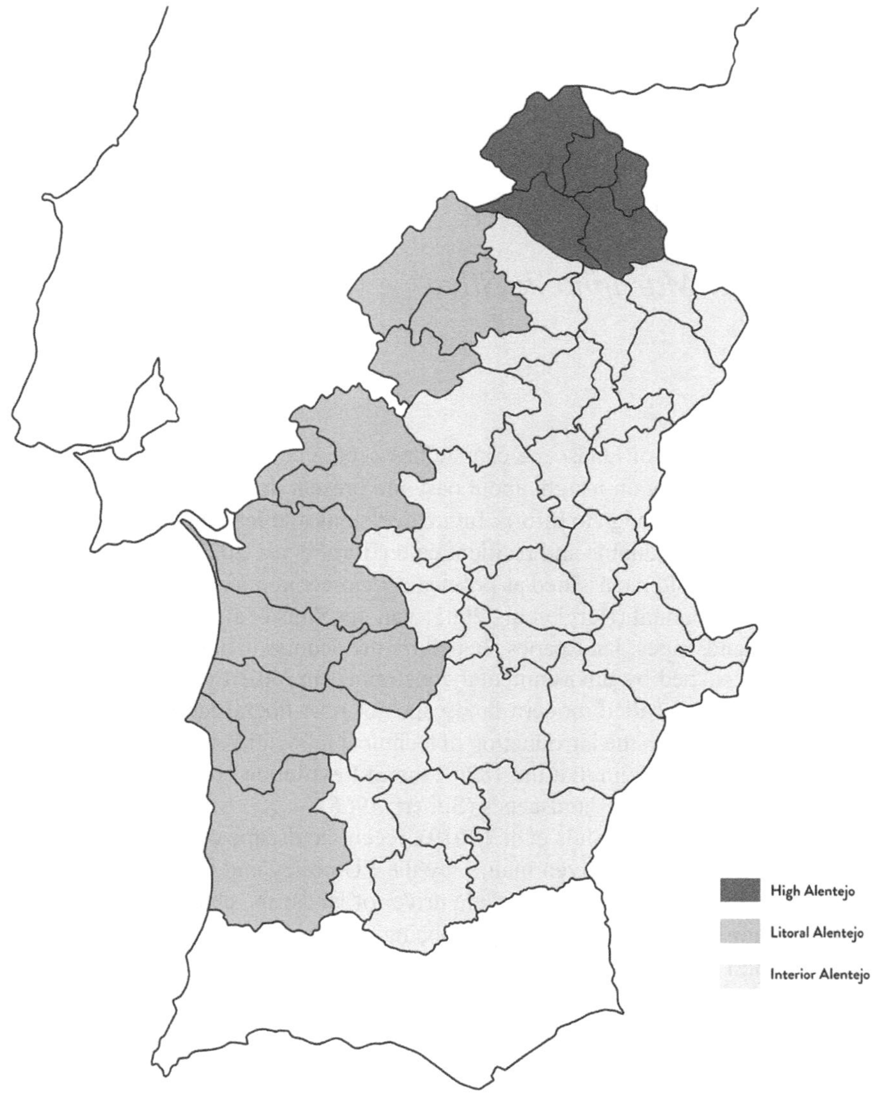

Figure 15.1 Location of Interior Alentejo.

Source: Image by the author.

continental Mediterranean climate and a native flora dominated by the holm oak (*Quercus rotundifolia*). Soil fertility varies; more fertile areas are located on clay soils and less fertile areas on schist soils.

The study was made based on an extensive literature review, extracting and decoding information from publications from neighbouring disciplines, such as history, sociology, and geography, and on the analysis of land use and agricultural census statistical data to build a coherent landscape narrative and identify and classify drivers of change.

The evolution of Interior Alentejo landscape in between revolutions

The Liberal Reforms

The 1820 Liberal Revolution terminated the Ancient Regime in Portugal. Liberal reforms brought major changes to the IA landscape, arising mainly from three political drivers: land redistribution, fiscal charge reduction, and the abolition of collective rights. To make land available to those cultivating it or wishing to cultivate it, the Liberal Government abolished the indivisibility and inalienability of land, expropriated former landowners (Crown, Military Orders, Church, aristocracy), and sold at auction confiscated lands. This sale sought to provide wider access to land to promote agriculture development. The second goal was achieved, but the sale didn't correct asymmetries in land distribution, as the bourgeoisie bought the largest share (Caldas, 1991). To allow the capitalization needed to develop agriculture, several feudal tributes were abolished (Caldas, 1991; Vaquinhas & Neto, 1993), and the 1867 Civil Code ended common grazing for being incompatible with the new definition of land ownership as an absolute, individual, and exclusive right and stimulated the privatization of common lands. As agrarian collectivism had been long blamed for the underdevelopment of IA agriculture, most of its common lands became private property. The privatization of commons and the end of common grazing reduced the region's poor peasants to the condition of journeymen (Caldas, 1991; Vaquinhas & Neto, 1993). This, together with the new class of landowners, changed the social structure of the region. The post Liberal Revolution social structure was a fundamental (social) driver of landscape change in the region.

While the liberal reforms hadn't resulted in a wider access to the land, they substantially increased the area of regularly cultivated land in IA.[2] The new class of landowners carried out large land clearances and adopted new technologies to increase productivity. Landscape intensification was also aided by the market expansion allowed by the construction of roads and railways and an increase of labour offer caused by the demographic growth and the abolishment of common rights. Grain fields with shorter rotations and modern *montados* (an enclosed holm oak forest with evenly spaced trees, regular cereal cultivation and regular maintenance[3]) substituted for scrub and forest (Balabanian, 1980; Caldas, 1991; Vaquinhas & Neto, 1993). However, landscape intensification, varied with soil fertility and population density. Despite some montado intensification scrub continued to dominate the landscape of low fertility and low population areas.

The milling industry

Until the 1870s, the expansion of cultivated lands resulted mainly from an increase of wheat fields. The milling industry, arising in the 1850s, substituted the cheaper

American wheat for national wheat, due to a free trade policy, causing a substitution of wheat fields for pastures and montados, as the market was favourable to livestock production. The milling industry led also to the decay and loss of many wind and water mills settled in the IA landscape (Cutileiro, 1971; Caldas, 1991; Vaquinhas & Neto, 1993).

Wheat protectionism

In 1899 a fixed price for wheat was established, 25 to 30% higher than the former price, and any import of this grain prior to the selling of national production was prohibited. These measures increased the consumer price of flour and bread and became known as the Hunger Law (Feio, 1949; Ribeiro et al., 1987; Caldas, 1991; Vaquinhas & Neto, 1993). The Hunger Law led to the clearance of large areas of scrub and to an increase of cultivated areas: open fields, montados, and olive tree co-planted fields. While in fertile soils trees were felled to maximize wheat production, in poor soils oaks and olive trees were maintained to supplement the small grain productivity. In low fertility areas tree-covered fields came to dominate the landscape (Ribeiro et al., 1987; Santos, 2015).

Wheat protectionism decreased during the First World War due to the need to decrease the price of bread (Caldas, 1991), but it resumed in 1929 when the first wheat campaign was launched. This campaign aimed to achieve wheat self-sufficiency in the country. Policies included a fixed price for wheat, technical support (selected seeds, chemical fertilizers, machinery, and technical advice to farmers), a low interest public credit, and incentives for land clearances (Cutileiro, 1971; Caldas, 1991). The subsidy of 200 escudos (1 euro) per hectare of land cleared eradicated the last tracts of scrub. Everything was cultivated, even the steepest slopes and thinnest soils that erosion soon carried away (Gomes et al., 1945; Feio, 1949; Cutileiro, 1971; Caldas, 1991). In fertile soils, open fields of grain became the dominant landscape. In poor soils, grain fields increased in a first phase, but the exhaustion of soil reserves and erosion soon led the landscape to be dominated by grazing land (long rotations) (Cutileiro, 1971; Feio, 1949; Ribeiro et al., 1987).

If the wheat rush destroyed some tree-covered landscapes, soon trees had to be replanted to guarantee economic (income complement to livestock and cereal production) and ecological (erosion control) sustainability, leading to an increase of montados and olive tree co-planted fields (Cutileiro, 1971; Ribeiro et al., 1987).

The rural exodus and African swine fever

Together with wheat protectionism, labour surplus was fundamental to the expansion of open fields of wheat in IA. This surplus resulted from a demographic increase, the abolition of common rights, and the lack of jobs outside agriculture. As grain farming didn't provide for a constant labour demand over the year, unemployment was the rule after the summer harvest and when soaked soil prevented weeding during the winter. Without access to common resources, the IA population endured severe difficulties (Feio, 1949; Baptista, 1978; Rosas, 1994). The central government proposed several measures to overcome the employment crisis in the region: industrialization, distribution of lands, and irrigation projects; all fiercely opposed by the large farmers as precarious work favoured their activity. To prevent emigration, they lobbied for a programme of public works, financed by the local and central governments, to absorb labour surplus in periods of low farming work demand. With wages below those of agriculture, public works were interrupted in times of intense labour in the fields (Cutileiro, 1971; Caldas, 1991; Rosas, 1994). These strategies became ineffective after the 1950s due to the halt of work migrations from other regions of the country and emigration to the industrial and urban centres in the litoral or to other European countries.

Labour shortage had a great impact on IA agricultural landscapes. Changes arose from mechanization, some private irrigation projects financed by the 1946 Farm Improvement Policy, and the construction of some public irrigation structures.[4] But despite some expansion of irrigated crops, wheat continued to dominate the IA fertile lands, as landowners were able to maintain wheat subsidies regardless of the imports increase resulting from the non-conversion of national agriculture to new consumer demands (Rosas, 1994).

The landscape of poor soils became more extensive, producing livestock, olive oil, and cork (Baptista, 1978). Despite the extensification, the area of holm oak montado continued to decrease. Several adverse factors led this landscape to almost disappear in the 1960s: i) the increase of management costs; ii) changes in consumer preferences; and iii) the outbreak of African swine fever. Until the 1950s, holm oak montado maintenance was self-supporting. Charcoal men did all the pruning, first for half and later for all the charcoal produced, and sharecroppers did the necessary cultivation to control shrubland and regenerate pastures. The substitution of bottled gas for charcoal transferred to montado owners the costs of the expensive pruning operation (Balabanian, 1980). To this added the better contract conditions obtained by sharecroppers, in face of the labour deficit.

New consumer preferences towards lean pork meat and the 1957 outbreak of African swine fever deprived the holm oak montado of its main production, the Iberian pig, leading to a long-lasting decline of this agro-sylvo-pastoral system (Baptista, 1978; Balabanian, 1980; Correia, 1993; Santos, 2015). There was some substitution of cattle and sheep for black pigs, but many holm oak montados were abandoned or transformed into hunting reserves,[5] a major landscape change arising from the extensification process (Baptista, 1978).

Table 15.1 and Figure 15.2 show land uses in 14[6] IA counties in the 1880s (data source: Feio, 1998), 1950s (data source: Feio, 1998), and 1968 (data source: Agricultural Census, 1968[7]). Between the 1880s and the 1950s, the area of open fields increased in all the counties; montado area only decreased in Almodovar, Moura, and Mourão; and the area of scrub decreased in all the counties, Mourão exempted (with an 10% increase). In the 1950s, open fields occupied more than half of the total area of 11 counties. Montados tended to dominate the landscape in counties dominated by poor soils, such as Barrancos, Portel, Almodovar, and Moura.[8] Scrub almost disappeared in all counties.

Table 15.1 Land use cover in 14 IA counties (% of total area) in the 1880s, 1950s, and 1968.

	Open Fields			***Permanent Crops***			***Montado***			***Forest***			***Scrub***		
	(1880s)	***(1950s)***	***(1968)***	***(1880s)***	***(1950s)***	***(1968)***	***(1880s)***	***(1950s)***	***(1968)***	***(1880s)***	***(1950s)***	***(1968)***	***(1880s)***	***(1950s)***	***(1968)***
Aljustrel	38.0	74.3	47.4	2.71	5.99	3.80	10.42	18.48	6.34	0.02	0.08	0.02	47.74	0.25	NA
Almodovar	15.8	57.6	42.7	0.20	0.87	1.10	57.21	44.04	34.43	0.01	0.01	0.01	26.74	2.93	NA
Alvito	21.3	57.5	44.7	4.57	9.38	7.68	23.19	32.15	25.05	0.00	0.16	0.00	50.35	0.06	NA
Barrancos	12.6	13.5	5.3	1.51	3.43	4.80	70.36	70.50	39.27	0.05	0.07	0.05	14.02	12.04	NA
Beja	40.0	65.0	62.8	4.87	6.99	14.67	16.70	26.26	10.08	0.01	0.05	0.01	37.15	0.89	NA
Cuba	40.5	60.4	41.7	12.56	10.12	7.44	24.91	27.66	6.29	0.50	0.02	0.49	20.11	0.69	NA
Évora	56.8	61.1	48.1	6.46	5.95	3.80	28.44	31.44	22.26	0.15	0.63	0.13	7.42	0.11	NA
Mértola	62.9	91.7	48.1	0.17	0.54	0.48	7.13	7.66	3.96	0.00	0.08	0.00	29.17	1.26	NA
Moura	23.9	34.9	32.2	9.47	18.51	17.14	53.15	40.19	26.73	0.01	0.04	0.01	12.33	5.53	NA
Mourão	55.3	62.7	35.5	5.20	8.38	9.24	32.12	20.98	3.32	0.00	0.09	0.00	5.81	6.28	NA
Portel	20.8	32.5	23.1	3.19	7.44	9.96	52.99	58.15	32.80	0.31	0.32	0.31	22.06	1.00	NA
Serpa	22.7	51.8	32.1	9.58	20.35	14.29	23.39	26.13	14.36	0.00	0.03	0.00	43.69	0.78	NA
Viana	31.1	59.8	35.0	5.23	4.60	4.83	25.24	33.54	25.94	0.11	0.45	0.11	37.74	1.05	NA
Vidigueira	45.5	52.6	27.7	9.83	12.31	10.32	23.66	32.62	12.58	0.08	0.10	0.08	19.46	0.89	NA

Source: Author.

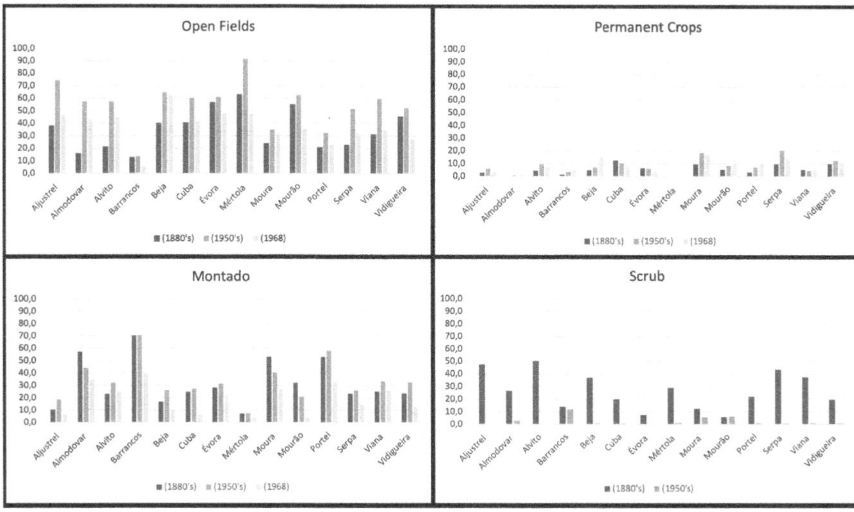

Figure 15.2 Landscape change in 14 IA counties (% of total area).

Source: Author.

Unlike previous data, the 1968 Agricultural Census has several deficiencies. It excludes large county areas and undermeasures several land uses, leading land use decreases to be unconclusive. However, area increases, despite all the possible errors, may be indicative of a trend. Permanent crops increased in several poor counties: Almodovar, Barrancos, Mourão, and Portel. This increase, together with the increase of forest[9] area in all counties, is probably a consequence of the decline of the holm oak montado. One cannot evaluate the evolution of scrub, as the 1968 census didn't assess this land use.

Discussion and conclusions

As aforementioned, this research aimed to describe the evolution of IA landscape and identify its main drivers of change between 1820 and 1974. Table 15.2 provides a summary of this evolution for open fields (OF), permanent crops (PC), montados (M), and scrub (S) in areas of poor soils and rich soils, while identifying the main drivers of change by category. Until the late 19th century, changes resulted mainly from the liberal reforms (political drivers with socio-economic implications). A new class of capitalist landowners, dominated by the bourgeoisie, substituted for a class of absentee landowners, and a land structure based on private property substituted for a quite complex and hierarchical social structure largely supported by common rights. This transformed the IA ancient landscape, described in detail by Silbert (1966), into a modern landscape. The new farmers, aided by a labour surplus arising from a demographic increase and the abolishment of common rights, entailed a notorious expansion of cultivated lands at the expense of

Table 15.2 Drivers of landscape change in IA (1820–1974).

Date	***Chronology*** ***Drivers of change (category)***	***Poor Soils***				***Rich Soils***			
		OF	***PC***	***M***	***S***	***OF***	***PC***	***M***	***S***
1820	**Liberal Revolution**					+/i	(–+	+	–
1821	*Feudal tributes abolished (P)*								
1832	**Agrarian Revolution Law**								
	Donations of Crown lands revoked (P)								
	Tithe abolishment (P)								
1834	**Male religious Orders extinct**								
	Lands confiscated by the State (P)								
1835	*Sale in auction of national lands begins (P)*								
1850	*Milling industry (begin) (T)*								
1853	*Construction of roads and railroads (begin)(P/T)*					+			
1857	*Liberalization of grain trade (P/SE)*					+	+	+	–
1867	**Civil Code**	–		+	+	–	+	+/i	–
	Common grazing abolished (P)								
	Privatization of common lands (P)								
1885	*Introduction of chemical fertilizers (T)*					+	+	+	–
1889	*Wheat Protectionism (P)*					+/i			
1899	*Hunger Law (P)*	+	+	+/i	–	+/i	+	+/i	–
1914–1918	**World War I**								
	Decline in wheat protectionism (P)								
	Boom in charcoal price (SE)								
1929	**First Wheat Campaign**	–/e	–	–	+	–/e	–	–	+
	Wheat protectionism (P)								
	Low interest public credit (P)								
	Technical support (T)								
	Subsidy to land clearances (P)								
1939–1945	**World War II**	+/i	+/i	+/i	–	+/i	+/i	–	–
	Boom in charcoal price (SE)								
1950	*Bottled gas (T)*			–	+			–	+
1957	*African Swine Fever (E/T)*	–/e	+	–	+	–/i	+	–	+
1960	*Rural exodus (SE)*								
	Mechanization (T)								
	New consumer preferences (SE)								
1974	**April Revolution**								
	(P) political; (SE) socio-economic; (E) environmental; (T) technological	(+) area increase; (–) area decrease; (i) intensification; (e) extensification							

Source: Author.

scrub and forest. This result is supported by Van Vliet et al. (2015), which identifies the farmer as a key driver of agricultural land use change.

Economic drivers, such as the liberalization of grain trade and the volatility of livestock prices, brought some instability to this evolution trend. But technical innovations (chemical fertilizers, selected seeds, and machinery) together with wheat protectionism led the expansion of cultivated lands to unprecedented levels and to some ecologically unsustainable situations, due to soil depletion and erosion. Ecological factors (soil fertility and erosion risk) became determinant in poor soils and led to an expansion of the holm oak montado, olive tree co-planted fields, and grazing lands.

Wheat protectionism continued to shape the IA agricultural landscape in fertile soils until the 1974 revolution, despite the rural exodus occurring after the 1950s. However, the area of holm oak montados suffered a sudden decline due to a conjuncture of adverse technological and social drivers: the substitution of bottled gas for charcoal, the outburst of African swine fever, rural exodus, and new consumer preferences. This decline resulted in landscape extensification.

Acquired insight about the drivers and actors of landscape change in IA in the studied period, when modern landscapes were shaped, allows us to predict future evolutions and inform planning and political decisions regarding current challenges affecting the region's landscapes (Bürgi et al., 2005, 2022). The region is currently facing an intensification and simplification of its landscape around the Alqueva Dam and in other irrigation areas, due to the cultivation of intensive monocultures, namely intensive olive groves. This fact is already aggravating the extreme drought affecting the region, due to climate changes. As suggested by Bürgi et al. (2005), regulations and subsidies, namely in the context of CAP, can be enacted to counter these forces of change, as occurred in the past with the wheat campaigns, and lead IA landscapes to a more resilient and sustainable evolution. Sustainable tourism, exploring the potential of montado and other traditional landscapes, well planned solar parks, and agricultural practices and products adapted to the ecological conditions of the region can be solutions to protect the landscape heritage of the region and to finally achieve the social justice sought by the Liberal Revolution and April Revolution.

Notes

1 According to Silbert (1966) the leading role of agriculture in the economic life is what defines the "economic ancient regime" (p. 121).
2 In Alentejo, untilled lands had been reduced to 50% of the region's area in the 1870s (Pery, 1875).
3 The literature acknowledges three types of montado: cork oak montados, holm oak montados, and mixed montados (composed by cork oaks and holm oaks). In IA, the holm oak montado is dominant.
4 Areas benefited by public irrigation structures were to be expropriated, parceled, and distributed among the landless population, but it was the class of large farmers that benefited from the public investment (Rosas, 1994).
5 In 1974, the area of hunting reserves totaled about 15% of the total Alentejo area.
6 Land use evolution was provided for only 14 counties as those were the counties with data available for the 1880s and 1950s.
7 Inquérito às explorações agrícolas do continente (distritos de Beja, Évora, e Portalegre). Lisboa: INE.
8 Curiously, Mértola, one of the poorest IA counties, had the lowest montado area.
9 Given the 1968 Census deficiencies, under the forest category are considered the forests composed by species other than the holm oak and cork oak and the item "untilled lands with forest trees" (Martinho da Silva, 2001).

References

Balabanian, O. (1980). *Les Exploitations et les Problemes de l'Agriculture en Estremadure Espagnole et dans le Haut-Alentejo: Contribuition a l'Etude de Campagnes Mediterraneenes. Thèse présentée en vue de l'obtention du Doctorat d'Etat à l'Université de Clermont-Ferrant, le 22 Juin 1979.* Braga: Tipografia Barbosa & Xavier.

Baptista, F.O. (1978). *Portugal 1975—Os Campos.* Porto: Edições Afrontamento.

Bürgi, M., *et al.* (2022). Advancing the study of driving forces of landscape change. *Journal of Land Use Science*, 17(1). https://doi.org/10.1080/1747423X.2022.2029599 [Accessed 10th July 2023].

Bürgi, M., Hersperger, A.M., & Schneeberger, N. (2005). Driving forces of landscape change-current and new directions. *Landscape Ecology*, 19. https://doi.org/10.1007/s10980-005-0245-3 [Accessed 10th July 2023].

Caldas, E.C. (1991). *A Agricultura Portuguesa Através dos Tempos.* Lisboa: Instituto Nacional de Investigação Científica.

Correia, T.P. (1993). Threatened landscapes in Alentejo, Portugal: the "montado" and other "agro-silvo-pastoral" systems. *Landscape and Urban Planning*, 24. https://doi.org/10.1016/0169-2046(93)90081-N 43-48.

Cutileiro, J. (1971). *Ricos e Pobres no Alentejo (Uma Sociedade Rural Portuguesa).* Lisboa: Livraria Sá da Costa Editora.

Feio, M. (1949). *Le Bas Alentejo et L'Algarve.* Lisboa: Centro de Estudos Geográficos.

Feio, M. (1998). *A Evolução da Agricultura do Alentejo Meridional. As Cartas Agrícolas de Pery. As difíceis perspetivas atuais na Comunidade Europeia.* Lisboa: Edições Colibri.

Gomes, M.A., Barros, H., & Caldas, E.C. (1945). Traços principais da evolução da agricultura portuguesa entre as duas guerras mundiais. *Revista do Centro de Estudos Económicos*, 1, 22–203.

Martinho da Silva, I. (2001). *Historic Anthropogenic Factors Shaping the Rural Landscape of Portugal's Interior Alentejo.* PhD dissertation, University of Arizona, Tucson. Available at: https://repository.arizona.edu/bitstream/handle/10150/289759/azu_td_3040148_sip1_m.pdf?sequence=1&isAllowed=y.

Pery, G. (1875). *Geografia e Estatística Geral de Portugal e Colónias.* Lisboa: Imprensa Nacional.

Ribeiro, O., Lautensach, H., & Daveau, S. (1987). *Geografia de Portugal.* Lisboa: Edições João Sá da Costa.

Rosas, F. (1994). Several chapters. In Mattoso, J. (ed), *História de Portugal.* Lisboa: Círculo dos Leitores.

Santos, H.P. (2015). *Evolução da Paisagem Rural do Continente Português no século XX.* Tese de Doutoramento. Porto: Repositório da Universidade do Porto. Available at: https://hdl.handle.net/10216/80814 [Accessed 8th July 2023].

Silbert, A. (1966). *Le Portugal Méditerranéen à la fin de l'Ancien Régime-XVIIIe—Début du XIXe siècle. Contribution à l'histoire agraire comparé.* Paris: SEVPEN.

Van der Sluis, T. *et al.* (2019). Drivers of European landscape change: Stakeholders' perspectives through Fuzzy Cognitive Mapping. *Landscape Research*, 44(4). https://doi.org/10.1080/01426397.2018.1446074 [Accessed 10th July 2023].

Van Vliet, J. *et al.* (2015). Manifestations and underlying drivers of agricultural land use change in Europe. *Landscape and Urban Planning*, 133. https://doi.org/10.1016/j.landurbplan.2014.09.001 [Accessed 10th July 2023].

Vaquinhas, I.M., & Neto, M. (1993). Agricultura e mundo rural: tradicionalismos e inovações. In Mattoso, J. (ed), *História de Portugal.* Lisboa: Editorial Estampa, Vol. V, pp. 325–337.

16 Urban adaptation strategy for the Tinto riverscape, at watershed scale, towards its multifunctionality and hydrological resilience—impacts and governance challenges

Diana T. Fernandes, Maria José Curado and Rodrigo Maia

Introduction

Landscape policy can contribute to a better understanding of the complex issues involved in riverscape planning and management and to find more valid future proposals. Thus, it is believed that strengthening the knowledge base within the scope of integrated water resources management and landscape policy governance will facilitate and ensure the systematic improvement of biophysical processes associated with river restoration and rainwater drainage efficiency at watershed scale.

This research study aimed to demonstrate which set of measures promotes urban riverscape multifunctionality and hydrological resilience, at watershed scale, and how its operationalization could be facilitated, in short- and medium-long terms, regarding intermunicipal policy formulation and decision-making processes (Fernandes, 2021). Assuming that analytical tools selection must be determined by the decision's context and problem characteristics and criteria considered important by decision-makers, the investigation was developed around a case study.

Case study

This study focuses on the Tinto riverscape, a corridor 110 meters wide to each side of the river, from its spring to its mouth at Douro's right margin (Figure 16.1). The Tinto watershed has an area of approximately 23 km^2 and crosses four municipalities of the Portugal North region, namely, Valongo, Maia, Gondomar and Porto.

The Tinto riverscape develops in an open, wavy and low altitude valley where the steepest slopes are predominantly located on the river's downstream section, being divided into two landscape units, the watershed's middle and upper sectors (A) and its lower sector (B). Both evolved following the abandonment of agricultural activity, but physiographic features, urban poles' proximity and different demographic conditions led to two completely different realities in the Tinto riverscape and watershed (Figure 16.2).

Unit A–Tinto North corresponds to upstream and middle section of the Tinto River corridor, mostly artificial and impermeable as it crosses two major urban

DOI: 10.4324/9781003454007-21

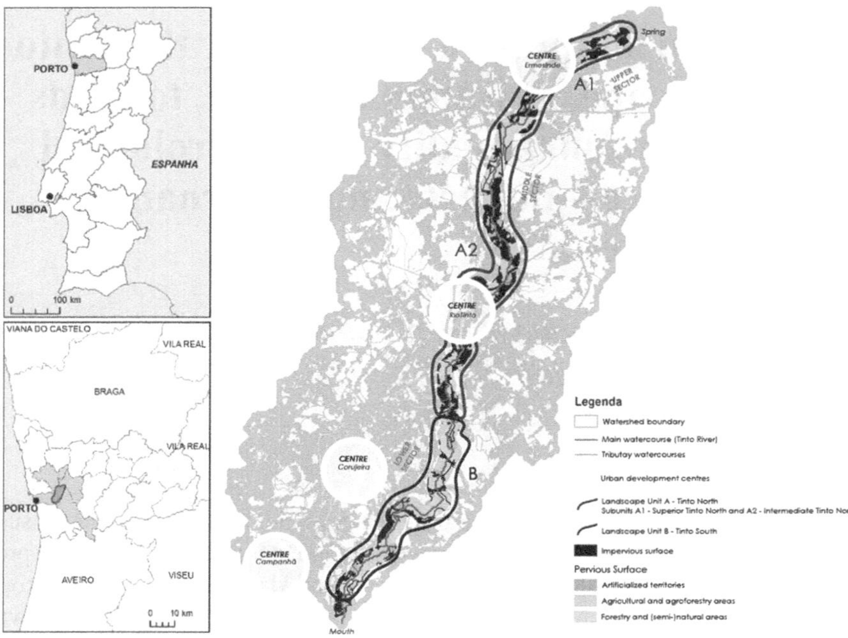

Figure 16.1 Tinto watershed and corridor map.

poles, with higher and increasing population density. Here, a pattern of suburban occupation is revealed, with agricultural and forest plots occurring between large housing blocks or simple houses, industries, warehouses and dense network of traffic routes. However, the landscape unit is divided into two subunits due to the difference in channel's width (lower in A1 and higher in A2) since it directly influences the riverscape's visual field breadth, as well as the relationship between watercourse and its users, despite similar surrounding conditions and the same diffuse and fragmented landscape.

Unit B–Tinto South is more naturalized and coincides with the downstream section of the Tinto's River corridor. This landscape unit has a lower and decreasing population density and only about 25% of impervious surface. The remaining territory is made up of agricultural fields that, over time, were abandoned and gave way to bush woods and forests. Two urban poles with a historically strong dynamic, both belonging to Porto's municipality area, developed on the valley slopes of this Tinto unit, imprinting a unique identity to this zone now reinforced with the creation of a riverside park.

Despite the different character, there are weaknesses set along the entire Tinto River corridor, which with more or less expression are common to both landscape units, mostly related to the health and safety of people and goods, natural resources protection and material heritage preservation, including frequent point water pollution events; high damage to infrastructure following major flood events; linear, impervious and regularized watercourse stretches which increases water flow and

Figure 16.2 Tinto riverscape units, registered between 2015 and 2017.

Source: Photographs by the author.

erosion risk; banks and margins without infiltration capacity; absence of a continuous riverine forest buffer to guarantee its stability and ecological connectivity and small rural settlements near watercourse with poor state of conservation.

Methodology

The methodology applied in this study was developed around the Tinto riverscape's case study and followed two main research lines regarding planning and governance issues (Figure 16.3).

Within planning scope, the research started with a case study's characterization, through literature review, followed by two assessment methods to demonstrate

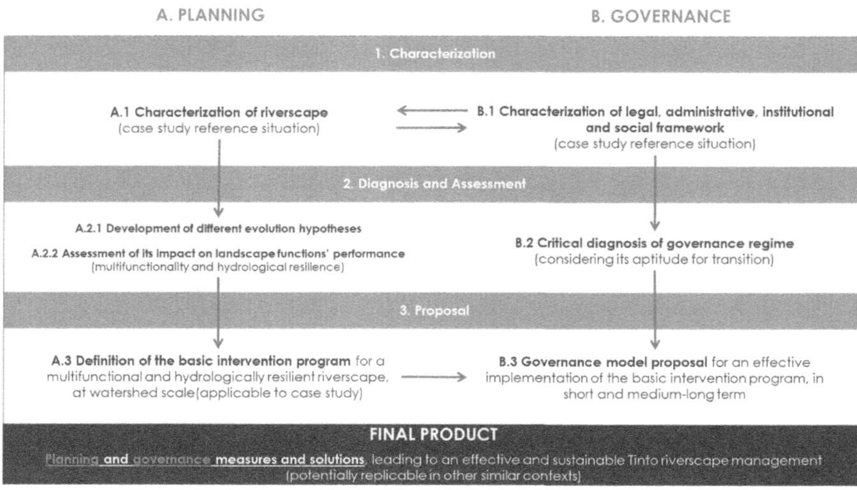

Figure 16.3 Research methodology applied.

which set of measures and coverage level is needed to guarantee a multifunctional riverscape with hydrological resilience capacity, within an urban context. Both methods were applied based on construction and comparison of three urban evolution hypotheses—growth, requalification and adaptation—with same base scenario (reference situation in 2012) and one alternative scenario, being each of them a simulated result of different sets of measures theoretically applied until 2030, the internationally established deadline for the achievement of water policy goals, at local scale, including by Porto municipality, under Urban Water Agenda 2030.

Within the scope of multifunctionality assessment, a comparative analysis was carried out qualitatively, through expert appraisal, and it focused on the riverscape's functions performance at economic (provision), environmental (regulation and habitat) and social (information) dimensions, considering the sustainable development principles related with multifunctional landscape concept (Lovell & Johnston, 2009) and integrated water resources management (Hassing et al., 2009). The selection of landscape's (sub-)functions under assessment were inspired by other classifications (Bastian, 1997, 1999; De Groot, 2006; Castelo-Branco & Coito, 2011; Vallés-Planells et al., 2014; Haines-Young & Potschin, 2018) and according to their applicability to local reality and policy objectives arising from European directives and strategies on water (Water Framework Directive and Floods Directive), biodiversity (EC, 2020) and climate change (EC, 2021).

Additionally, to assess hydrological resilience's capacity, each evolution hypothesis' potential biophysical impacts in the performance of water regulation subfunctions related with water flow (natural soil recharge and flood regulation) and water quality (pollutants' filtration) were quantitatively modelled and compared, under two periods with different extreme rainfall conditions, using the software i-Tree Hydro, which is based on an upgraded TOPMODEL framework (Wang et al., 2008).

Based on these results, an action programme and its territorial scope was defined for the Tinto riverscape improvement at watershed scale and a second research line was conducted to understand and improve the governance model's aspects under its operationalization, towards practical efficiency of the measures proposed, in short- and medium-long terms. Considering the current legal, administrative, institutional and social frameworks of the Tinto watershed, previously characterized, an analysis of the governance model's level aptitude for transition was carried out and supported by Rijke's governance pattern (Rijke et al., 2013), followed by a critical diagnostic, inspired by the "Transition Governance Framework" of Farrelly et al. (2012), to identify structural and procedural governance factors lacking improvement, regarding intermunicipal policy formulation and decision-making processes, both based on actual case studies with verifiable practical results ("Room for the River in The Netherlands" and "Water Sensitive Cities in Australia"). This exercise led the proposal to a set of solutions focused on increasing the governance model's level aptitude for transition, aiming to better suit the implementation of the urban adaptation action programme. Here, aptitude for transition is understood to be the capacity or ability of the governance regime to transform its mechanisms to suit the specificities of each phase of the transition process.

Results

Following first research line methodology, three alternative scenarios of each urban evolution hypothesis were developed according to different territorial development trends of Tinto watershed and riverscape, namely: growth—alternative scenario (2030-C) reflecting the status quo maintenance regarding the evolution of its land uses and water management, between 1995 and 2010; requalification—alternative scenario (2030-R) reflecting the implementation of most structuring actions which have been or are being implemented in the Tinto watershed since 2012 and, adaptation—alternative scenario (2030-A) reflecting the application of the latest European guidelines for water, biodiversity and climate change.

The Tinto riverscape's multifunctionality assessment results were presented through a three-level colour scale (Figure 16.4), regarding the presence and level performance of landscape (sub-)functions in a base scenario (2012-B) and each alternative scenario. In general, the main weaknesses in the 2012-B scenario are related to null or low performance of environmental and social functions. Since 2030-A scenario's assessment results present the riverscape with a superior performance of most of the landscape (sub-)functions, including in environmental and social dimensions, the set of measures under the 'adaptation' evolution hypothesis was considered the most suitable for the Tinto riverscape's multifunctionality improvement.

In turn, the hydrological resilience capacity assessment confirms the previous conclusion regarding the impact of the 'adaptation' evolution hypothesis in water regulation function by presenting the most positive results on the three indicators modelling exercise related to water flow and water quality subfunctions' performances (Table 16.1), for both annual periods with climatic values like 2005

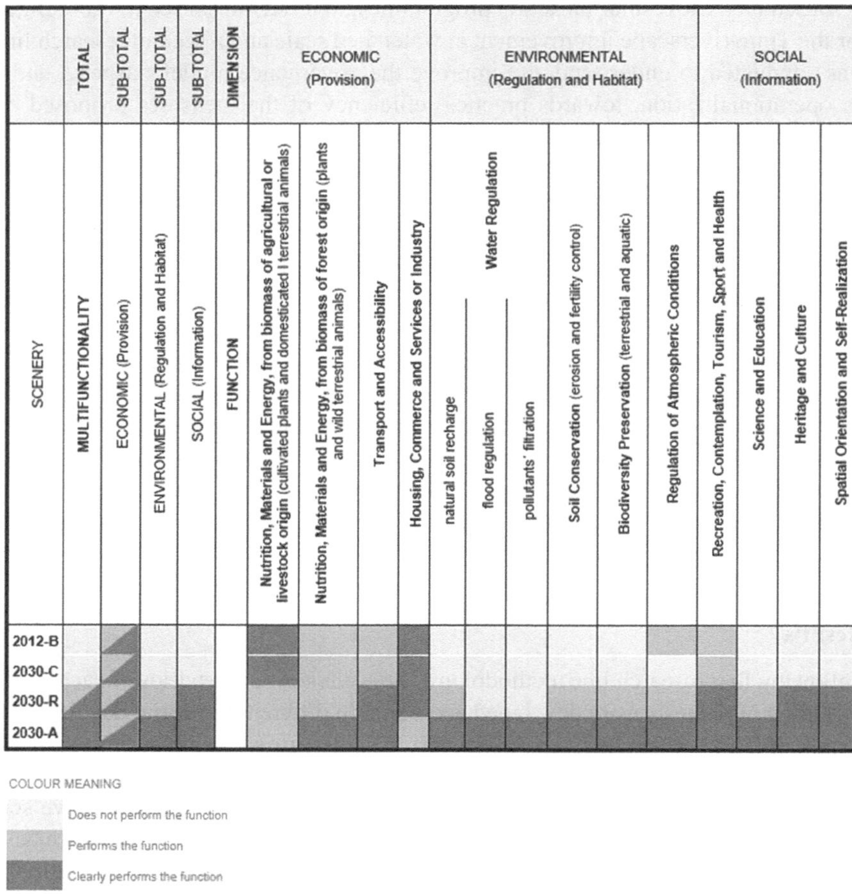

Figure 16.4 Results of the multifunctionality assessment of the Tinto riverscape.

Table 16.1 Results of the resilience capacity assessment of the Tinto riverscape.

Landscape Subfunction	***Natural soil recharge***	***Flood regulation***	***Pollutants' filtration***
Indicator	**Variation of total baseflow volume**	**Variation of maximum hourly surface run-off flow over impervious surface**	**Variation of average concentration of total pollutant annual load, depending on total surface run-off volume**
Unity	%	%	%
1. Evolution Hypothesis "Urban Growth" (2012-B → 2030-C)			
Extremely Dry Year (2005)	–23,1	24,8	10,4
Very Rainy Year (2014)	–23,4	24,8	7,4

(*Continued*)

Table 16.1 (Continued)

Landscape Subfunction	***Natural soil recharge***	***Flood regulation***	***Pollutants' filtration***
2. Evolution Hypothesis "Urban Requalification" (2012-B → 2030-R)			
Extremely Dry Year (2005)	−10,7	10,5	4,1
Very Rainy Year (2014)	−10,8	10,5	2,8
3. Evolution Hypothesis "Urban Adaptation" (2012-B → 2030-A)			
Extremely Dry Year (2005)	2,0	−15,1	−1,1
Very Rainy Year (2014)	−5,1	−15,1	−1,2

('extremely dry') and 2014 ('very rainy'), as recorded by IPMA (2006, 2015). These results are mainly due to limitation of impervious surface expansion, appliance of natural water retention measures in drainage system and an increase of tree cover. However, the reduced variation between the 2012-B and 2030-A scenarios (in terms of relative values, in %) of total volumes of baseflow and of average annual concentration of the pollutant load shows that, at least regarding subfunctions of natural soil recharge and pollutants' filtration, the action measures proposed will only be effective in terms of limiting, mitigating or compensating the pressure and effects of soil sealing on water and other natural resources if they are complemented with other measures and actions.

Summing up, in both assessments, the 'adaptation' measures set had the most positive impact, although requiring the adoption of complementary ones. Following these results, an action programme was designed towards a long-term effective urban adaptation of Tinto riverscape, with specific measures (Table 16.2) to be applied inside and outside the river corridor area, at watershed scale (Figure 16.5), and based on a step-by-step process with different challenges in its development stages, requiring a transitional governance model.

Within the governance scope (second line of research), and with reference to Rijke's governance pattern (Rijke et al., 2013; Rijke, 2014), the transition process of an action programme focused on the socio-ecological system's adaptation cycle is developed in four main steps (Figure 16.6). It is believed that the current Tinto watershed governance model already has some pre-established elements or mechanisms—listed in Table 16.3 as current facts—that can facilitate the development of each step of the Tinto riverscape's action programme.

However, considering that the programme's success requires a continuous step-by-step transition process, it appears that the current governance model fails in responding to the different challenges and needs of each step, mainly in adaptation measures' transposition into public and private entities' current practices, towards a specific long-term goal. These elements and mechanisms are not enough to guarantee the robustness and flexibility required for the transition process' practical effectiveness, revealing the urgent need to introduce new resources and approaches, thus

Table 16.2 Action programme for the Tinto riverscape's urban adaptation, at watershed scale.

Main Goal	*General Action Measures*	*Territorial Zones (Figure 5)*		
		River Corridor		*Tinto Watershed*
		Tinto River	*Tributary Streams*	
A. Reinforcement of riverscape's social functions	A1. Creation of recreation and leisure opportunities along watercourses, through new open collective spaces (e.g., parks and gardens)			
	A2. Improved accessibility in soft modes, through the creation of an intermunicipal pedestrian and cycle paths network along watercourses			
	A3. Rehabilitation of open and built spaces, including rural centres and other material heritage elements			
B. Reinforcement of riverscape's environmental functions	B1. Improvement of support conditions to autochthonous biodiversity, in artificialized territories and agricultural areas, through the provision of spaces dedicated to their natural habitat			
	B2. Changing the way of agricultural spaces production for integrated or organic methods			
	B3. Conversion of forest production spaces for conservation with native species			
	B4. Improved efficiency of the wastewater system (up to 10% infiltration into the sanitation network)			
	B5. Construction of an outfall to discharge the sanitation network into Douro River			
C. Increase of riverscape's hydrological resilience	C1. Application of natural water retention measures (NWRM) in the hydrographic network and riparian zones, to improve their hydromorphological conditions (including bank stabilization and river forest buffers)			
	C2. Application of natural water retention measures (NWRM) in rainwater drainage system (e.g., green roofs, rainwater harvesting, permeable surfaces, swales, rain gardens), to improve its efficiency and sustainability, both in artificial territories and in agricultural and forestry areas			
D. Assurance of people and goods' safety	D1. Planned removal of the building (remove, relocate or retrofit), in river corridors and wet systems, followed by their renaturation			
	D2. Conditioning the artificialization of permeable areas (e.g., for building and road network) in sensitive areas (river corridors, wet systems, bollards and slopes greater than 25%)			

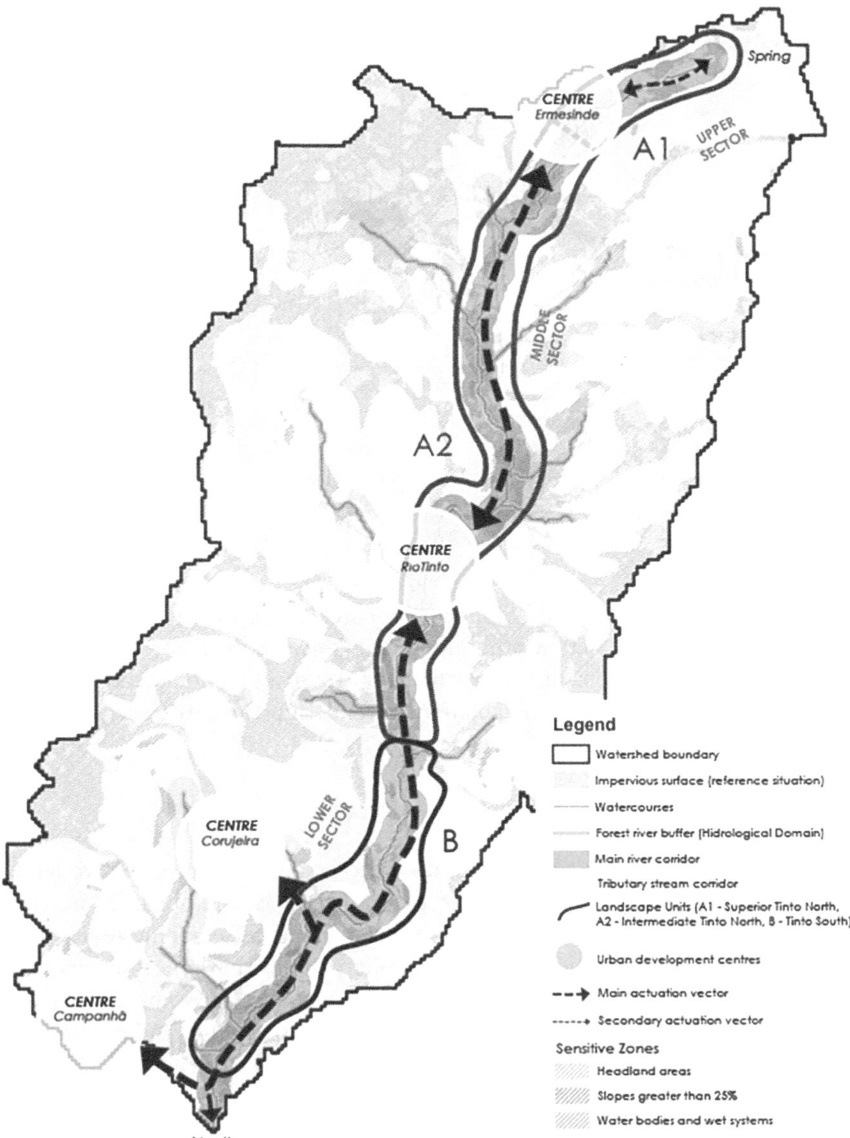

Figure 16.5 Map of areas and action vectors regarding the Tinto watershed's action programme.

reinforcing existing procedural and structural governance factors' performances for long-term adaptive riverscape planning and management.

By applying the transition governance framework (Farrelly et al., 2012; Rijke et al., 2012), several solutions were proposed (Figure 16.7), that can be introduced in the Tinto watershed's current governance model in a systemic way. The

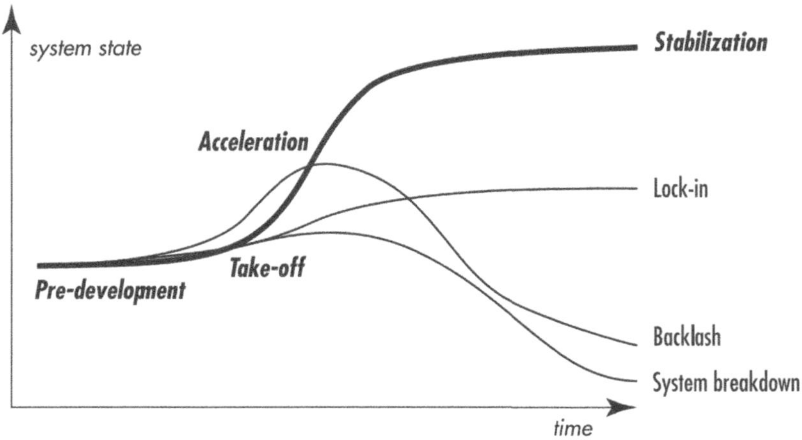

Figure 16.6 The multi-phase concept of transition theory.
Source: Rotmans, 2005, p. 24

aim is to make it more suitable for transition according to the challenges inherent to each step of the action programme appliance process and thus guarantee its coherent development and consistent long-term operationalization towards the Tinto riverscape's urban adaptation, considering the 2012 scenario as the reference.

Discussion

All multifunctionality assessment values were assigned in an objective-tending way, although through an individual exercise, based on a reasonable knowledge of the studied riverscape, either by bibliographic and cartographic review or by observation and participation in events, public participation processes and informal conversations. However, in a real context, it would be advisable to subject his assessment to discussion and validation by different specialists and eventually by local population, through collaborative planning workshops.

Under resilience capacity's assessment, the results obtained through modelling are generally in line with the most recent European and international guidelines on water management, revealing the utility of i-Tree Hydro programme as an expedited tool and a first base of analysis to discuss and justify some options in riverscape planning. Nevertheless, its application still has some restrictions in the Portuguese context, due to programme's own limitations and the unavailability of some local data needed, which led to the use of the model in uncalibrated mode and with some values from the North American reality, drawing conclusions only by variations and relative values. This, along with the programme's inability to simulate some specific measures with the potential impact on water regulation and quality, are the reasons why any option that is more technical or focused on a specific

Table 16.3 Transition governance process for the Tinto riverscape's urban adaptation, adapted from Rijke's governance pattern (Rijke et al., 2013).

LANDSCAPE		TRANSITION GOVERNANCE				
ADAPTATION CYCLE (socio-ecological system)		**TRANSITION PROCESS** (governance regime systems)		**GOVERNANCE MODE**	***TYPES OF ACTION OR INITIATIVE***	**CURRENT FACTS (TINTO WATERSHED)**
Launch Step	⟷	EARLY STAGE	Pre-development Step	**Decentralized and informal:** looking for alternative solutions, testing innovations and sharing experiences, establishing, and stimulating new relationships	*Network formation* *Experimentation and learning*	Intermunicipal Intervention Plan (AdP, 2011) Tinto River Monitoring Project (UFP-PAAS and LIPOR, 2011-2020) H2PORTO Technological Platform (AdP, 2016) LIPOR's Ecological Pathway (2018)
Reorganization / Renewal Step	⟷	EARLY STAGE	Take-off Step	**Hybrid:** formal policy decisions to catalyze and/or coordinate actions, and informal and decentralized learning for greater innovation	*Political decision in response to a crisis*	Definition of measures in River Basin Management Plan and Flood Risk Management Plan of Douro hydrographic European and national funds
Growth / Exploration Step	⟷	INTERMEDIATE STAGE	Acceleration Step	**Hybrid:** centralized policy to promote initiatives or actions, decentralized implementation, informal network for demonstration and dissemination of empirical knowledge, coordinated capacity building to create synergies and avoid inefficient use of resources	*Implementation of innovative solutions*	Hybrid approach at the decision-making power level in Tinto river's depollution and requalification (promoted by APA and implemented by local actors)
Conservation Step	⟷	FINAL STAGE	Stabilization Step	**Centralized and formal:** adjust or establish legal frameworks and coordinate technical capacity building to convince and promote the adoption of innovative approaches in current practice and safeguard the new *status quo*	*Creation of regulation and legislation to maintain and popularize the status quo*	Existence of laws and programs that provide the required legal framework Individuals' binding dependent on the transposition of programs to the territorial plans and respective municipal regulations

action should be complemented with specialized studies to provide a more precise basis for decision-making.

Finally, regarding governance solutions—and given the difficulties in evaluating its impacts and effectiveness and so the impossibility of objectively proving that they work for the Tinto riverscape's urban adaptation—the proposal was developed focusing on aptitude level improvement of the Tinto watershed's governance regime for transition towards long-term effectiveness of planning and management processes. In this sense, the work constitutes an example for application, at local scale, of concepts and approaches that have been developed and recommended by scientific

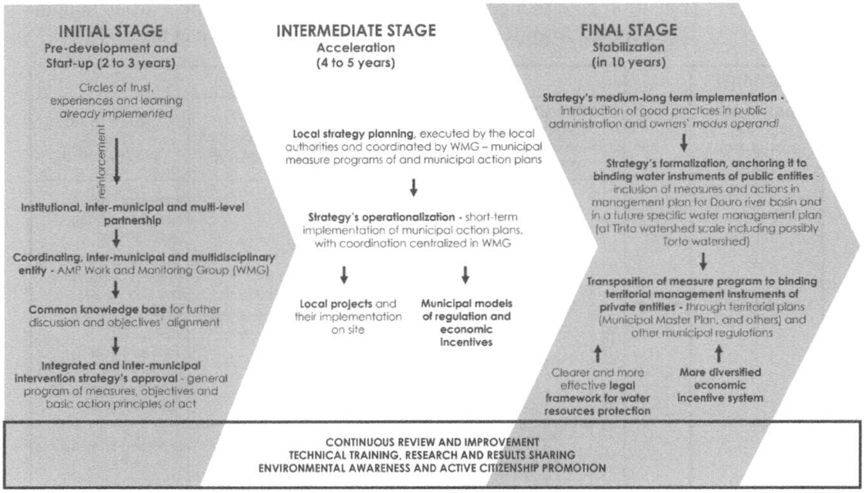

Figure 16.7 Main solutions proposed for the governance model's transition aptitude improvement towards a long-term effective urban adaptation of the Tinto riverscape.

literature and several international organizations, including the European Environment Agency (Geels et al., 2019), within water and environment policy scope.

Conclusion

The results of the Tinto riverscape's assessments regarding its multifunctionality and hydrological resilience's capacity revealed the need to apply different types of nature-base solutions, under an intermunicipal urban adaptation action programme at watershed scale.

Given the multiple actors and authorities potentially involved in the Tinto watershed governance regime and the changeable and unpredictable condition of its context variables, the proposed programme will be more effectively implemented in the long-term through a robust and flexible governance model, with a high-level aptitude for transition. To this end, context-specific procedural and structural solutions have been proposed in this research that can potentially be replicated in other identical Portuguese watersheds, if local additional studies are developed for adjustments.

This work was financially supported by Fundação para a Ciência e Tecnologia, under the Individual PhD Scholarship SFRH/BD/92367/2013.

Bibliography

Bastian, O. (1997). *Gedanken zur Bewertung von Landschaftsfunktionen: Unter besonderer Berucksichtigung der Habitatfunktion*. Schnevedingen: Alfred Toepfer Akademie für Naturschutz.

Bastian, O. (1999). Leitbilder fur Naturraume auf der Basis von Landschaftsfunktionen (Ecological goals for (natural) landscape units on the basis of landscape functions.). *Natur und Landschaft*, 74, 361–373.

Castelo-Branco, M., & Coito, A. (2011). *Servidões e Restrições de Utilidade Pública.* Lisboa: Direcção-Geral do Ordenamento do Território e Desenvolvimento Urbano.
De Groot, R. (2006). Function-analysis and valuation as a tool to assess land use conflicts in planning for sustainable, multi-functional landscapes. *Landscape and Urban Planning*, 75(3–4), 175–186. https://doi.org/10.1016/j.landurbplan.2005.02.016.
EC. (2020). *Our life insurance, our natural capital: An EU biodiversity strategy to 2020.* Communication from the Commission to the European Parliament, the Council, the Economic and Social Committee and the Committee of the Regions (COM(2011) 244 final). Brussels: European Commission.
EC. (2021). *Forging a climate-resilient Europe—The new EU strategy on adaptation to climate change.* Communication from the Commission to the European Parliament, the Council, the Economic and Social Committee and the Committee of the Regions (COM(2021) 82 final). Brussels: European Commission.
Farrelly, M., Rijke, J., & Brown, R. (2012). Exploring operational attributes of governance for change. *WSUD2012–7th International Conference on Water Sensitive Urban Design: Building the Water Sensitive Community.* Melbourne, 21–24 February.
Fernandes, D. (2021). *A Paisagem ribeirinha em contexto urbano. Um modelo de planeamento e governação para a bacia hidrográfica do rio Tinto.* Tese de Doutoramento. Porto: Faculdade de Ciências da Universidade do Porto.
Geels, F., Turnheim, B., Asquith, M., Kern, F., & Kivimaa, P. (2019). *Sustainability Transitions: Policy and Practice, EEA Report No 9/2019.* Luxembourg: European Environment Agency.
Haines-Young, R., & Potschin, M. (2018). *Common International Classification of Ecosystem Services (CICES) V5.1—Guidance on the Application of the Revised Structure.* United Kingdom: Fabis Consulting Ltd.
Hassing, J., Ipsen, N., Clausen, T.J., Larsen, H., & Lindgaard-Jorgensen, P. (2009). *Integrated Water Resources Management in Action.* DHI Water Policy, UNEP-DHI Centre for Water and Environment, UNESCO World Water Assessment Programme. Paris: United Nations Educational, Scientific and Cultural Organization.
IPMA. (2006). *Relatório de Caracterização Climática do Ano 2005.* Lisboa: Instituto Português do Mar e da Atmosfera.
IPMA. (2015). *Boletim Climatológico Anual—2014 Portugal Continental.* Lisboa: Instituto Português do Mar e da Atmosfera.
Lovell, S.T., & Johnston, D. (2009). Designing landscapes for performance based on emerging principles in landscape ecology. *Ecology and Society*, 14(1).
Rijke, J. (2014). *Delivering Change: Towards fit-for-Purpose Governance of Adaptation to Flooding and Drought.* Dissertation, UNESCO-IHE Institute for Water Education, Delft University of Technology, The Netherlands.
Rijke, J., Farrelly, M., Brown, R., & Zevenbergen, C. (2013). Configuring transformative governance to enhance resilient urban water systems. *Environmental Science & Policy*, 25(2013), 62–72.
Rijke, J., van Herk, S., Zevenbergen, C., & Ashley, R. (2012). Room for the river: Delivering integrated river basin management in the Netherlands. *International Journal of River Basin Management*, 10(4), 369–382.
Rotmans, J. (2005). Societal innovation: Between dream and reality lies complexity. In *ERIM Inaugural Address Series Research in Management 7293.* Rotterdam: Erasmus Research Institute of Management, Rotterdam School of Management, Erasmus University, Erasmus School of Economics.
Vallés-Planells, M., Galiana, F., & van Eetvelde, V. (2014). A classification of landscape services to support local landscape planning. *Ecology and Society*, 19(1), 44.
Wang, J., Endreny, T., & Nowak, D. (2008). Mechanistic simulation of tree effects in an urban water balance model. *Journal of the American Water Resources Association*, 44(1), 75–85.

17 Diversity of street trees in Lisbon, towards city sustainability

Ana Luísa Soares, Leónia Nunes, Inês Duarte, Ana Júlia Francisco, Davide Gaião, Francisco Castro Rego and Susana Dias

Introduction

The urban ecological framework promotes urban resilience by minimising the negative effects of urbanisation (McLean et al., 2020). The urban forest is an important asset for the quality of life in a city, giving the idea of proximity to the countryside, with positive impacts on improving human health and well-being and valuing heritage and landscape (Konijnendijk et al., 2023).

Street trees are a vital component of the urban forest, and managing its composition and structure has a significant effect on the provision of ecosystem services, as they contribute to a milder climate, reduce air pollution, provide aesthetic urban settings and promote biodiversity (e.g., McLean et al., 2020).

Urban planning and the managing of public street trees according to social, economic and climate scenario impacts is challenging. Legislation on these topics stresses the need to conduct and update tree inventories in order to foster a city's sustainability. In Portugal this subject has been strengthened with Law no. 59/2021, of 18 August, on the legal regime for the management of urban trees.

Street trees in Portugal have been the subject of relevant works, such as a compilation of historical installation in Lisbon green spaces (Andresen, 1982) and the quantification of the benefits of urban trees in Lisbon and Porto (e.g., Soares et al., 2011; Graça et al., 2017, 2018).

The transfer of knowledge by academics and other professionals has contributed to the increase of abundance and richness in Lisbon street trees by encouraging the selection of species that best adapt to each setting (Castel-Branco, 2000).

Knowing the current abundance and diversity of tree species, their geographic origin, phenology and health is the first step towards quantifying the benefits provided by Lisbon's green structures.

The present study aims to characterise the public street trees in Lisbon from 1929 to 2021. We used data from the Lisbon Municipal Council (CML) archives for the first period (CML, 1929) and data collected in 2021, under an agreement between Lisbon University's School of Agriculture (ISA/ULisboa) and CML, with the aim of updating the CML database and estimating the ecosystem services (ES) provided by the street trees.

DOI: 10.4324/9781003454007-22

Methods

This study focuses on street trees, defined as any public arboreal elements (trees and shrubs with height higher than 2.5m) included in a buffer of 15m from the axis of the road (Figure 17.1a). In order to understand the changes in the city's street tree composition, a comparative analysis was carried out between the data from the 1929 (CML, 1929; Andresen, 1982) and the data from the 2019–2021 inventory, based on the 10 most representative genera and their tree abundance.

The city parishes were clustered by Soares et al. (2011) through multivariate analysis, into four management zones (stratum 1 to 4), taking into account the composition, abundance and locations of the arboreal species. The recent parish of *Parque das Nações* was added to stratum 2, according to the previously established criteria (Figure 17.1b).

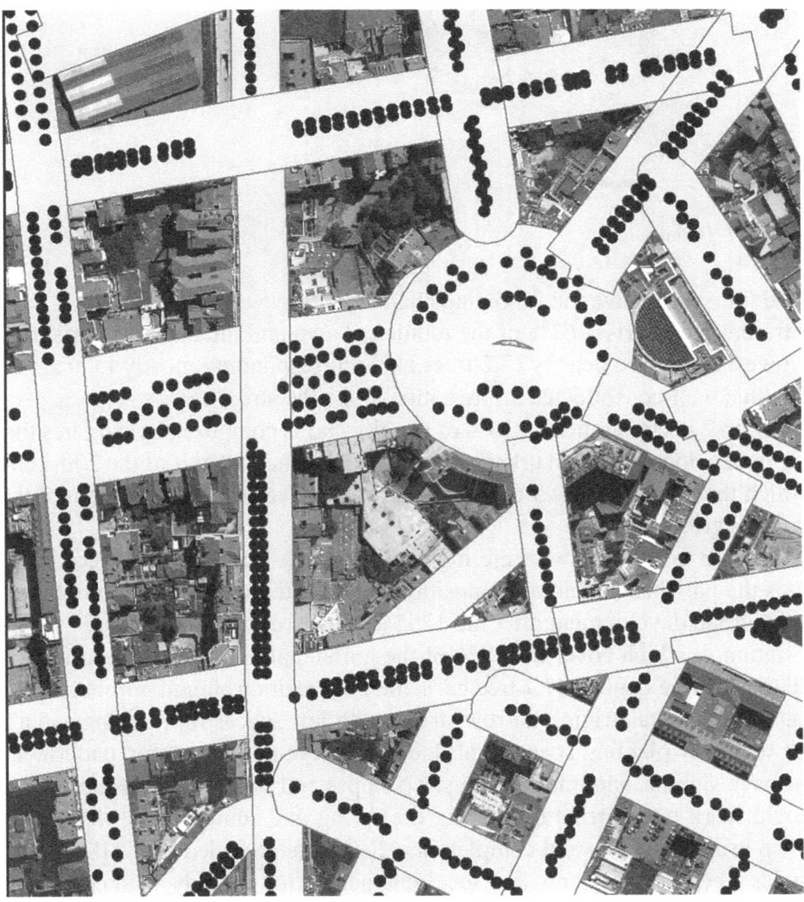

Figure 17.1 Street trees selection (a) and their distribution (black dots) in Lisbon city strata 1 to 4 (b).

Figure 17.1 (Continued)

The four strata have the following characteristics:

Stratum 1 comprises 63% of the municipal area and includes the parishes with the lowest street tree density (7.2 trees.ha^{-1}), corresponding mostly to areas of the city with a well-consolidated afforestation along the streets.

Stratum 2 occupies around 30% of the city and is composed by parishes that, in general, developed from an urbanisation system in the 2nd half of the 20th century, in which the street trees (9.5 trees.ha^{-1}) appear embedded in the residential fabric, with large green areas.

Stratum 3 is the one with the highest street tree density (15.5 trees.ha^{-1}) and groups the parishes that are the transition between the old city pattern (stratum 4) and the urban development after the 1755 earthquake.

Stratum 4, which cover only 2% of the municipal area, includes parishes with similarity of tree density (7.7 tree.ha^{-1}) and composition, found within an organic layout (of medieval origin), narrow streets with few spaces for planting and a tight mesh with few planting spaces that does not obeys any geometric pattern. These areas have similar topography, with steep slopes and southern exposure.

Field work was carried out during the spring and summer of 2019, 2020 and 2021 in order to update and complete the 2019 data provided by CML. Field data includes tree measurements and local characteristics, namely diameter at breast height (DBH), tree height, crown dimension and condition.

Street trees data were used to evaluate species origin (exotic or native), phenology (evergreen or deciduous), compute species richness and abundance, measure diversity (Shannon index, H'), dominance and similarity between strata (Magurran, 1988).

A sample of street trees (6928), distributed by the four strata, was used to quantify ecosystems services (ES) with i-Tree Eco-V6 (www.itreetools.org), taking in consideration tree leaf period in Lisbon, atmospheric and meteorological hourly data (in particular temperature, precipitation and winds). The i-Tree estimates the effects of urban forests and trees on carbon sequestration and storage, air pollutant removal, building energy use, hydrology, volatile organic compounds and oxygen production (Nowak, 2021).

The next section highlights the results regarding replacement values, carbon storage and total benefit values. Replacement value is the value of a tree based on the physical resource itself e.g., the cost of having to replace a tree with a similar tree, following CTLA (1992) procedures. Trunk area and species are used to determine the basic value, which is then multiplied by condition and location ratings (0 to 1) to determine the final tree compensatory value (Nowak, D. et al., 2002). Condition factors are based on percent crown dieback (Nowak et al., 2008).

Carbon storage is the amount of carbon bound up in the aboveground and belowground parts of woody vegetation. Carbon estimations were made using specific allometric equations for each taxonomic entity (e.g., McPherson et al., 2016). The annual benefit value includes the economic value of the environmental benefits provided by all of the trees in the study area. These values reflect the amount of carbon storage, avoided run-off and air pollutant removal. We compare the annual benefits with the total leaf area, as many tree benefits directly equate to the amount of healthy leaf surface area of the plant.

Results

Species composition

The composition of Lisbon's public street trees increased from 21822 individuals and 33 genera in 1929 to 70792 individuals and 140 genera in 2021 (Table 17.1), not considering palm trees. This increase resulted from a combination of factors such as a greater willingness of technicians to use a wider spectrum of species that have become more available in the nurseries. The selection of species for the EXPO'98 event contributed as well to the diversity and number of trees in the city (Castel-Branco, 2000; Almeida, 2006).

Street trees in Lisbon in 2021 are comprised 365 taxa, including genera, species, subspecies and cultivars. *Celtis* is the most abundant genera in both periods, with a similar representativeness (≈14.2%). The *Ulmus* genera disappears from the list of the top 10 due to diseases and pests (Andresen, 1982), whereas the *Pinus* genera appears in the top 10 in 2021. *Platanus* and *Jacaranda* genera have increased in abundance in recent decades.

Celtis australis is the most abundant tree (13.8%). *Platanus hispanica* and *Jacaranda mimosifolia* are the other two most common species in strata 1 to 3, while

Table 17.1 Composition of public street trees in Lisbon (the 10 genera most abundant) comparing 1929 to 2021.

Order	**1929**			**2021**		
	Genera	**Number**	**%**	**Genera**	**Number**	**%**
1	*Celtis*	3099	14.2	*Celtis*	10150	14.3
2	*Ulmus*	2976	13.6	*Platanus*	5271	7.4
3	*Platanus*	2057	9.4	*Tilia*	5131	7.2
4	*Styphnolobium*	1566	7.2	*Populus*	4109	5.8
5	*Robinia*	1081	4.9	*Jacaranda*	3863	5.5
6	*Ligustrum*	907	4.2	*Acer*	3542	5.0
7	*Olea*	850	3.9	*Fraxinus*	3083	4.4
8	*Jacaranda*	767	3.5	*Prunus*	2846	4.0
9	*Cercis*	662	3.0	*Cercis*	2303	3.3
10	*Broussonetia*	621	2.8	*Pinus*	1872	2.6
	Other taxa	7236	33.2		22622	40.4
Total		21822	100.0		70792	100.0

Table 17.2 Number of individuals (N), richness (S), diversity (Shannon Index H') and evenness (index E) for the strata in Lisbon in 2021.

Stratum	*N*	*S*	*% Exotic*	*% Native*	*Shannon, H'*	*E*
S1	39077	365	69.4	23.6	4.1	0.7
S2	24686	316	71.7	25.4	4.3	0.9
S3	5702	113	61.2	6.0	3.1	0.6
S4	1317	96	54.4	20.0	3.2	0.7

Fraxinus sp. is also very important in stratum 4. This stratum has one of the lowest tree densities and some remarkable trees that reveal great longevity, such as *Celtis australis* representing ca. 26% of the total number of older trees.

More than 78% of the trees are deciduous, with perennials representing only 6% in stratum 3 (Table 17.2). Only around 30% of the specimens are native. Exotic species have a higher representation in stratum 2 (≈72%).

There is an asymmetry in terms of the values of richness, diversity and dominance between parishes, with the lowest number of trees, lowest richness and diversity occurring in the older areas (stratum 3 and 4) and the high diversity corresponding to new urban sprawl (stratum 2; Table 17.2).

Diversity and its variability between strata (expressed as 95% confidence intervals) and between parishes in each stratum are summarised in Figure 17.2. Diversity variability is lowest among the 20 parishes included in stratum 1 and highest in stratum 3. In general, the diversity of the street tree community is at around 80% of the maximum potential, with the high dominance of species such as *Celtis australis* contributing to low diversity levels, particularly in stratum 3.

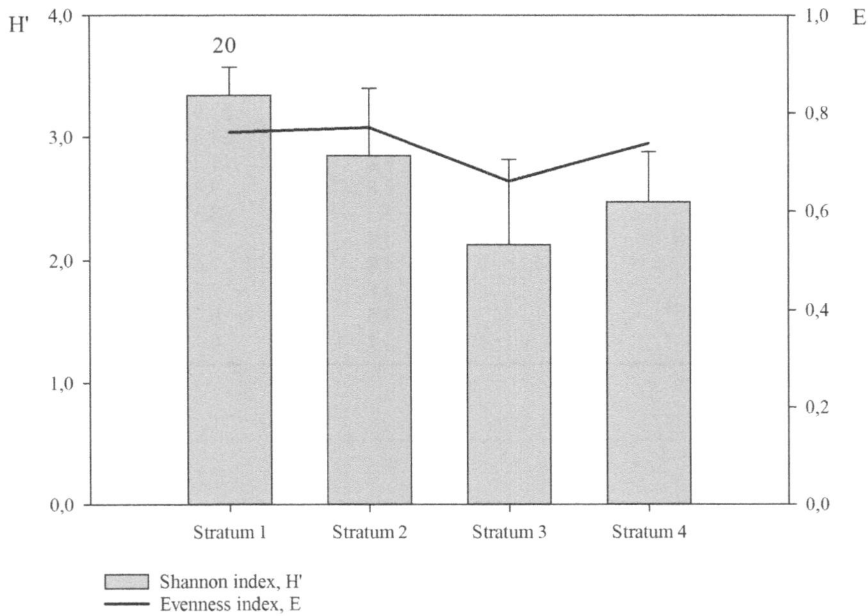

Figure 17.2 Values of street tree diversity (mean and 95% IC for Shannon diversity index, H') and evenness (Evenness index, E) obtained for the four strata comprising a variable number of parishes (number in the top of the column).

Relationship between tree characteristics and i-Tree estimations

The relationship between tree characteristics and i-Tree estimations was evaluated with a subsample of 6928 street trees.

The species with higher leaf area are *Celtis australis*, *Platanus* x *hispanica* and *Jacaranda mimosifolia*. The 10 species with the highest importance values are listed in Table 17.3. Importance values (IV) are calculated as the sum of percent population and percent leaf area. Although with different percentage to the population, *Platanus* x *hispanica* and *Jacaranda mimosifolia* have similar IV due to different leaf shape and subsequent area.

High importance values do not mean that these trees should necessarily be encouraged in the future; rather it means that these species currently dominate the urban forest structure.

High values of DBH, crown size and leaf area positively affect the provision of ES by trees. Figure 17.3 shows the relationship between the replacement values estimated by i-Tree for each tree species and the DBH measured during the field work. It is possible to observe that DBH has a significant importance on the evaluation of the replacement value ($R^2 = 0.815$), as the DBH itself is also an indicator of the size and age of a tree.

Table 17.3 Most important 10 species in sample of 6928 individuals of street trees in Lisbon.

Species Name	***% Population***	***% Leaf Area***	***% Importance Value (IV)***
Celtis australis	16.7	21.2	37.9
Jacaranda mimosifolia	7.7	6.8	14.5
Platanus x *hispanica*	4.4	8.8	13.2
Tilia cordata	4.5	4.8	9.4
Tilia tomentosa	3.0	5.3	8.3
Fraxinus angustifolia	4.2	3.4	7.7
Tipuana tipu	2.8	4.8	7.6
Platanus sp.	2.3	4.8	7.0
Acer negundo	3.2	3.5	6.7
Melia azedarach	3.0	3.1	6.1

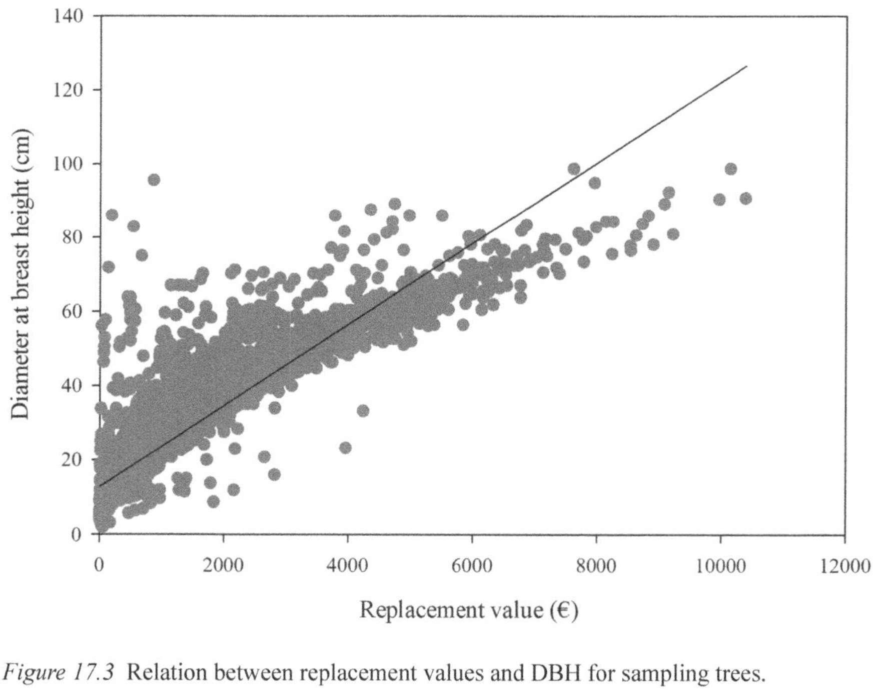

Figure 17.3 Relation between replacement values and DBH for sampling trees.

Trees reduce the amount of carbon in the atmosphere by sequestering carbon in new growth every year. The amount of carbon sequestered annually increases with the size and health of the trees. Lisbon's street trees are estimated to store around 1600 tons of carbon (Soares et al., 2022).

Figure 17.4 shows the relationship between carbon storage and DBH for the 10 most important species in the sample. Tree diameter is a variable that gives an indication of the stem size. Some studies have shown that the stem is the tree component with more biomass and productivity, compared with branches and leaves: the component where carbon sequestration is retained longer (Nunes et al., 2013).

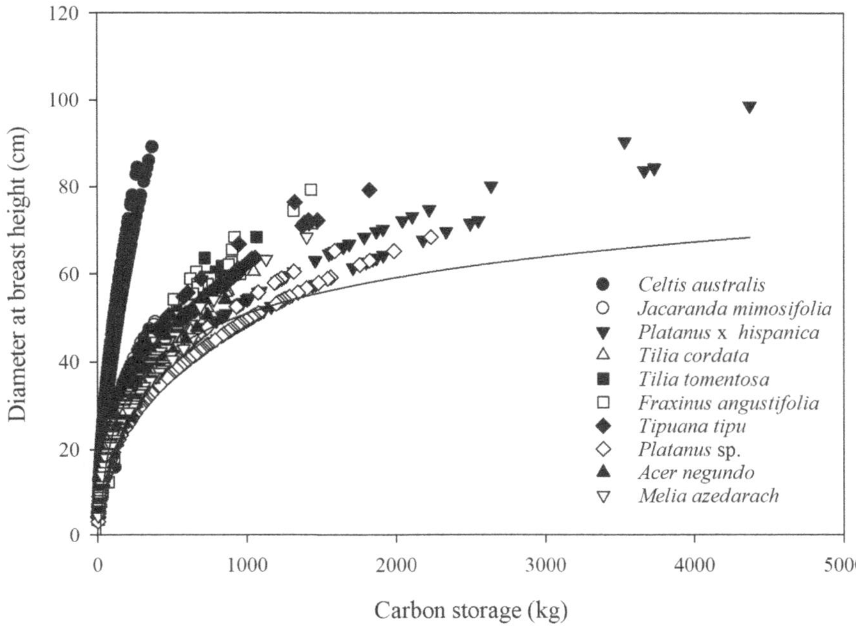

Figure 17.4 Relation between carbon storage and DBH for the most important 10 species.

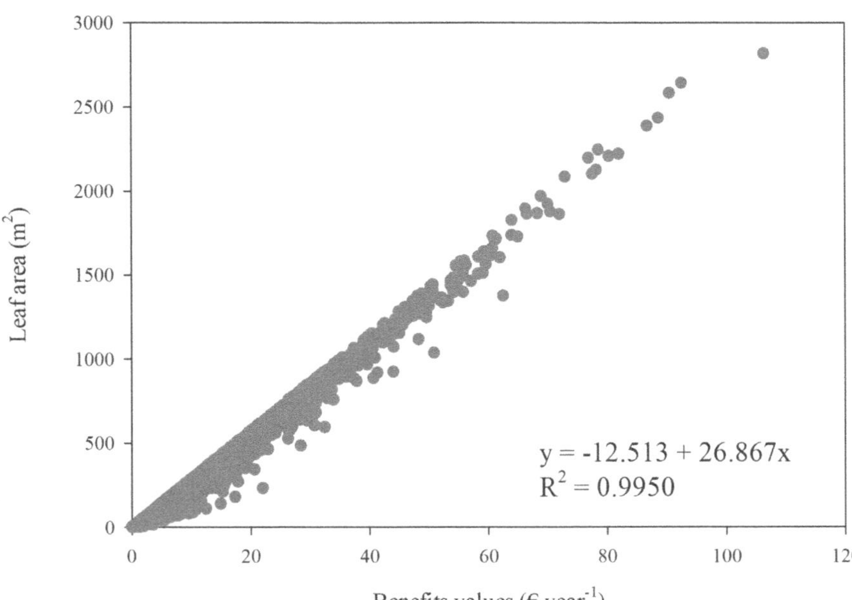

Figure 17.5 Relationship between benefits values and leaf areas for sampled trees.

Figure 17.5 shows the relationship between the benefits value and the leaf area. As expected, there is a strong relationship between the quantity of the leaves in a crown and the benefits that a tree can provide. Larger crowns provide more ecosystem services.

Discussion and conclusions

Accurate measures of urban forest structure and diversity are critical for proper urban forest planning to maintain or improve environmental quality, human health and well-being in cities (Nowak et al., 2008), particularly when is expected an increase in the global urban population and a need for climate change adaptation (Measham et al., 2011; UN, 2019).

From 1929 to 2021 there was a significant increase in the number and composition of street trees in Lisbon, in line with the expansion of urban planning and management of the municipality over the last century. These green structures, either with native, exotic and/or cultivars species, seem to be well adapted to the environmental city conditions (Almeida, 2006).

Celtis sp. can be considered Lisbon's preferred tree as it has been the most common species in the city during the last decades (around 14% of total street trees). This could be explaining by its ecological attributes: a fast growth species, native to the warm valleys of Tagus and Douro rivers and well adapted to Lisbon soil and climate (Martínez-Varea et al., 2023). This species has a large canopy that provides excellent shade, although it suffers from the excessive heat that causes the leaves to curl; shows resistance to insect attacks and possesses robust, deep roots and is capable of growing in the interstices between rocky layers (Magni & Caudullo, 2016).

Deciduous species are dominant on the streets of Lisbon. The most relevant to providing environmental benefits are *Celtis australis*, *Jacaranda mimosifolia*, *Platanus* x *hispanica*, *Tilia cordata*, *T. tomentosa*, *Fraxinus angustifolia*, *Tipuana tipu*, *Platanus* sp., *Acer negundo* and *Melia azedarach*. Slow growing species are important for carbon storage while the broadleaf ones are important for pollutant removal and other ES. Nevertheless, during winter, they make a limited contribution to lessening the impacts of flash floods and improving infiltration. Along with leaf type, the tree age is also an important factor to consider. As a tree grows, it stores more carbon by retaining it in the tissues. When a tree dies and decays, it releases much of the stored carbon back into the atmosphere. Thus, carbon storage is an indication of the amount of carbon that can be released if trees are allowed to die and decompose. Maintaining healthy trees will keep the carbon stored in trees, but tree maintenance can contribute to carbon emissions (Nowak, D.J. et al., 2002). The presence of trees with their shade also contribute to decrease the air temperature (Soares et al., 2011).

This study highlighted the usefulness of software, such as i-Tree, to estimate ES that could support urban planning. A sample of 6928 street trees provides ES valued at 62000 euros annually, which extrapolated to all street trees in Lisbon estimates a value around 447100 euros annually (Soares et al., 2022). This evaluation could be improved by integrating other components, such as the aesthetic value.

Quantifying the degree to which street trees are serving the urban ecosystem proved to be a useful tool for supporting decisions on tree diversity in cities (i.e., Graça et al., 2017), both at specific, age and dimension levels. Currently, municipalities can use this approach for the establishment of priorities and the optimisation in the cost-benefit ratio in green spaces public investment and to foster public awareness about the importance of urban trees to support sustainable urban planning and development.

Acknowledgments

The authors thank the Lisbon Municipal Council and all the ISA/ULisboa's team collaborators.

References

Almeida, A.L. (2006). *O Valor das árvores, árvores e floresta urbana de Lisboa (Tree value assessment, Lisbon urban forest)*. Lisboa: Instituto Superior de Agronomia, Universidade Técnica de Lisboa.

Andresen, M.T. (1982). *Árvores de arruamento de Lisboa, contribuição para a sua classificação*. Lisboa, Portugal: Relatório Curso Arquitectura Paisagista, ISA, UTL.

Castel-Branco, C. (2000). *A Celebration of Architecture Contribution That Landscape Architects Have Made to the History of Expo 98*. United Kingdom: Landscape Design 295.

CML. (1929). *Inventário do arvoredo existente nos vários logradouros públicos da cidade de Lisboa e nos viveiros municipais (relativamente a Setembro de 1929)*. Tipografia Municipal, Lisboa: Câmara Municipal de Lisboa.

CTLA. (1992). *Guide for Plant Appraisal*. Savoy, IL: International Society of Arboriculture. Council of Tree and Landscape Appraisers.

Graça, M. *et al.* (2018). Assessing how green space types affect ecosystem services delivery in Porto, Portugal. *Landscape and Urban Planning*, 170(November 2017), 195–208. https://doi.org/10.1016/j.landurbplan.2017.10.007.

Graça, M.S. *et al.* (2017). Assessing mismatches in ecosystem services proficiency across the urban fabric of Porto (Portugal): The influence of structural and socioeconomic variables. *Ecosystem Services*, 23(November 2016), 82–93. https://doi.org/10.1016/j.ecoser.2016.11.015.

Konijnendijk, C. *et al.* (2023). *Forests and Trees for Human Health: Pathways, Impacts, Challenges and Response Options*. A Global Assessment Report. IUFRO World Series Volume 41. Vienna. 232 p.Magni, D., & Caudullo, G. (2016). Celtis australis in Europe: Distribution, habitat, usage and threats. In San-Miguel-Ayanz J., de Rigo D., Caudullo, G., Durrant, T.H., & Mauri, A. (ed.), *European Atlas of Forest Tree Species*. Luxembourg: Publication Office of the European Union.

Magurran, A.E. (1988). *Ecological diversity and its measurement*. Princeton, NJ: Princeton University Press.

Martínez-Varea, C.M. *et al.* (2023). Redrawing the History of Celtis Australis in the Mediterranean Basin under Pleistocene—Holocene Climate Shifts. *Forests*, 14(4), 1–28. https://doi.org/10.3390/f14040779.

McLean, D.C. *et al.* (2020). Florida's urban forest: A valuation of benefits. *Edis*, 2020(6), 1–13. https://doi.org/10.32473/edis-ep595-2020.

Mcpherson, E., Van Doorn, N., & Peper, N.S. (2016). *Urban tree database and allometric equations*. General Technical Report PSW-GTR-235. Albany, CA: U.S. Department of Agriculture, Forest Service, Pacific Southwest Research Station. https://doi.org/10.13140/RG.2.2.35769.98405.

Measham, T.G. *et al.* (2011). Adapting to climate change through local municipal planning: Barriers and challenges. *Mitigation and Adaptation Strategies for Global Change*, 16(8), 889–909. https://doi.org/10.1007/s11027-011-9301-2.

Nowak, D., Crane, D., & Dwyer, J. (2002). Compensatory value of urban trees in the United States. *Journal of Arboriculture*, 28(4), 194–199.

Nowak, D. *et al.* (2008). A ground-based method of assessing urban forest structure and ecosystem services. *Arboriculture & Urban Forestry*, 34. https://doi.org/10.48044/jauf.2008.048.

Nowak, D.J. (2021). *Understanding i-Tree: 2021 Summary of Programs and Methods*. General Technical Report NRS-200–2021. Madison, WI: U.S. Department of Agriculture, Forest Service, Northern Research Station, 100 p., 200–2021(December), 1–100. Available at: www.nrs.fs.fed.us/pubs/.

Nowak, D.J. *et al.* (2002). Effects of urban tree management and species selection on atmospheric carbon dioxide. *Journal of Arboriculture*, 28(3), 113–122.

Nunes, L. *et al.* (2013). Aboveground biomass and net primary production of pine, oak and mixed pine-oak forests on the Vila Real district, Portugal. *Forest Ecology and Management*, 305. https://doi.org/10.1016/j.foreco.2013.05.034.

Soares, A.L. *et al.* (2011). Benefits and costs of street trees in Lisbon, Portugal. *Urban Forestry and Urban Greening*, 10(2), 69–78. https://doi.org/10.1016/j.ufug.2010.12.001.

Soares, A.L. *et al.* (2022). *Estudos especializados para a quantificação dos serviços de ecossistema que proporcionam as árvores de arruamento da cidade de Lisboa*. Relatório—Serviço dos ecossistemas. Projecto LX-Tree. Instituto Superior de Agronomia. Lisboa, Portugal.

UN. (2019). *United Nations. Sustainable Cities*. Available at: www.un.org/en/sustainablefuture/.

18 Contributions to the integration of socio-cultural landscape assessment methods in the learning process

Lígia Vaz de Figueiredo, Isabel Loupa Ramos, Maria da Graça Saraiva and Fátima Bernardo

Introduction

The relationship between society and nature is complex but crucial to understand the subjective character of the landscape. An in-depth understanding of the socio-ecological dynamics of the landscape is relevant to dealing with the main challenges in urban areas. Thus, the integration of landscape assessment methodologies into landscape planning and learning enables the establishment of a conceptual framework that allows a common language between the actors. According to the European Landscape Convention (ELC), landscape assessment must consider the specific values attributed by stakeholders and social actors. The landscape services framework (Termorshuizen & Opdam, 2009) considers landscape as a socio-ecological system, meaning that citizens' well-being depends on the services that landscape provides. In this sense, landscape assessment is based on identifying and measuring these services.

The ELC defines landscape as "an area, as perceived by people, whose character is the result of the action and interaction of natural and/or human factors" and recommends "taking into account the particular values assigned to them by the interested parties and the population concerned" in the landscape assessment processes, which means considering values that people attribute to places.

The relevance of values for landscape research and learning has been pointed out since the early 80s by several researchers, highlighting that people's values are a fundamental component of the so-called "human dimension" of landscape (Buchecker, 2007). The studies developed by Kaplan and Kaplan (1989) were already based on the approach that the natural environment is experienced as a preferred or aesthetic environment, and Nassauer (1997) had concluded that landscape policies and strategies should be designed to align aesthetic expectations with the ecological health. This emphasizes people's perception, preferences, and desires of the landscape, highlighting that the landscape is socially constructed (Antrop, 2005).

Landscape assessment is a subjective process from which a specific valuation is expressed in different ways, dependent on the subject. In this process, people assign values to landscapes based on what they consider important (Antrop & Van Eetvelde, 2017). These values pertain to human preferences and are strongly

DOI: 10.4324/9781003454007-23

influenced by the interactions that occur between ecological and social systems. Socio-cultural assessment is the process through which it is possible to collect, synthesize, and communicate these values.

The concept of value has different meanings depending on the lens through which we analyze it, considering different disciplines, such as economics, philosophy, psychology, and ecology. Figure 18.1 highlights the conceptual changes through time, based on Ueda et al. (2009), considering several scientific frameworks, the main researchers, and the interconnections between the different typologies.

The first studies on values attributed to Ancient Greece focused on how people could live well. In the 17th century, Hobbes reflected on the subjective idea of goods, explaining, "scientifically" the logical relationships between social phenomena and their causes. During the 18th century, Kant reflected on subjective values. At the beginning of the 20th century, with behaviourism, values were studied from the learning point of view. The discussion of the concept in economic science occurred in the mid-18th century (Ueda et al., 2009). With the emergence of environmental sustainability, the concept is also explored in the field of ecology. Under the Rio Conference in 1992, sustainable development places humans as the centre, living in harmony with nature and the environmental value as fundamental for the development of societies.

At the beginning of the 20th century, the idea that nature is somehow "valuable" became stronger since society derives benefits from ecosystems that are essential for its survival. The Millennium Ecosystem Assessment (MEA, 2005) emerged in 1999 to assess how ecosystem changes impact human well-being. Since then, the values associated with nature, ecosystems, or landscape have been based on the services they provide.

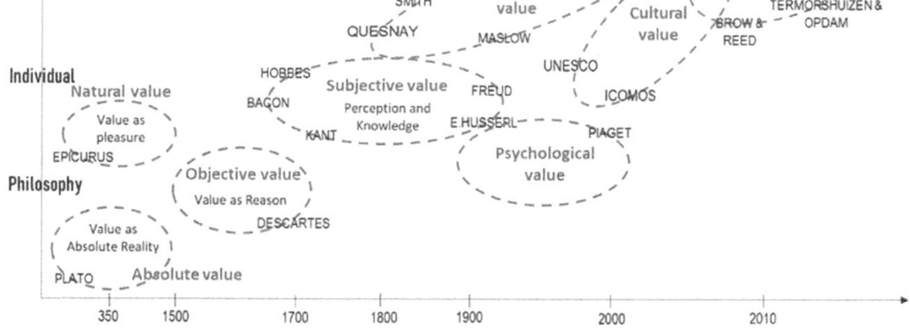

Figure 18.1 Concept of value through time, main references, and interconnections.

Source: Adapted from Ueda et al., 2009.

The subjective component of the landscape can be assessed through socio-cultural methods that are the process of collecting, synthesizing, and communicating knowledge about how people assign importance and meaning to nature's contributions to well-being (Pascual et al., 2017).

The biggest challenge is to give value to aspects that do not, directly, or indirectly, translate into material benefits (Chan et al., 2012) and to develop appropriate methodologies to achieve this goal, integrating people's needs and perspectives of nature.

Methods

There are several methods of socio-cultural landscape assessment. In this study, a systematic review was made using an online database "Isi Web of Knowledge". The search was carried out in 2019, including the words "socio-cultural" and "valuation" and "socio-cultural" and "assessment". Only peer-reviewed articles were considered leading to a total of 153 articles.

In the first phase, all the articles were critically analyzed based on title, keywords, and abstracts to identify which ones explicitly integrated socio-cultural assessment methods in landscape studies and operationalized the method with a practical application.

After we excluded articles focused only on the economic or ecological assessment, the approach was mostly theoretical, and the practical application was not identified. Finally, we identified 62 articles that met the established criteria.

Those articles were grouped based on Harrison et al. (2018), according to the following categories: participatory mapping (24), narrative analysis (5), deliberative valuation (2), preference assessment (15), photo-series analysis (3), photo-elicitation (3), multi-method (8), values suitability analysis (1), and time use studies (1). A comparative analysis was performed to discuss the main differences between the existing methods, based on the following questions:

- What is the specific aim of the method?
- What kind of audience is engaged or mobilized?
- Which instruments or tools are used?
- Which aspect is the assessment focused on: benefits, services, or values?
- What kind of output is obtained?

Results

Results show several methods for socio-cultural assessment which can be grouped according to aim, target community, tools, and outputs.

Participatory mapping (24) and preference assessment (15) are the most used methods, probably due to the variety of tools available and the diversity of results. Participatory mapping consists of mapping and describing (what exists and where), measuring (how much of something exists in a place), and valuing (how much it is valuable in quantitative terms) (e.g., Pietrzyk-Kaszyńska et al., 2017). Its main goal is the identification of potential conflict areas in land use planning through the study of landscape values identified by people. It is based on participatory data and GIS technologies to obtain non-expert spatial data through direct consultative processes. This allows us to integrate people's perspectives, combining theoretical and scientific knowledge with the people's experiences and values. Is widely used within the ecosystem services framework to obtain qualitative and quantitative data, spatially explicit, such as ecosystem services mapping. This method usually uses face-to-face interviews, online surveys, or collective interviews, like focus group sessions or workshops. With the development of GIS technology and online platforms, new practical and simpler tools have emerged, such as Maptionaire. Maps are used as interactive vehicles for spatial learning, analysis, discussion, and decision-making to communicate results with the public in a spatially explicit way.

Preference assessment is a direct consultative process for the analysis of people's preferences for a particular landscape or place (e.g., Schmidt et al., 2017). This method usually uses surveys to capture individual perceptions through free listening exercises, ranking of landscape typologies, identification of landscape elements and features, or visual stimulation with illustrations of different landscapes. This method has been widely used to evaluate cultural services, namely to study which service is perceived as more vulnerable or which contributes more to human well-being and, together with scenario integration, to study people's preferences concerning land use change.

Narrative analysis methods aim to capture the values people attach to nature through their stories, communicated in their own words (e.g., Bieling et al., 2014). These methods allow the collection of qualitative data and are also suitable for measuring some aspects of the Society–Nature relationship in quantitative or semi-quantitative terms. Thus, it is a flexible methodology, as it allows participants to express their emotions, feelings, and perceptions. Data analysis is therefore more complex and subjective.

Deliberative valuation is rather a valuation paradigm that provides a framework combining instruments and techniques that integrate population, stakeholders, and experts from different disciplines. The aim is to assess their preferences about the landscape through open and inclusive dialogues to debate ideas regarding their relationship with nature and promote consensus, based on panels, surveys, or debate forums (e.g., Raymond & Alexander, 2014).

The use of photographs of a specific landscape is a way to evaluate landscape through methods such as photo-analysis and photo-elicitation.

In photo-analysis, photographs published on photo-sharing websites such as Flickr, Panoramio, or Instagram are analyzed based on several criteria, such as the place that is photographed and the number of times a certain feature is highlighted, among others, to identify social preferences related to a particular landscape. This is a widely used method to identify cultural ecosystem services through people's perception based on photos and producing a map of their distribution (e.g., Gliozzo et al., 2016).

Another example of the usage of photographs is photo-elicitation, which translates visual experiences and perceptions of the landscape. Participants are asked to take a photo, in a non-directed way, of a particular landscape and then analyze which aspects are valued. Regarding the ecosystem services framework, this method has been used to identify the key ecosystem services provided by a particular landscape, using a series of photographs representing several landscape features (e.g., García-Llorente et al., 2012).

Given the complexity associated with socio-cultural assessment processes, often the use of a single method is not sufficient. Thus, 8 of the 62 case articles analyzed use multi-method approaches, combining two or more methods to integrate values perception data, meaning values that the public attributes to nature (qualitative information) with territorial and spatial information (land use analysis) or ecological information (indicators) (e.g., Plieninger et al., 2018).

The development of land use change scenarios is not in itself a method used in the socio-cultural landscape assessment; however, it can be a useful tool to integrate in participatory mapping (Vaz de Figueiredo, 2022).

The values suitability analysis (e.g., Reed & Brown, 2003) is a method to identify and integrate public perceptions of ecosystem values, which consists of empirical observations of what and where people value in nature.

Discussion

The concept of landscape values

The analysis of Figure 18.1 demonstrates that the concept of value evolves from a philosophical perspective to a social and environmental perspective.

In the MEA framework, the concept of value is associated with the utilitarian and anthropocentric perspective and is based on social preferences of nature and ecosystems connected to human well-being. With this framework, the socio-cultural value domain emerged and was defined as "non-material benefits that people obtain from ecosystems" (MEA, 2005) and contribute to human well-being.

After that, several developments occurred, and other frameworks appeared, such as landscape services, which allowed us to consider, in addition to the services that ecosystems provide to people, the services that result from the interactions between them, integrating the heterogeneity of the landscape.

The landscape services framework, within the concept of landscape values (Antrop, 2005; Termorshuizen & Opdam, 2009) reflects "the contributions of landscapes and landscape elements to human well-being". This approach is interdisciplinary, decentring the focus from ecology and economy to a more comprehensive and holistic assessment perspective. Society is not only considered as a "receptor of services", but also as part of the whole process. Landscape values are intrinsically related to cultural landscape services, which are understood as the social representations of nature, corresponding to the images developed in people's "encounters" with places, which allow them to understand those places (Muhar et al., 2018). This perspective reinforces the importance of the society-nature relationship to understand the urban landscape (Scholte et al., 2015). In this process, people, individually and socially, attribute meaning and value to places, as they are "place-makers".

Socio-cultural assessment methods

The chapter's analysis intended to provide answers to the questions raised before. Regarding the assessment aim, we concluded that the integration of local knowledge, spatialization of values, and decision-making support are the goals most identified. In this sense, these methods aim to capture local knowledge, through preferences, perceptions, and values assigned to nature and places. Participatory mapping is the method that allows the integration of qualitative data with spatial location, although this technique can also be used in photo-analysis through the georeferencing of photographs. We also identify the study of the relationship between landscape and human well-being and quality of life and land use conflict management as a frequent aim of the landscape assessment.

The target public varies according to the method under consideration. The population is the common target of all methods, and there are specific methods for approaching different publics. For example, stakeholders are more involved in narrative analysis, preference assessment, and photo-series analysis, and experts and decision-makers in deliberative valuation. Participatory mapping is a method addressed to all types of publics.

Surveys are the most used instrument, as in preference assessment and participatory mapping. These can be combined with other tools, such as maps and photographs, as they can reach a larger number of participants. Interviews, focus groups, or debate panels are commonly used in methods such as deliberative valuation and narrative analysis. Photo-elicitation and photo-analysis are based on the use of photographs through online platforms or photographs taken by the participants themselves.

Deliberative assessment, photo-analysis, and photo-elicitation focus on service evaluation and participatory mapping on landscape values identification.

The results are essentially qualitative, although in some methods quantitative data is obtained, such as preference assessment, photo-elicitation, and multi-methods. Participatory mapping results in maps of landscape or service value distribution. Multi-methods may also provide maps, indices, or indicators depending on the combination of methods used.

Conclusion

The process of valuing nature is an ongoing human concern, however, some relevant developments in recent times have led to new approaches. In the landscape context, monetary valuation, stressed by the ecosystem services framework revealed limitations on valuing non-tangible aspects, and therefore the development of approaches focusing on socio-cultural approaches and landscape values emerged as a valid option for subjective landscape assessment.

In general, there are two ways of measuring and quantifying landscape services (Brown & Fagerholm, 2015): the traditional economic approach that highlights the importance of natural systems and resources in providing benefits that ensure human well-being, and the place-based approach, supported in non-monetary valuations, producing guidelines for decision-making, within landscape planning and management.

Landscape assessment requires an integrated approach, considering different, complementary, and diverse methods (Jacobs et al., 2018), ensuring the engagement of stakeholders through participatory methods to identify society's needs, expectations, and preferences. This process promotes society's commitment to ecological issues and its engagement in landscape regeneration processes.

Landscape planning should integrate landscape assessment methods as it enables the qualitative and quantitative valuation of landscape services, and this knowledge contributes to informed decision-making. The subjectivity underlying social processes requires that its assessment integrates knowledge from social sciences and methods that include perception, attitudes, and behaviours. The socio-cultural assessment is a qualitative approach based on the perspective of stakeholders, experts, or citizens and should be used as a complement to the economic and ecological assessment, ensuring a holistic landscape approach.

As we demonstrate, there are several methods with different aims, target communities, tools, and different outputs. In the context of landscape assessment, the most appropriate method is often the result of a combination of methods that better enable the achievement of the goals.

As a conclusion, to ensure the mainstreaming of the methods into landscape planning practices, this knowledge should be integrated and be trained during the learning process. The subjective component approach of the landscape should complement the assessment methodologies and be integrated into the landscape management tools.

Bibliography

Antrop, M. (2005). From holistic landscape synthesis to transdisciplinary landscape management. In Tress, B., Tress, G., Fry, G., & Opdam (eds.), *From Landscape Research to Landscape Planning: Aspects of Integration, Education, and Application*, Vol. 12. The Netherlands: Springer, pp. 27–50. http://hdl.handle.net/1854/LU-339574

Antrop, M., & Van Eetvelde, V. (2017). *Landscape Perspectives. The Holistic Nature of Landscape*, Vol. 23 (Landscape). The Netherlands: Springer. doi:10.1007/978-94-024-1183-6.

Bieling, C., Plieninger, T., Pirker, H., & Vogl, C.R. (2014). Linkages between landscapes and human well-being: An empirical exploration with short interviews. *Ecological Economics*, 105, 19–30.

Brown, G., & Fagerholm, N. (2015). Empirical PPGIS/PGIS mapping of ecosystem services: A review and evaluation. *Ecosystem Services*, 13, 119–133.

Buchecker, M., Kianicka, S., & Junker, B. (2007). Value systems: Drivers of human-landscape interactions. In Kienast, F., Ghosh, S., & Wildi, O. (eds.), *A Changing World: Challenges for Landscape Research*. Switzerland: Springer Landscape Series, pp. 7–26. ISBN: 978-1-4020-4434-2.

Chan, Kai M.A. *et al.* (2012). Where are cultural and social in ecosystem services? A framework for constructive engagement. *BioScience*, 62(8), 744–756.

García-Llorente, M. *et al.* (2012). The role of multi-functionality in social preferences toward semi-arid rural landscapes: An ecosystem service approach. *Environmental Science and Policy*, 19–20, 136–146.

Gliozzo, G., Pettorelli, N., & Haklay, M. (2016). Using crowdsourced imagery to detect cultural ecosystem services: A case study in South Wales, UK. *Ecology and Society*, 21(3).

Harrison, P.A. *et al.* (2018). Selecting methods for ecosystem service assessment: A decision tree approach. *Ecosystem Services*, 29, 481–498.

Jacobs, S. *et al.* (2018). The means determine the end—pursuing integrated valuation in practice. *Ecosystem Services*, 29, 515–528.

Kaplan, R., & Kaplan, S. (1989). *The Experience of Nature: A Psychological Perspective.* Cambridge: Cambridge University Press.

Landscape European Convention. (2000). Council of Europe (ETS No. 176).

MEA. (2005). *Ecosystems and Human Well-being: Synthesis. Millennium Ecosystem Assessment.* Washington, DC: Island Press.

Muhar, A. *et al.* (2018). A model integrating social-cultural concepts of nature into frameworks of interaction between social and natural systems. *Journal of Environmental Planning and Management*, 61(5–6), 756–777.

Nassauer, J.I. (1997). Cultural sustainability: Aligning aesthetics and ecology. In Nassauer, J.I. (ed.), *Placing Nature: Culture and Landscape Ecology*. Washington, DC: Island Press, pp. 65–83.

Pascual, U. *et al.* (2017). Valuing nature' s contributions to people: The IPBES approach. *Current Opinion in Environmental Sustainability*, (26–27), 7–16.

Pietrzyk-Kaszyńska, A., Czepkiewicz, M., & Kronenberg, J. (2017). Eliciting non-monetary values of formal and informal urban green spaces using public participation GIS. *Landscape and Urban Planning*, 160, 85–95.

Plieninger, T. *et al.* (2018). Identifying and assessing the potential for conflict between landscape values and development preferences on the Faroe Islands. *Global Environmental Change*, 52, 162–180.

Raymond, C.M. *et al.* (2014). Comparing instrumental and deliberative paradigms underpinning the assessment of social values for cultural ecosystem services. *Ecological Economics*, 107, 145–156.

Reed, P., & Brown, G. (2003). Values suitability analysis: A methodology for identifying and integrating public perceptions of ecosystem values in forest planning. *Journal of Environmental Planning and Management*, 46(5), 643–658.

Saraiva, M.G. (Coord). (2009). Cidades e Rios: Perspectivas para uma relação sustentável. *Expoentes*, 9. ParqueExpo.

Schmidt, K., Walz, A., Martín-López, B., & Sachse, R. (2017). Testing socio-cultural valuation methods of ecosystem services to explain land use preferences. *Ecosystem Services*, 270–288.

Scholte, S.S.K., van Teeffelen, A.J.A., & Verburg, P.H. (2015). Integrating socio-cultural perspectives into ecosystem service valuation: A review of concepts and methods. *Ecological Economics*, 114, 67–78.

Termorshuizen, J.W., & Opdam, P. (2009). Landscape services as a bridge between landscape ecology and sustainable development. *Landscape Ecology*, 24(8), 1037–1052.

Ueda, K., Takenaka, T., Váncza, J., & Monostori, L. (2009). Value creation and decision-making in sustainable society. *CIRP Annals—Manufacturing Technology*, 58(2), 681–700.

Vaz de Figueiredo, L. (2022). *Metodologia de avaliação da paisagem fluvial: perspetivar o rio no espaço urbano. Tese de doutoramento.* Lisboa: Faculdade de Arquitetura da Universidade de Lisboa. http://hdl.handle.net/10400.5/28037.

19 Multiscalarity, interdisciplinarity, and research-by-design towards a metropolitan landscape

Ana Beja da Costa, Wim Wambecq and João Rafael Santos

Introduction

'Studio Lisbon' was an intensive 6-week design course of the postgraduate Advanced Master's of Human Settlements (MaHS) and Urbanism, Landscape, and Planning (MaULP), organized by the Department of Architecture, KU Leuven, as one of the 21/22 Spring semester's design studios working in a territorial context outside Belgium. Nineteen international students from the two postgraduate master's programs worked together to study and respond to emerging territorial problematics, having the Trancão floodplains site, in the Loures Municipality, Portugal, as a territorial scale case study.

The design studio organization took advantage of a collaboration protocol established between the KU Leuven and the School of Architecture of the University of Lisbon to foster synergies between the two universities. In this way, the studio took place at the School of Architecture in Lisbon and was developed under the framework of the ongoing FCT-funded research project 'MetroPublicNet–Building the foundations of a Metropolitan Public Space Network to support the robust, low-carbon and cohesive city: Projects, lessons, and prospects in Lisbon' in partnership with the Lisbon Metropolitan Area Authority. MetroPublicNet has as its main goal mapping, decoding, assessing, and discussing the combined result of public space improvements in Lisbon Metropolitan Area (LMA) delivered in the past two decades (1998–2020) and builds upon the hypothesis that a public space network at a metropolitan scale can offer possibilities to interconnect and integrate various fields for synergic responses to today's societal and urban challenges (Santos & Matos Silva, 2021). Within this frame, Studio Lisbon paid particular attention to public space networks that can be sustained and/or proposed by the students' design strategies, contributing to the research goals.

Studio Lisbon ambitioned to explore MetroPublicNet's research rationale as a fundamental way of city-making. It puts forward the hypothesis that a designed network of qualified public space can be shaped by acknowledging and integrating existing open space and urban segments into a larger-scale and consistent system that synergically articulates city and landscape as part of a complex metropolitan figure (Clos, 2016). Pedagogically, the studio was devised to immerse the next generation of professionals in a truly collaborative and interdisciplinary setup, necessary

DOI: 10.4324/9781003454007-24

to respond to the city's challenges, giving a strong emphasis on the landscape as the structural entity (Magalhães, 2001; Ribeiro Telles, 1994; Waldheim, 2006).

The territory was investigated through design at the scale of the Trancão River and its floodplain, a tributary of Tagus, and the surrounding urban patchwork as an outstanding landscape entity and ecological corridor. The studio explored networks, flows, and inhabitation modes that are to become coherent with the specificity of its context within the LMA and embedded in the broader resilient and dynamic estuarine landscape systems of the Tagus River. It aimed to combine fieldwork, critical survey, and interpretive mapping to understand the site's complexity and a strategic approach to large-scale public space networking and tactical urban design to reimagine this territorial entity as (part of) the systemic functioning of the metropolis.

Theoretical framework and interdisciplinary collaboration

Taking further on the ambitions of the renowned quality and tradition in teaching urbanism of the OSA department at KUL, landscape urbanism became the means for essential contributions to the achievement of the studio's objective, as it proposes to think the city (and at the core, its urbanism) through the medium of landscape (Waldheim, 2006). The discipline strongly anchors the definition of urban (or rural) design interventions on the characteristics of the landscape, canalizing development through the landscape structures that intertwine nature and cultural elements and addressing the territory's resourcefulness. Landscape urbanism is, by definition, appropriate to deal with climate change and water and natural resources management, also, as a design discipline that uses design as a medium of mediation (Corner, 1991; Pellegrino et al., 2015), develops visions and explores futures for the territories by integrating stakeholders' different interests (cross-disciplinary). In this regard, landscape urbanism is simultaneously able to deal with the needs of local communities, which have a strong link to the resourcefulness of the landscape, by developing strategies to guide interventions at the local scale and can systematize these in workable visions on a large scale (cross-scale) (De Meulder & Shannon, 2010).

It can be discussed that landscape urbanism appropriates Landscape Architecture's core theory and practice by a broader constituency. Landscape urbanism stems from Landscape Architecture's core body of knowledge, as defined by Cabral (1993) as 'the art of ordering outdoor space in relation to man' and the 'global landscape' advocated by Ribeiro Telles (1992, 1997). Landscape Architecture has evolved throughout the last century into what James Corner proposes as the need to 'recover' landscape through the repurposing of its role from a 'passive product of culture' into an 'active and strategic agent of culture' (Swaffield, 2002). Landscape urbanism is also strongly related to the 'hermeneutic tradition of interpretation, which means the relevance of mapping as a way of understanding the territory' (Corner, 1991) and can also be framed under Swaffield's (2002) 'theories of process', translated into site planning, ecological design, and the aesthetics of sustainability, relatable to the 'landscape-system' (Ribeiro Telles, 1994) and the 'ecological land suitability' (Magalhães et al., 2011) concepts. Design-wise, this argument is reinforced in KUL Urbanism's long-standing tradition in Belgium, where landscape architects are mostly dedicated to smaller-scale garden design, leaving a gap (and opportunity) for architects and urbanists to embrace large parks and wider scale landscape strategies design thinking. With Studio Lisbon taking place *in situ*, it took advantage of the Landscape Architecture core knowledge being developed in Portugal since 1940s, as a cross-scale discipline to tackle the Trancão floodplains design challenges.

The Trancão floodplains case study

The Tagus estuary is the largest estuary in Europe, and its vast water body is at the centre of the LMA. It is the common ground that originated many urban settlements with distinct genesis, economic activities, and social structures, justified by its ecological richness and by its sheltered geography that favoured the port development and related activities. Lisbon, the capital city, sits on the north bank, but many more urban centres punctuate the margins of the estuary. Throughout the 20th century, shifts in economic activities, abundant rural-urban migration, and car-oriented expansion resulted in cities dispersing away from the estuarine dynamics. As sprawl became the main mode of urbanization, its quality and specificity in relation to the geographical context diluted, increasing mobility inefficiency, social exclusion, and ecological ruptures (Wambecq & Beja da Costa, 2022). The LMA was created in 1991 to establish strategic spatial and social cohesion for the metropolis. It includes 18 municipalities and houses a growing 2.9 million inhabitants, almost 30% of Portugal's population.

Located in a prominent and central location in the metropolitan territory, River Trancão's floodplain was chosen as a case study. It forms a clear landscape figure in the Lisbon metropolis and crosses two municipalities: Lisbon and Loures. Historically regarded as the city (Lisbon) and its productive hinterland (Loures), they are now communicating barrels on the housing market. As Lisbon is rapidly turning unaffordable for a large stretch of the population, the expanding housing market in Loures remains more accessible, resulting in the projection of a space-consuming, generic, suburban housing production atop of a traditional productive landscape, gradually reducing its ecological richness and identity. The hypothesis is that the Trancão floodplain could serve as an intermediate landscape figure that strengthens the identity of the region (Ribeiro Telles, 1992), reactivating its systemic and ecological role, and allowing accessible inhabitation through new urban typologies.

The design's social and ecological challenges

For decades urbanization turned its back on the industrialized and consequently polluted floodplains of the Trancão River, which are also prone to extreme flood episodes (Leal & Ramos, 2013), such as the devastating ones that took place in 1967. With special decontamination efforts around the Expo 98 construction period, soils and water have generally been remediated. Nevertheless, flooding is likely to be exacerbated by climate change and the possibility of extreme high tides that enter the Trancão floodplain. At the same time, drought events might be more problematic as the need for water renders the natural wetlands dry, pushing unique fish and plant species towards extinction. Besides the ecological challenges, its role in the social tissue needs to be revisited. With changing perspectives on the role of such landscapes in the metropolis, the Trancão floodplain has the potential to become a qualitative place for urban resilience.

In this context, the students had to tackle three challenges strongly related to the MetroPublicNet research hypothesis: 1) the one of resilience and environmental robustness, that questioned how a recalibration of a multifunctional landscape can allow for the strengthening of ecological processes and biodiversity, the remediation of climate change impact as floods and drought, while maintaining the identity of a rural, productive landscape with restored, diverse ecological assets, through the improvement of waterlines riparian galleries, afforestation of the hillside slopes, preservation of infiltration areas on the plateaus; 2) the one of sustainable and low-carbon mobility, that questions what a consistent soft mobility network would work, namely between existing public transportation hubs, and if it is possible to recover some ancient boat transportation routes, etc.; 3) the one of inclusion and territorial cohesion, questioning how the elaboration of a new urban tissue and innovative public space, grafted on the landscape's structures, can improve urban cohesion and neighbourhood connectivity around the Trancão floodplain and streamline development speculation to a frame coherent with the aforementioned challenges. Various examples from well accomplished case studies of river restoration and design interventions were exposed, such as the Qian'na Sanhihe Greenway, Hebei Provice in China, (Kongjian You/Turenscape & Beijing University College of Architecture and Landscape, 2010); the Wijkerroogpark, Velsen in the Netherlands (Bureau B+B & Erick de Lyon, 2006); the Linear Park of Ribeira das Jardas in Cacém, Portugal (NPK Arquitectos Paisagistas, 2008); and the Emscher River Reconversion Master Plan 1990–2020 in the Ruhr Valley, Germany. These respond to similar challenges, in diverse scales and geographical contexts, sharing the common conceptual frame of recovering water related degraded and/or vulnerable spaces, with the consequent biodiversity recovery and improvement of hydrological performance in heavily or progressively urbanized contexts (McDowell, 2017; Prominski et al., 2017; Rossano, 2021; Sijmons et al., 2017).

Methodology

The studio functioned as a 6-week intensive design workshop, an immersion experience of living and working in Lisbon and in the broader LMA territory, with weekly organized outputs designed to streamline the students' workflow and help construct critical design research within the time frame. The stages of the workshop included the following:

- Stage 1: critical understanding; analytic mapping through the landscape-system method (Cunha & Magalhães, 2019; Magalhães, 2001; Magalhães et al., 2007); delayering topography, slopes, aspect, geology, soils, vegetation cover; interpretative mapping (Alfaiate & Ribeiro, 2021; Cavalieri & Viganò, 2019; Orff, 2016); adding the mobility systems flows in relation to urban tissue, historical evolution, flood scenarios; and fieldwork.
- Stage 2: based on the previous, design explorations on the landscape-system through the redefinition of the water cycle were carried out, testing of urban typologies and densities in relation to the landscape-system, hovering in scale between the floodplain as a system, and the cases as places that play specific roles in the system—in the Trancão river mouth—Tejo confluence, in the

Figure 19.1 The Trancão floodplain and the six case study sites.

Source: Map by W. Wambecq.

Trancão hillsides (Sacavém and S. João da Talha), and in the floodplain (Frielas, Infantado and Sto. Antão do Tojal).

- Stage 3: detailed design exercises on the six case studies and elaboration of urban typologies. This stage included production of plans, hand-drawn sections, sketches, 3D visualizations and collages of proposal's masterplan, and detailed design of the housing typologies.
- Transversal inputs were organized to build contextual and thematic knowledge that assisted the design work: Field trips were organized around projects of MetroPublicNet, the Tagus estuary, and the floodplain of Trancão; Open Lecture series: publicly organized lecture series on MetroPublicNet, water in public space, housing, Landscape Architecture theory and practice, and landscape urbanism; MetroPublicNet case study investigations: students were assigned to critically evaluate a case study of MetroPublicNet according to the three rationales of the research project.

Three geomorphological features of Trancão formed the subject of design explorations: the Trancão river mouth, the steep Trancão hillsides (*costeiras de Loures*), and the Trancão floodplains to respond to the three challenges mentioned earlier. Across the three features, six case studies were defined (Figure 19.1). The final output was organized as an exhibition and presented to the public and a jury, at the end of the six-week period.

Results and discussion

Students from various academic, professional, and cultural backgrounds showed a remarkable intake of core concepts of Landscape Architecture, taught according

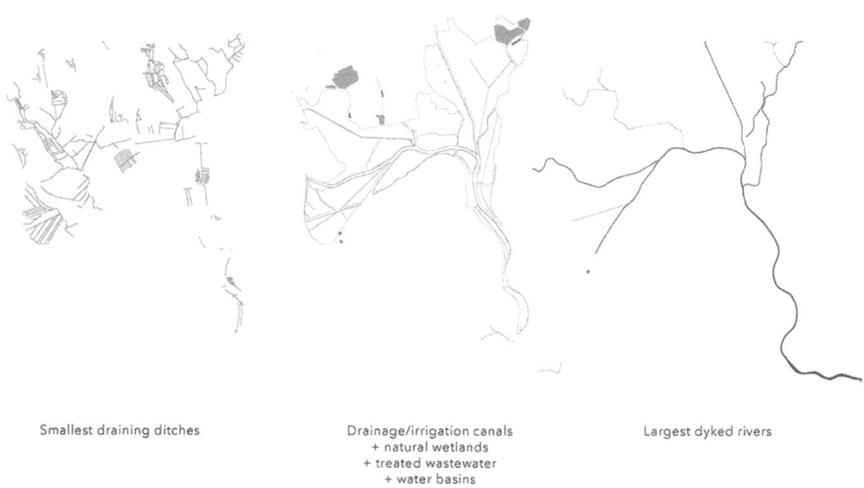

Figure 19.2 Understanding of the current water cycles and its components to rebalance water drainage, retention, and tidal movement.

Source: Maps by W. Wambecq.

to the Portuguese Landscape Architecture methodologies, such as the 'system-landscape' mapping method, delayering the landscape through a set of themes. Learning from this, the ecological landscape suitability (Cunha & Magalhães, 2019; Magalhães et al., 2011) was introduced on the scale of the floodplain, (re) introducing former land uses and reimagining the water cycle, designed in a way that installed a balance between the natural ecology of the wetland (attempting to retain water on site) and the productive landscape (essentially draining water out of the floodplain to accommodate agricultural production). A system of wetland parks was imagined where main river courses enter the floodplain, often grafted on former river courses while maintaining the interior of the floodplain available for agricultural production (Figure 19.2). This rebalancing was realized by inverting the meaning and functioning of the dikes along the water that created a network of soft mobility lines and riparian vegetation galleries to reinforce the *continuum naturale*. In strategic locations, dykes are removed to allow flooding and the creation of wetlands. In other areas, they are maintained to protect the productive fields. Water is no longer guided to be drained as fast as possible towards the Tagus

Figure 19.3 Rethinking urbanity through a sustainable water cycle in Frielas.
Source: Authors: E. Karanja, A. Van Lint, O. Vé-Réveillac; Studio Lisbon; the Trancão floodplain.

Figure 19.4 The re-establishment of a tidal landscape as a generator of public space and new urban typologies at Sacavém.

Source: Authors: M.R. Armoutaki, A. Tsouvara, L. Sudhakar; Studio Lisbon; the Trancão floodplain.

estuary. In a drought-prone country with increasing problems of water accessibility, water should be kept on the terrain, rather than flushed away. The imagined water cycle was accompanied by natural and cultural (productive) landscape systems and actively sought to enrich the relation with the surrounding urban structures.

Students were able, on the scale of group case, to incorporate the landscape processes into a hybrid design of landscape, public space, and urban development, intertwining novel practices of city-making that were coherent with and contributed to reinforce the systemic functioning of the landscape: the housing typologies became part of a water retention structure; sports facilities were embedded in the constructed wetland and feeding a new water body for leisure activities (Figure 19.3); the renewal of a ruined industrial site allowed for a physical connection between the floodplain and the upper plateaus overseeing a re-naturalized, meandering floodscape of the Trancão (Figure 19.4).

The students showed the capacity to build and use the knowledge of landscape systems to inform urban design, dealing with the three thematic challenges, and the floodplain as a large-scale landscape figure, transversal to the different design groups. The trans-scalar approach demanded flexible collaboration methods in an effort to make a collective project.

Conclusions

Studio Lisbon was a 6-week intensive collaboration between the Department of Architecture (KU Leuven) and the Lisbon School of Architecture (ULisboa). The setup of local context immersion allowed students to be in constant contact with the object of study, gather research and information from local sources, and to experience cultural habits throughout the workshop. Teaching was organized as a collaborative effort with complementary expertise from both universities between urbanism, architecture, and Landscape Architecture. Pedagogically, the 6-week intensive workshop drew from two significantly different design cultures, both in terms of fields of expertise, as methodologies, pushing both students and teachers beyond their usual frame of reference. Landscape analysis methodologies, based on the fields of knowledge of ISA, formed the backbone for the urban design explorations, as explored at KU Leuven, that incorporated systemic territorial functioning, its inhabitation conform urban norms of development, and the new imaginary that sprouted from such new models of design. Local stakeholders welcomed the exploration of alternative development scenarios that spring from locally embedded knowledge yet also draw from the creativity of the 'outsider' view. In addition, their continuous involvement raises the potential relevance of the design explorations and the possible uptake by the local authorities (Figure 19.5).

The collaborative workshop proved to be an interesting vehicle to explore the hypothesis of MetroPublicNet, to understand its potential as a projective ambition, and especially to underpin the importance of the landscape—the notion of resilience and environmental robustness—as the leading rationale in searching for a continuity of public space. The Trancão floodplain is an impressive landscape within the LMA. Its ecological refitting—based on a broad inclusion of uses, people, and practices—forms the base of a new form of living, of producing, and of moving in the metropolis.

Figure 19.5 The final review and ideas exchange moment between universities and stakeholders, based on the students' research-by-design presentations.

Source: Photograph by W. Wambecq.

The workshop has proven to be a successful exchange model that can hopefully motivate further internationalization of the Landscape Architecture course at the Higher Institute of Agronomy–ISA, both through the possibility for intensive workshops design studios setup and the use of English as main teaching language. The collaboration between KU Leuven and the Lisbon School of Architecture—the organizational flow, knowledge exchange, professors and students' interaction, and open lectures with staff from both universities—was a mutually enriching experience.

Acknowledgements

The authors would like to acknowledge the support of KU Leuven, FA-ULisboa, and CIAUD. MetroPublicNet research project was supported by Portuguese national funds through the Fundação para a Ciência e a Tecnologia (FCT, I.P.) under the grant PTDC/ART-DAQ/0919/2020 (MetroPublicNet: Building the foundations of a Metropolitan Public Space Network to support the robust, low-carbon and cohesive city: Projects, lessons and prospects in Lisbon).

Reference list

Alfaiate, T., & Ribeiro, J.M. (2021). *Coreografias da água—Pensar o sítio de Lisboa (Water Choregraphies—Thinking the Lisbon site)*. Lisboa: Margem Arquitectura Paisagista.

Caldeira Cabral, F. (1993). *Fundamentos da Arquitectura Paisagista (Landscape Architecture Fundamentals*). 2nd Edn. Lisbon: Instituto da Conservação da Natureza.

Cavalieri, C., & Viganò, P. (eds.). (2019). *The Horizontal Metropolis a Radical Project*. Zurich: Park Books.

Clos, O. (2016). *Espace(s) public(s) metropolitains(s)*. Lille, l'Agence de dévelopment e d'urbanisme de Lille Métropole. www.adu-lille-metropole.org/wp-content/uploads/2021/05/SyntheseEPM.pdf.

Corner, J. (1991). The hermeneutic landscape. In Swaffield, S. (ed.), *Theory in Landscape Architecture: A Reader*. Philadelphia: Pennsylvania University Press, pp. 130–131.

Cunha, N.S., & Magalhães, M.R. (2019). Methodology for mapping the national ecological network to mainland Portugal: A planning tool towards a green infrastructure. *Ecological Indicators*, 104, 802–818. https://doi.org/10.1016/j.ecolind.2019.04.050.

De Meulder, B., & Shannon, K. (2010). Traditions of landscape urbanism. *Topos Magazine*, (71), 68–73.

Leal, M., & Ramos, C. (2013). Susceptibility to flooding in northern Lisbon metropolitan area. Natural factors and impacts of changes in land use. [Susceptibilidade às cheias na área metropolitana de Lisboa norte factores de predisposição e impa ctes das mudanças de uso do solo]. *Finisterra*, 48(95), 17–40. https://doi.org/10.18055/Finis3129.

Magalhães, M. (2001). *A Arquitectura Paisagista—morfologia e complexidade*. Lisboa: Editorial Estampa Lda.

Magalhães, M.R., Abreu, M.M., Cortez, N., & Lousã, M. (2007). *Estrutura Ecologica da Paisagem, conceitos e delimitação—escalas regional e municipal (Ecological Network—Landscape, concepts and delimitation—regional and municipal scales)*. Lisbon: ISAPress.

Magalhães, M.R., Cunha, N.S., & Pena, S. (2011). The ecological land suitability in the land-use plan. In *7VCT Conference, 11–13 October 2011, Lisbon.*

McDowell, S. (2017). *Water Index—Design Strategies for Drought, Flooding and Contamination*. Actar.

Orff, K. (2016). Introduction. Revive, Cohabit, Engage, Scale essays. In Orff, K. (ed.), *Toward and Urban Ecology*. New York, NY: The Monacelli Press, pp. 7–16.

Pellegrino, P., Ahern, J., & Becker, N. (2015). Green infrastructure: Performance, appearance, economy and working method. In Czechowski, D., Hauck, T., & Hausladen, G. (eds.), *Revising Green Infrastructure Concepts Between Nature and Design*. Boca Raton: Taylor & Francis Group, pp. 385–403.

Prominski, M., Stokman, A., Stimberg, D., Voermanek, H., Zeller, S., & Bajc, K. (2017). *River. Space.Design, Planning Strategies, Methods and Projects for Urban Rivers*. Basel: Birkhäuser.

Ribeiro Telles, G. (1992). A Paisagem Global da Região de Lisboa (The Global Landscape of Lisbon's Region). *AGROS*, (2), 5–9. Available at: www.repository.utl.pt/bitstream/10400.5/24847/1/A%20paisagem%20global%20da%20Regi%C3%A3o%20de%20Lisboa.pdf.

Ribeiro Telles, G. (1994). Paisagem Global. Um conceito para o Futuro (Global landscape. A concept for the future). *Iniciativa*, (no. Special Issue).

Ribeiro Telles, G. (1997). *Plano Verde de Lisboa (The Green Plan of Lisbon)*. Lisboa: Edições Colibri.

Rossano, F. (2021). *Floodscapes Contemporary Landscape Strategies in Times of Climate Change*. Amsterdam: nai010.

Santos, J.R., & Matos Silva, M. (2021). Public space as a metropolitan network. Making the bridge between policy and design in Lisbon Metropolis. In Charytonowicz, J., Maciejko, A., & Falcão, C.S. (eds.), *Advances in Human Factors in Architecture, Sustainable Urban Planning and Infrastructure Proceedings of the AHFE 2021 Virtual Conference on Human Factors in Architecture, Sustainable Urban Planning and Infrastructure, 25–29 July 2021, USA*. Cham, Switzerland: Springer. pp. 310–317.

Sijmons, D., Feddes, Y., Feddes, F., Luiten, E., & Nolden, M. (2017). *Room for the River. Safe and Attractive Landscapes*. Netherland: blawdruk.

Swaffield, S. (ed.). (2002). *Theory in Landscape Architecture: A Reader*. Philadelphia: University of Pennsylvania Press.

Waldheim, C. (2006). Landscape as urbanism. In Waldheim, C. (ed.), *The Landscape Urbanism Reader*. 1st Edn. New York: Princeton Architectural Press, pp. 35–53.

Wambecq, W., & Beja da Costa, A. (eds.). (2022). *Studio Lisbon, the Trancão Floodplain*. Leuven: University of Leuven.

Part IV

Conclusion: Landscape Architecture and the practice of ecological thinking

Considerations for the next 80 years of history

Maria Matos Silva and João Ferreira Nunes

As a field of study and practice, Landscape Architecture began to emerge in the late 19th and early 20th centuries (Cabral, 2003[1993], p. 34) despite its remote origins. As a discipline forming part of higher education for over a century, Landscape Architecture is not only 1) the holder of precious accumulated knowledge, but it is also 2) an active knowledge producer where specific habits[1] are formed.

Regarding the discipline's accumulated knowledge, one may highlight emblematic milestones, of both theoretical reasoning or professional practice, from "L'Art des Jardins" (The Art of Gardens) by André Le Nôtre or "Design with Nature" by Ian McHarg (1969) to the "The Third Landscape" by Gilles Clément (2004), just to mention a few; or from Villa d'Este (Pirro Ligorio, modified by Cardinal Ippolito, 16th century), Boston's Emerald Necklace (Frederick Law Olmsted, late 19th century), to Qunli Stormwater Park (Turenscape, 2010), among many other examples extensively recognized and valued by society. In Portugal, Landscape Architecture education and practice dates back to 1941, when Cabral founded the first Autonomous Course in Landscape Architecture *(Curso Livre de Arquitectura Paisagista, CLAP)* (Cabral, 1968). Throughout these past 80 years, Portuguese Landscape Architecture has played an important role with unquestionably significant contributions in the face of societal challenges (Cf. "Heritage and History of Portuguese Landscape Architecture: A time frame for Landscape Architecture in Portugal." by Teresa Andresen), considering namely the seminal legal diplomas of the Landscape Planning Policy (namely the National Ecological Reserve (REN, DL 321/1983)); the participation in international executive boards or scientific committees of globally recognized institutions of great value to the field (such as IFLA, ICOMOS, Fondazione Benetton Studi Ricerche, Institut Européen des Jardins et Paysages, Europeam Route of Historic Gardens); the encompassing contributions as decision-makers in national state agencies such as the National Service for Parks and Reservations (currently the Institute for Nature Conservation and Forests); the contributions to the World Heritage List of UNESCO, namely in the category of

DOI: 10.4324/9781003454007-25

cultural landscapes such as the sites of Sintra Cultural Landscape, Douro Wine Region, and Sanctuary of Bom Jesus do Monte in Braga; the creation of cultural associations decisive in the valorisation and safeguarding of many Portuguese landscapes, such as the Liga para a Protecção da Natureza (1948) or the Portuguese Association of Historic Gardens (2003); the pioneering endeavours in computer-aided landscape planning (as mentioned by Ribeiro when referring to the OTAC Symposium (Cf. "Part II – A Portuguese context on education and pedagogy in Landscape Architecture: Heading to a comprehensive logical/rational, aesthetic, and ethic approach" by Luís Paulo Ribeiro); or the designs of emblematic parks and gardens such as the Garden of the Gulbenkian Foundation (António Viana Barreto & Gonçalo Ribeiro Telles, 1969), Parque da Cidade at Oporto (Sidónio Pardal, 1993), or Parque Tejo e Trancão (PROAP + Hargreaves Associates, 1998) in Lisbon, or even the recently awarded public space projects in Lisbon, such as the requalification of Praça do Martim Moniz (F|C Arquitectura Paisagista, 2003) or the Alcântara Valley Green Corridor, Avenida de Ceuta unit (BALDIOS Arquitectos Paisagistas), naming just a very small selection. A somewhat violent overview, so small and superficial when compared to the size and richness of the entire 80-year history, just to highlight how significant the role of Landscape Architecture in Portugal has been.

Yet research presented in this book highlights further examples of accumulated knowledge in Portuguese Landscape Architecture. More specifically, regarding historic gardens and cultural heritage, Marques and Curado develop upon "the way the restoration of historic gardens has evolved in Portugal from the first actions and writings by Cabral and some of his disciples to the present day", emphasizing the "knowledge and sensitivity of the senior tutors" in the "attentive and very careful orientation, directed towards the cultural significance of the gardens" (Chapter 1, "Historic gardens as cultural heritage"). Also Adagói et al. present the history of the Green and Blue Infrastructures Thematic Line (GBI-TL), which is part of the LEAF research centre (Linking Landscape Environment Agriculture and Food) at ISA-UL, emphasizing its 30 years contribution to the development of Portuguese Landscape Architecture education and research (Chapter 8, "Thirty years of research in green infrastructure and landscape planning"). On the other hand, using the particular example of three generations of landscape architects, who reconciled professional practice with teaching and research, Costa Pinto argues how accumulated expertise serves to be "reflected in a way of projecting, and in the transmission of knowledge, ideas, and values, between masters and disciples" (Chapter 11, "The legacy of three generations of architects-professors").

Concerning the production of knowledge through the creation of certain habits in conduct, in Landscape Architecture one can refer to ecological reasoning also as a "modus operandi", not as something that can be explicitly learned, but rather conveyed by example, by neighbouring references, by routine. A habit that is essentially intellectual, rational, non-emotional that is useful for managing complex problems: the habit of reflecting, talking, reasoning, as well as the habit of picking fights or of not being satisfied with certain justifications (Tamen, 2023). Having ecological thinking as a habit, it is possible not only to evolve and further mature on its reflexive processes but also to continuously contribute and serve education and the professional practice with new knowledge. An outstanding example of the concept is clearly emphasized in the teaching of landscape design by Manuel de Sousa da Câmara, who urged for "constant questioning and debate" helping students to "develop critical thinking towards the feasibility of their design solutions" (Chapter 10, "Nature-based solutions in the teaching of landscape design by Manuel de Sousa da Câmara"), as well as from the history of Lisbon city gardeners and their on-the-job experiences that, together with other more classical knowledge transfer methods such as training, publications or horticultural events, enabled their know-how to be passed along generations (Chapter 6, "Lisbon city gardeners, from horticulture to Landscape Architecture [1840–1960]").

Interpreting the research entailed in this book, it is possible to recognise not only its contribution towards Portuguese Landscape Architecture accumulated knowledge but also its contribution towards the enrichment of a specific way of thinking that is rooted in cultural, aesthetic (Cf. "Part III – Portuguese perspectives on theory and methods in Landscape Architecture. Theory versus circumstances: the help of methods in Landscape Architecture" by Cristina Castel-Branco) and ecological principles of action (cf. three principles of ecological thinking, see the Introduction: Three principles of ecological thinking for an interpretation of Portuguese Landscape Architecture education, heritage, and research). Two aspects of a disciplinary approach are here considered as particularly useful when facing today's global crises. A disciplinary approach that is namely highlighted in the following six views evidenced throughout the chapters:

(1) **Landscape Architecture as a discipline with an overall comprehensive outlook that is not specialized in a singular subject.** The ability to focus sharply towards distant and independent objectives, albeit being very much necessary to our communal scientific and social development, if applied to the analysis, diagnosis, or proposal of landscapes, runs the enormous risks of being myopic and not able to see the whole. As argued by Timothy Morton, "It is vital for us to think and act in more general, wider terms. . . . The reactionary response to wind farms in the United Kingdom, for instance, has tried to bog down environmentalists with the idea that birds will be caught in the blades of the windmills. Yes, we need to cultivate a more comprehensive view of 'humanity' and 'nature'" (Morton, 2007, p. 6). And the range of chapters presented in this book—which address themes that go from the role of trees in the city to planning policies, from the importance of cultural landscapes to the development of research-by-design processes—fits well with this generalist and comprehensive way of looking at landscape and its imbricated ecosystems. "Landscape architects are specialists in generalities" as stated by Cabral (2003 [1993]).

(2) **The preparation of Landscape Architecture to read and design realities from the human to the planetary scale.** From local to municipal, metropolitan, national, continental, and even planetary scales, Landscape Architecture has been exercising, both in theory and methods, a cognitive and practical elasticity between scales. There is no such thing as a small or large landscape.

Regarding the local scale, Cunha et al. demonstrate how tree abundance and species richness increased in Lisbon since the thirties (Chapter 7, "From historical research to urban ecology") while Soares et al. estimate the ecosystem services (ES) provided by the city's public street trees (Chapter 17, "Diversity of street trees in Lisbon, towards city sustainability"). On the watershed scale, Fernandes et al. propose an urban adaptation strategy for Tinto riverscape so that to "reverse the increasing destruction of the watersheds' natural structure, whose impacts are being aggravated by extreme climatic events and severe pollution" (Chapter 16, "Urban adaptation strategy for Tinto riverscape, at watershed scale, towards its multifunctionality and hydrological resilience—impacts and governance challenges"). On the regional scale, Martinho da Silva describes the landscape changes of Interior Alentejo (Portugal) comparing two important historical milestones, the 1820 Liberal Revolution and the 1974 April Revolution, arguing that "the region is currently facing an intensification and simplification of its landscape around the Alqueva dam and in other irrigation areas, due to the cultivation of intensive monocultures", a "fact is already aggravating the extreme drought affecting the region, due to climate changes" (Chapter 15, "In between revolutions"). On the planetary scale one must mention the research by Carl Steinitz, recently honoured with the title Honoris Causa by a Portuguese Landscape Architecture School (2023), reaffirming the disciplinary competences of Landscape Architecture to "change geography by design" (Steinitz, 2012).

(3) **The irreverence of Landscape Architecture towards the limitations associated with the idea of a confining "frontier";** that is resulting from the aforementioned generalist and comprehensive approach respecting landscape and the trained flexibility with its working scales. Indeed, today's greatest challenges (nuclear radiation, viruses, heat waves) are totally oblivious to man-made frontiers. All societies (almost) equally share the risk from the environment itself (Morton, 2007). In his latest book, Ulrich Beck proposes an interpretation of what kind of world we live in today, insisting on the idea of metamorphosis: that we live in a world in metamorphosis and how this is distinct from a world in transformation.[2] Compared to Galileo's hypothesis that it was not the Sun that revolved around the Earth but the Earth that revolves around the Sun, he suggests that "Climate risk teaches us that the nation is not the centre of the world. The world is not circulating around the nation, but the nations are circulating around the new fixed stars: 'world' and 'humanity'" (Beck, 2016, p. 11).

Among the presented research, two chapters stand out in their conscious and reflective approach towards existing frontiers. More specifically, Nardella compared the birth and evolution of Landscape Architecture education in Portugal and Italy through the personalities of Cabral and Pietro Porcinai, highlighting "how Cabral and Porcinai, both inextricably linked to their countries' unique natural and cultural contexts and very active within IFLA, have affected the development of innovative approaches to Landscape Architecture educational programmes in Italy and Portugal, albeit following different pathways and destinies" (Chapter 4, "Landscape Architecture education in Portugal and in Italy).

Focusing on a different continent, and developing from an established cooperation with the Architecture and Physical Planning Faculty of the Eduardo Mondlane University (FAPF–UEM) in Maputo, Mozambique, Beja da Costa aimed at "testing Landscape theory and methods as they have been developed in the past 80 years in Portugal, in a different geographical, climate and cultural context bring valuable insights on the methodologies efficiency and validity, showing a potential to be continued as a valuable academic and professional exchange" (Chapter 12, "Landscape Architecture and urban ecology research as means for knowledge transfer on urban resilience in Maputo, Mozambique").

(4) **The value that Landscape Architecture gives to experimental research and design processes**, which is particularly evident in Chapter 19, "Multiscalarity, interdisciplinarity, and research-by-design towards a metropolitan landscape" and Chapter 9, "Teaching restoration of historical gardens: Research through design experience". The first, built on a 6-week intensive design studio in which three academic institutions collaborated, aimed to combine "fieldwork, critical survey, and interpretive mapping to understand the site's complexity" so that to propose strategic approaches on "large-scale public space networking and tactical urban design" for the Trancão river basin in Lisbon's Metropolitan Area. Overall, it served as a successful means to explore and experiment on the research hypothesis launched by the MetroPublicNet research project (Chapter 19). The second article develops upon the different teaching methodologies on historic garden restoration that have been explored at the University of Lisbon, and how they have matured throughout the 33 years of education practice (Chapter 9). Both chapters evidence the contemporary concern for new pedagogical models that embrace the new generations, which are namely bringing new forms of interaction together with a straightforward digital practice.

(5) **The attentive and responsible social sensitivity of Landscape Architecture**. It is always worth recalling, in the words of the Portuguese Landscape Architecture "father figure", that "our mission . . . always requires no less knowledge of men than of plants and the physical environment", and "it is necessary that we have been able to guess the desires and wishes of those for whom we work for" (free translation, Cabral, 2003[1993], p. 47). And while this responsibility is consistently present in all chapters in this book, two research endeavours develop particularly interesting viewpoints. Vaz de Figueiredo et al. stress how values assigned by people are essential to urban landscape planning processes, systematizing a set of socio-cultural assessment methodologies that can be used to engage stakeholders, experts, or citizens in order to identify society's needs, expectations, and preferences. The research furthermore highlights how this qualitative approach is to be used "as a complement to the economic and ecological assessment, ensuring a holistic landscape approach" (Chapter 18, "Contributions to the integration of socio-cultural landscape assessment methods in the learning process"). Also, Marques and McIntosh propose new perspectives on how to embrace social and cultural aspects of sustainability by exploring recent trends in Landscape Architecture within New Zealand that

incorporate native knowledge. A reform in the current Portuguese Landscape Architecture academic curriculum and pedagogy is furthermore strongly encouraged, arguing that such change must namely entail "a deep and personal connection to the landscape that can ground scholars, students and practitioners to establish a more holistic and respectful framework" (Chapter 13, "Breaking siloes and embracing the future"). And last but not least,

(6) **The Landscape Architecture pursuit to understand the correlations between natural and man-made laws.** This seemingly clear delineation between natural and man-made laws gets more complex as society advances and our understanding of the world broadens. In essence, natural laws respond to the principles that govern the physical universe (such as gravity or thermodynamics), and man-made laws are designed by human societies, in the form of legal codes, to regulate behaviours or sustain justice. Yet it is evident how these categories have become more and more intertwined as humans wield increasing power to manipulate genetics, control climate, and reshape ecosystems. Indeed, the notion of a rigid distinction is being challenged, namely with the rise of fields such as environmental ethics or biotechnology (Kolbert, 2014). While Cabral was a pioneer and a key figure in the ecological awareness in Portugal (Andresen, 2001) it was Ribeiro Telles who laid the foundations of environmental policy, as thoroughly detailed by Vasconcelos et al. (Chapter 2, "Gonçalo Ribeiro Telles). Both had strong ethical principles, as recalled by Gomes when mentioning the misfortune on the remodelling project for the Avenida da Liberdade in Lisbon, stating how in the face of a strong "political backlash, they upheld their technical views, refusing to give in to external pressure." (Chapter 3, "Francisco Caldeira Cabral and the 'Panorama' magazine"). Cultural landscapes, which were always a matter of interest to both personalities, represent this paradox rather clearly, as a concept that, while acknowledging this entanglement between natural and human laws existing in landscape, allows us to think on how to "work with the natural, cultural, tangible and intangible elements and processes that reinforce landscape identity" (Chapter 14, "From the concept of cultural landscape to its application in conservation policies and higher education in Portugal")

But what are we not seeing, not engaging, or not enabling when considering this spectrum of research dedicated to the 80 years of Landscape Architecture in Portugal? Bluntly, one may argue about the lack of a socially diverse representation in the celebrated professionals and academics. Although the discipline did emerge in a social and political context that clearly privileged white males, today's portrait of Portuguese Landscape Architecture is expected to have evolved.[3] But in this regard, how has it progressed? For instance, what has been the role of women, or other less visible professionals, in the growth and recognition of the discipline?[4] And throughout the 80 years, was there no communal influence among Portuguese speaking countries and cities?[5] These are just some of many possible questions that are paramount in future research.

Indeed, bearing in mind this important milestone that celebrates Landscape Architecture in Portugal, it is important to critically reflect not only on the achievements and misadventures of the past but also, and just as importantly, on the future. Considering the 80 years of accumulated knowledge, what should the next steps be? Like in so many other areas, major concerns regarding the current global environmental, social, and mental crisis (as highlighted in the introductory text (Introduction) are challenging Landscape Architecture towards a certain degree of reinvention. Overall, it is here argued that a significant part of this disciplinary revisitation must entail the leadership on the promotion of cultural, aesthetic, and ecological principles of action in society and economics. A culture of action that is based on the accumulated knowledge of eight decades and that practises a habit of thinking and acting that is specific to the discipline of Landscape Architecture. A process that encompasses the investment on the creation and development of an agenda of values rooted in cultural, aesthetic, and ecological principles and that steps outside the strictly academic sphere. For a landscape architect is more than just a designer, planner, and manager of landscapes, it is also a social actor with particular causes reinforced with specific technical expertise. The ability of landscape architects to reinvent environments, enhancing the quality of life and emotional experiences; the knowledge on environmental justice, community participation, engagement, and inclusiveness; the holistic vision that enables an integral view and a general understanding of phenomena; working methods based on an elasticity of relationships between different times and territorial scales; the skills towards innovative and improved designed solutions inspired in natural processes (Chapter 10, "Nature-based solutions in the teaching of landscape design by Manuel de Sousa da Câmara"); the humble and un-romantic attitude towards nature; political perspectives on green economy—these are just a few of the much-needed competences of Landscape Architecture which call for a more preponderant role in future decision-making endeavours.

In light if this reality, together with the emergence of the millennium generation with new digital practices and ways of collaboration, Landscape Architecture pedagogical models and practices are also being rethought and adapted in terms of content and interaction inside and outside the academic community. Indeed, there is a strong tendency for teaching to be focused on learning, through applied and applicable models, interactive and attractive, inclusive and engaging with society. An education that emphasis active methodologies that develop students' autonomy and reasoning, through the promotion of critical thinking, problem-solving skills, teamwork, and flexibility. Overall, the practical approach is being reinforced in Landscape Architecture education, endorsing the will to explore and experiment, and to learn by researching, designing, and building possibilities. Arguably, Landscape Architecture education should therefore embrace surrounding disciplinary fields, such as architecture, geography, or agricultural engineering, evidencing how its disciplinary frontiers are richer and more resilient by being thick, permeable, and moldable. Elaborating on a course that is more profoundly linked to the land and its people, which, by seeking to respond to the new global challenges, will more likely attract the youth. In search for a revisited discipline that breaks pedestals and unwelcomes the fall "into the dogmatism of 'eternal values' or into an apology for a 'timelessness'"

(Certeau, 1992 [1975], p. 57, in Chapter 5, "Landscape Architecture in Portugal"). All this ambition seeks to ensure that the next 80 years will honour the 80 years that have passed.

Notes

1 Habits in the sense conveyed by Aristoteles, which are understood as part of the best of what universities can give to society, as in the learning of a practice, a certain way of doing things, certain way of responding to problems, or even a certain way of doing nothing and keeping quiet. In other words, habit as an action that is intentionally carried out and which, when repeated, manifests in behavior (ARISTÓTELES 2019. Ética a Eudemo. Volume I–Tomo III. *Aristóteles Obras completas*. Imprensa Nacional Casa da Moeda, Centro de Filosofia da Universidade de Lisboa).

2 "with 'metamorphosis,' a more extensive and lasting transformation occurs, utilizing the concept of change and moving beyond change exclusively. Drawing from geological processes found in the rock cycle or the evolution from a caterpillar to a butterfly, these changes are more complete and radical" HONEYBUN-ARNOLDA, E. 2017. The Metamorphosis of the World: How Climate Change is Transforming Our Concept of the World. *The AAG Review of Books*, 5, 177–179.

3 Remarkably, three of the five winners of the recent national award "Prémio Ribeiro Telles" are female (Teresa Andresen [2019], Aurora Carapinha [2020], and Manuela Raposo Magalhães [2022]), and of the 46 contributing authors (including editors) of this book, 32 are female.

4 Recent research has been developing upon such subject, albeit within a different and particular focus on second world war landscape architecture in Britain, "Women of the Welfare Landscape" (CSEPELY-KNORR, L. & ALLEN, C. 2022. Women Of The Welfare Landscape. University of Liverpool: Arts And Humanities Research Council (AHRC).

5 The current IFLA president recently mentioned that "Cabral was known as a multilingual, multi-talented, constant traveller and an eternal student. He would adventure himself throughout Europe and Africa with students 'en route' to IFLA Councils and Congresses and, with that, explore magnificent landscapes and use that as a living lab of learning" MARQUES, B. 2023. *Speech for the inauguration of the Alameda Francisco Caldeira Cabral at the Instituto Superior de Agronomia* [Online]. Available: https://80anosap.isa.ulisboa.pt/programacao/ [Accessed 29th November 2023]. Yet up to today, there are no official Landscape Architecture associations in former Portuguese colonies, and only Angola is known to have active Landscape Architecture offices. There might be a Portuguese influence in Macau, which is presently creating a Landscape Architecture association, and perhaps India where the profession is very much developed.

Reference list

Andresen, T. (2001). *Francisco Cadeira Cabral*. United Kingdom: LDT Monographs No. 3.

Aristóteles. (2019). *Ética a Eudemo. Volume I—Tomo III. Aristóteles Obras completas*. Lisboa: Imprensa Nacional Casa da Moeda, Centro de Filosofia da Universidade de Lisboa.

Beck, U. (2016). *The Metamosphosis of the World: How Climate Change is Transforming Our Concept of the World*. Cambridge, UK: Polity Press.

Caldeira Cabral, F. (1968). *Memorial do Curso Livre de Arquitetura Paisagista 1941–1970*. Lisboa: Instituto Superior de Agronomia, Centro de Estudos de Arquitectura Paisagista.

Caldeira Cabral, F. (2003 [1993]). *Fundamentos da Arquitectura Paisagista*. Lisboa: Instituto de Conservação da Natureza.

Csepely-Knorr, L., & Allen, C. (2022). *Women of the Welfare Landscape*. Liverpool: University of Liverpool, Arts and Humanities Research Council (AHRC).

Honeybun-Arnolda, E. (2017). The metamorphosis of the world: How climate change is transforming our concept of the world. *The AAG Review of Books*, 5, 177–179.
Kolbert, E. (2014). *The Sixth Extinction: An Unnatural History*. New York: Macmillan.
Marques, B. (2023). *Speech for the inauguration of the Alameda Francisco Caldeira Cabral at the Instituto Superior de Agronomia*. Available at: https://80anosap.isa.ulisboa.pt/programacao/ [Accessed 29th November 2023].
Morton, T. (2007). *Ecology Without Nature: Rethinking Environmental Aesthetics*. Cambridge, MA: Harvard University Press.
Steinitz, C. (2012). *A Framework for Geodesign: Changing Geography by Design*. Redlands, CA: Esri.
Tamen, M. (2023). *Eutopos: algumas vantagens das universidades* | Brotéria. [online video]. Available at: www.youtube.com/@Broteria.

Index

Note: Page numbers in *italic* indicate a figure and page numbers in **bold** indicate a table on the corresponding page.

For Product Safety Concerns and Information please contact our EU representative GPSR@taylorandfrancis.com Taylor & Francis Verlag GmbH, Kaufingerstraße 24, 80331 München, Germany